Advances in Functional Inorganic Materials Prepared by Wet Chemical Methods (Volume II)

Advances in Functional Inorganic Materials Prepared by Wet Chemical Methods (Volume II)

Editors

Aleksej Zarkov
Aivaras Kareiva
Loreta Tamasauskaite-Tamasiunaite

MDPI • Basel • Beijing • Wuhan • Barcelona • Belgrade • Manchester • Tokyo • Cluj • Tianjin

Editors
Aleksej Zarkov
Vilnius University
Vilnius
Lithuania

Aivaras Kareiva
Vilnius University
Vilnius
Lithuania

Loreta
Tamasauskaite-Tamasiunaite
Center for Physical Sciences
and Technology
Vilnius
Lithuania

Editorial Office
MDPI
St. Alban-Anlage 66
4052 Basel, Switzerland

This is a reprint of articles from the Special Issue published online in the open access journal *Crystals* (ISSN 2073-4352) (available at: https://www.mdpi.com/journal/crystals/special_issues/wet_chemical_methods).

For citation purposes, cite each article independently as indicated on the article page online and as indicated below:

LastName, A.A.; LastName, B.B.; LastName, C.C. Article Title. *Journal Name* **Year**, *Volume Number*, Page Range.

ISBN 978-3-0365-6958-1 (Hbk)
ISBN 978-3-0365-6959-8 (PDF)

Contents

About the Editors

Aleksej Zarkov

Prof. Aleksej Zarkov currently holds a Full Professor position at the Institute of Chemistry of Vilnius University, as of 2023. He obtained a PhD in 2016 at Vilnius University, where he conducted research on the development of synthetic approaches for the preparation of oxide-based solid electrolytes. His research interests encompass the synthesis and characterization of functional inorganic materials in the form of bulk materials, nanoparticles and thin films. Recently, A. Zarkov has focused on the synthesis and investigation of calcium phosphates for medical, optical and environmental applications. He is a co-author of 82 articles and numerous conference announcements and presentations. During his scientific career, A. Zarkov has held long-term internships at various prestigious scientific institutions, including Osaka University (Osaka, Japan), Georgetown University (Washington D.C., USA), University of Cologne (Cologne, Germany), University of Aveiro (Aveiro, Portugal) and others. For his scientific achievements, A. Zarkov was awarded with several awards and scholarships from Vilnius University Rector and the Lithuanian Academy of Sciences.

Aivaras Kareiva

Prof. Aivaras Kareiva was promoted to the position of Full Professor at the Department of Inorganic Chemistry of Vilnius University in 1998. He is an expert in the sol–gel synthesis of different oxide materials. Since then, the "sol–gel chemistry" group has "prepared" 40 PhDs and 6 postdocs who have used the sol–gel method to synthesize superconductors, optical and magnetic materials, bioceramic materials, and nanostructured materials. The sol–gel method has also been used successfully to preserve cultural heritage. He cooperated for many years with scientists from Stockholm University, the University of Tuebingen, Masaryk University Brno, Muenster University of Applied Sciences, the University of Cologne, Kyushu University, the University of Venice, the University of Strasbourg, the National Taipei University of Technology, and others. Together with the co-authors, he has published more than 350 scientific articles and has participated in many research projects, grants, and contracts. In addition to scientific work, he also had some administrative positions. He was elected to the Dean position of the Faculty of Chemistry (2006–2016) at Vilnius University, and from 2017 to 2022, he was also the Director of the Institute of Chemistry, and in 2022, he was elected as the Dean of the Faculty of Chemistry and Geosciences at Vilnius University. He is a Member of the Lithuanian Academy of Sciences, and has received five Vilnius University Rector's Research Awards (2003, 2007, 2011, 2016, 2020), two Lithuanian Republic Research Awards (2004, 2020), Lithuanian State Last Degree Fellowship (2007–2008), and Lithuanian Academy of Sciences' Juozas Matulis Award (2009). He is laureate of the Medal SAPIENTI SAT of the Association and Chapter of A. Sniadecki, K. Olszewski and Z. Wrublewski (Poland) (2016), and was awarded the Medal of the Lithuanian Academy of Sciences (2021) and the Theodor von Grotthuss Medal.

Loreta Tamasauskaite-Tamasiunaite

Dr. Loreta Tamasauskaite-Tamasiunaite. is currently working as a chief research associate at the Department of Catalysis at the Center for Physical Sciences and Technology (FTMC) in Vilnius (Lithuania). She has interests ranging from physical chemistry, catalysis, and electrochemistry to developing catalysts for low-temperature polymer membrane fuel cells, water splitting, and hydrogen generation from aqueous hydride solutions. She has published 124 scientific articles and is a co-author of 19 patents (https://orcid.org/0000-0001-7555-4399).

crystals

Editorial

Advances in Functional Inorganic Materials Prepared by Wet Chemical Methods (Volume II)

Aleksej Zarkov [1,*], Aivaras Kareiva [1] and Loreta Tamasauskaite-Tamasiunaite [2]

1 Institute of Chemistry, Vilnius University, Naugarduko 24, LT-03225 Vilnius, Lithuania
2 Center for Physical Sciences and Technology, Sauletekio Ave. 3, LT-10257 Vilnius, Lithuania
* Correspondence: aleksej.zarkov@chf.vu.lt

Citation: Zarkov, A.; Kareiva, A.; Tamasauskaite-Tamasiunaite, L. Advances in Functional Inorganic Materials Prepared by Wet Chemical Methods (Volume II). *Crystals* **2023**, *13*, 324. https://doi.org/10.3390/cryst13020324

Received: 13 February 2023
Accepted: 14 February 2023
Published: 15 February 2023

Functional inorganic materials are an indispensable part of innovative technologies, which are essential for development in many fields of industry. The use of new materials, nanostructures, and multicomponent composites with specific chemical or physical properties promotes technological progress in electronics, optoelectronics, catalysis, biomedicine, and many other areas that are concerned with aspects of human life. Due to the broad and diverse range of potential applications of functional inorganic materials, the development of superior synthesis pathways, reliable characterization, and a deep understanding of the structure–property relationships in materials are rightfully considered to be fundamentally important scientific issues. Only synergetic efforts of scientists dealing with the synthesis, functionalization, and characterization of materials will lead to the development of future technologies. This Special Issue on "Advances in Functional Inorganic Materials Prepared by Wet Chemical Methods (Volume II)" covers a broad range of preparation routes, the characterization and application of functional inorganic materials, and metal–organic materials that are important in the fields of catalysis, electronics, optics, biomedicine, and others.

Metal phosphates are a family of inorganic compounds that are widely used in different areas such as medicine, catalysis, sensors, water treatment, the development of optical materials, etc. Calcium hydroxyapatite ($Ca_{10}(PO_4)_6(OH)_2$) is probably the most popular member of the calcium phosphate (CP) subfamily, whose solubility, specific surface area, surface wettability, and adsorption characteristics strongly depend on the morphology and crystallinity. Karalkeviciene et al. [1] investigated the effects of different organic solvents on the formation of hydroxyapatite via the hydrolysis of α-tricalcium phosphate under solvothermal conditions. The syntheses were performed with different proportions of water–organic solvents: ethylene glycol, methyl alcohol, ethyl alcohol, isopropyl alcohol, and butyl alcohol. The influence of the solvothermal conditions, the nature of the solvent, and the effect of the organic solvent fraction in the solution on the phase purity and morphological features of HA were investigated. It was demonstrated that methyl alcohol and ethylene glycol had a stronger inhibitory effect on the hydrolysis process. The morphology of the obtained samples varied from plate-shaped to rod-shaped depending on the reaction medium and synthesis conditions. Out of all the solvents analyzed, ethylene glycol had the highest impact on the sample morphology. A different efficient way to modify the morphology of hydroxyapatite considers the usage of various organic additives. A comprehensive experimental study was carried out to compare the effects of sodium lauryl sulfate and three amino acids (DL-aspartic acid, dodecanedioic acid, and suberic acid) on the formation of calcium-deficient hydroxyapatite via the hydrolysis of α-tricalcium phosphate [2]. The study revealed that higher concentrations of sodium lauryl sulfate and dodecanedioic acid induced the formation of impurities in addition to hydroxyapatite, whereas aspartic and suberic acids did not affect the phase purity. The morphology of the samples varied from plate- to rod-like depending on the concentration of the specific organic additive.

Another member of the CP group is brushite ($CaHPO_4 \cdot 2H_2O$), which is a material that can be synthesized in aqueous medium by using simple wet precipitation. Alshaaer et al. [3] studied the feasibility of full-scale substitution of Ca^{2+} by Ni^{2+} ions and its influence on the crystal structure of brushite. The authors concluded that the substitution level is limited, and that the replacement of 20% of Ca^{2+} leads to the formation of a mixture of compounds.

Rare-earth phosphates are another group of phosphate-based materials that recently have gained much scientific interest, especially in the field of luminescent materials. Ezerskyte et al. [4] reported the hydrothermal synthesis of $GdPO_4$:Eu^{3+} nanoparticles by varying different synthesis parameters: pH, <Gd>:<P> molar ratio, and Eu^{3+} concentration. It was shown that the nanoparticle length can be adjusted from hundreds to tens of nanometers by changing the pH. Increasing the <Gd>:<P> molar ratio resulted in a decrease in nanoparticle length and an increase in its width. The temperature-dependent measurements in the 77–500 K range revealed that the $GdPO_4$:50%Eu^{3+} sample maintains half of its emission intensity, even at room temperature.

Silver phosphate (Ag_3PO_4) can be used as an efficient photocatalyst, which shows a very strong oxidation capacity and high visible-light catalytic activity. However, Ag_3PO_4 is prone to photoelectron corrosion and becomes unstable during photocatalysis, which severely limits its application in this field. A tubelike g-C_3N_4/Ag_3PO_4 heterojunction was constructed with the chemical precipitation method by Yan et al. [5]. The photocatalytic activity of the photocatalyst was evaluated based on the degradation of RhB under visible-light irradiation. The degradation rate of the RhB was 90% under irradiation for 40 min; moreover, the recycling experiment showed that there was no apparent decrease in the activity of the heterojunction after five consecutive runs.

Precious metal-based catalysts are the most widely studied and are currently regarded as state-of-the-art electrocatalysts for formic acid/formate oxidation because of their high catalytic activity and good anti-poisoning properties. In a study by Nacys et al. [6], a platinum-coated Ni foam catalyst (PtNi/Ni) was investigated for the oxidation of the formate reaction in an alkaline medium. The catalyst was fabricated via a two-step procedure, which involved an electroless deposition of the Ni layer and the subsequent electrodeposition of the platinum layer. The obtained composite demonstrated enhanced electrocatalytic activity compared to the Ni/Ni foam catalyst and pure Pt electrode.

Another area of application of precious metals is molecular analysis by SERS spectroscopy. Zavatski et al. [7] fabricated gold nanostructures on two different types of substrates, namely, meso- and macroporous silicon, via an electroless chemical deposition method for different time periods. SERS measurements were performed for the R6G molecules adsorbed on different Au-coated samples at a 10^{-6} M concentration. The authors demonstrated the correlation of the R6G SERS spectra intensity with the morphology and structural properties of Au coverage on the silicon surface.

Carbon-based materials comprise another wide and diverse class of functional inorganic compounds, which can be modified in a variety of ways. A special subgroup of these materials, known as graphite intercalation compounds (GICs), are especially interesting due to their use in Li-ion batteries, superconductivity, catalytic activity, etc. Rimkute et al. [8] demonstrated the synthesis of GICs using three different oxidizers: $(NH_4)_2S_2O_8$, $K_2S_2O_8$, and CrO_3, with and without P_2O_5 as a water-binding agent. The use of CrO_3 resulted in a much higher structural disorder, whereas a minimal defect concentration was reached by using $K_2S_2O_8$. The use of P_2O_5 can help to achieve more effective intercalation and has a positive effect on the formation of the stage I GIC phase.

Graphene-polypyrrole nanocomposites were synthesized via a wet-way protocol using a graphite bisulfate precursor, which was prepared in the presence of concentrated sulfuric acid and potassium periodate [9]. Three different types of graphite precursors with particle sizes of <50 μm, ≥ 150, ≤ 830 μm, and ≤ 2000 μm were used for this purpose. The nanocomposites prepared from exfoliated graphene and polypyrrole had a core–shell structure, where the particle core is composed of exfoliated graphene covered by a polypyrrole shell. The most relevant areas of application for these obtained nanocomposites, including energy storage and (bio)sensing, were considered.

Synthetic conditions for the preparation of different types of spherical amine-holding silica particles were proposed by Melnyk et al. [10]. The Stöber version of a sol-gel method of co-condensation of two alkoxysilanes was used for the one-pot synthesis of silica particles with $\equiv Si(CH_2)_3NH_2$, $\equiv Si(CH_2)_3NH(CH_2)_2NH_2$, and $\equiv [Si(CH_2)_3NH]_2$ functional groups that had an available group content of 1.3–2.3 mmol/g. The sorption properties of these hybrid materials for nickel(II) and copper(II) ions, as well as for bovine serum albumin, were found to be dependent on the structure of the materials and the type of incorporated functional group. The maximal static sorption capacity values toward the targeted adsorbates were shown by the samples with 3-aminopropyl groups (1.27 mmol Ni/g), diamine groups (1.09 mmol Cu/g), and secondary amine groups (204.6 mg BSA/g).

The study of Pt and Pd complexes with various organic compounds is of fundamental and practical interest due to a wide range of applications, including biosensors, DNA visualization, and catalysis. Novikov et al. [11] reported the synthesis of new $[PdCl_2(HGua)_2]Cl_2\cdot H_2O$ and $[PtCl_5(HGua)]\cdot 2H_2O$ compounds using wet chemistry methods. It was shown that the complexes have an M-N bond, and the organic ligand is included in the first coordination sphere. In the palladium complex, π-halogen and π-stacking interactions were found, whereas in analogues, such interactions were not found. On the other hand, π-halogen and halogen interactions were found in the structure of a platinum complex and its analogues. The same group of authors reported the synthesis of new platinum and uranium compounds with nicotinic acid [12]. The authors described the synthesis of two new anionic complexes, $(HNic)_2[PtCl_6]$ and $(HNic)_2[UO_2Cl_4]$, using wet chemistry methods. It was determined that hydrogen bonds (H···Hal/Hal···H and O···H/H···O) are the main contributors to the intermolecular interactions in both compounds. Other important contacts in cations are H···H, C···H/H···C, and C···Hal/Hal···C; in anions, it is H···Hal/Hal···H. The Pt-containing complex has a halogen-π interaction and halogen bonds, but this type of interaction was not observed in the U-containing compound.

Another successful synthesis of metal–organic compounds was demonstrated by Zhang et al. [13]. In this study, two different phase-pure cobalt(II) cyanoguanidine bromide coordination compounds, $Co(C_2N_4H_4)_2(H_2O)_4\cdot 2Br\cdot 2H_2O$ (**1**) and $Co_3(C_2N_4H_4)_8(H_2O)_8\cdot 6Br$ (**2**), were precipitated from aqueous and methanol solutions, respectively. It was shown that the magnetic susceptibility of **1** is consistent with either isolated or very weakly interacting Co^{2+} centers, whereas the magnetic susceptibility of **2** is evidence of the potential weak antiferromagnetic exchange interactions that may arise from the superexchange within the binuclear clusters.

Layered double hydroxides or hydrotalcite-like compounds are layered basic salts, whose structural features give them a number of specific properties which can be tuned by the modification of the chemical composition and partial substitution of metal ions. Nestroinaia et al. [14] employed three different wet chemical methods (co-precipitation at variable pH, co-precipitation followed by hydrothermal treatment, and co-precipitation with microwave treatment) to synthesize Mg/Ni–Al layered double hydroxides containing triply charged nickel cations. Despite the variety of synthetic approaches, the properties of the samples synthesized by the different methods were shown to be quite similar.

Author Contributions: Conceptualization, A.Z., A.K. and L.T.-T.; writing—original draft preparation, A.Z., A.K. and L.T.-T.; writing—review and editing, A.Z., A.K. and L.T.-T.; funding acquisition, A.Z. All authors have read and agreed to the published version of the manuscript.

Funding: This work was funded by the grant WHITCERAM (No. S-LJB-22-1) from the Research Council of Lithuania.

Conflicts of Interest: The authors declare no conflict of interest.

References

1. Karalkeviciene, R.; Raudonyte-Svirbutaviciene, E.; Gaidukevic, J.; Zarkov, A.; Kareiva, A. Solvothermal Synthesis of Calcium-Deficient Hydroxyapatite via Hydrolysis of α-Tricalcium Phosphate in Different Aqueous-Organic Media. *Crystals* **2022**, *12*, 253. [CrossRef]
2. Karalkeviciene, R.; Raudonyte-Svirbutaviciene, E.; Zarkov, A.; Yang, J.-C.; Popov, A.I.; Kareiva, A. Solvothermal Synthesis of Calcium Hydroxyapatite via Hydrolysis of Alpha-Tricalcium Phosphate in the Presence of Different Organic Additives. *Crystals* **2023**, *13*, 265. [CrossRef]
3. Alshaaer, M.; Issa, K.; Afify, A.S.; Moustapha, M.E.; Alanazi, A.A.; Elsanousi, A.; Qahtan, T.F. The Impact of Full-Scale Substitution of Ca^{2+} with Ni^{2+} Ions on Brushite&rsquos Crystal Structure and Phase Composition. *Crystals* **2022**, *12*, 940.
4. Ezerskyte, E.; Zarkov, A.; Klimkevicius, V.; Katelnikovas, A. Hydrothermal Synthesis of Well-Defined Red-Emitting Eu-Doped $GdPO_4$ Nanophosphors and Investigation of Their Morphology and Optical Properties. *Crystals* **2023**, *13*, 174. [CrossRef]
5. Yan, X.; Wang, Y.; Kang, B.; Li, Z.; Niu, Y. Preparation and Characterization of Tubelike g-C_3N_4/Ag_3PO_4 Heterojunction with Enhanced Visible-Light Photocatalytic Activity. *Crystals* **2021**, *11*, 1373. [CrossRef]
6. Nacys, A.; Šimkūnaitė, D.; Balčiūnaitė, A.; Zabielaitė, A.; Upskuvienė, D.; Šebeka, B.; Jasulaitienė, V.; Kovalevskij, V.; Norkus, E.; Tamašauskaitė-Tamašiūnaitė, L. An Enhanced Oxidation of Formate on PtNi/Ni Foam Catalyst in an Alkaline Medium. *Crystals* **2022**, *12*, 362. [CrossRef]
7. Zavatski, S.; Popov, A.I.; Chemenev, A.; Dauletbekova, A.; Bandarenka, H. Wet Chemical Synthesis and Characterization of Au Coatings on Meso- and Macroporous Si for Molecular Analysis by SERS Spectroscopy. *Crystals* **2022**, *12*, 1656. [CrossRef]
8. Rimkute, G.; Gudaitis, M.; Barkauskas, J.; Zarkov, A.; Niaura, G.; Gaidukevic, J. Synthesis and Characterization of Graphite Intercalation Compounds with Sulfuric Acid. *Crystals* **2022**, *12*, 421. [CrossRef]
9. Rimkutė, G.; Niaura, G.; Pauliukaitė, R.; Gaidukevič, J.; Barkauskas, J. Wet Synthesis of Graphene-Polypyrrole Nanocomposites via Graphite Intercalation Compounds. *Crystals* **2022**, *12*, 1793. [CrossRef]
10. Melnyk, I.; Tomina, V.; Stolyarchuk, N. Synthetic Conditions for Obtaining Different Types of Amine-Holding Silica Particles and Their Sorption Behavior. *Crystals* **2023**, *13*, 190. [CrossRef]
11. Novikov, A.P.; Volkov, M.A.; Safonov, A.V.; Grigoriev, M.S.; Abkhalimov, E.V. Synthesis and Characterization of New Guanine Complexes of Pt(IV) and Pd(II) by X-ray Diffraction and Hirshfeld Surface Analysis. *Crystals* **2021**, *11*, 1417. [CrossRef]
12. Novikov, A.P.; Volkov, M.A.; Safonov, A.V.; Grigoriev, M.S. Synthesis, Crystal Structure, and Hirshfeld Surface Analysis of Hexachloroplatinate and Tetraclorouranylate of 3-Carboxypyridinium-Halogen Bonds and π-Interactions vs. Hydrogen Bonds. *Crystals* **2022**, *12*, 271. [CrossRef]
13. Zhang, J.; Corkett, A.J.; van Leusen, J.; Englert, U.; Dronskowski, R. Solvent-Induced Cobalt(II) Cyanoguanidine Bromides: Syntheses, Crystal Structure, Optical and Magnetic Properties. *Crystals* **2022**, *12*, 1377. [CrossRef]
14. Nestroinaia, O.V.; Ryltsova, I.G.; Lebedeva, O.E. Effect of Synthesis Method on Properties of Layered Double Hydroxides Containing Ni(III). *Crystals* **2021**, *11*, 1429. [CrossRef]

crystals

MDPI

Article

Preparation and Characterization of Tubelike g-C_3N_4/Ag_3PO_4 Heterojunction with Enhanced Visible-Light Photocatalytic Activity

Xin Yan *, Yuanyuan Wang, Bingbing Kang, Zhuo Li and Yanhui Niu *

School of Materials Science and Engineering, Chang'an University, Xi'an 710064, China; 2019131030@chd.edu.cn (Y.W.); 2019231022@chd.edu.cn (B.K.); lizhuo@chd.edu.cn (Z.L.)
* Correspondence: xinyan@chd.edu.cn (X.Y.); niuyh@chd.edu.cn (Y.N.); Tel.: +86-(0)29-8233-7345 (X.Y.); +86-(0)29-8233-7340 (Y.N.)

Abstract: Water pollution caused by dye wastewater is a potential threat to human health. Using photocatalysis technology to deal with dye wastewater has the advantages of strong purification and no secondary pollution, so it is greatly significant to look for new visible-light photocatalysts with high photocatalytic ability for dye wastewater degradation. Semiconductor photocatalyst silver phosphate (Ag_3PO_4) has high quantum efficiency and photocatalytic degradation activity. However, Ag_3PO_4 is prone to photoelectron corrosion and becomes unstable during photocatalysis, which severely limits its application in this field. In this study, a tubelike g-C_3N_4/Ag_3PO_4 heterojunction was constructed by the chemical precipitation method. An Ag_3PO_4 nanoparticle was loaded onto the surface of the tubelike g-C_3N_4, forming close contact. The photocatalytic activity of the photocatalyst was evaluated by the degradation of RhB under visible-light irradiation. The tubelike g-C_3N_4/Ag_3PO_4-5% heterojunction exhibited optimal photocatalytic performance. In an optimal process, the degradation rate of the RhB is 90% under visible-light irradiation for 40 min. The recycling experiment showed that there was no apparent decrease in the activity of tubelike g-C_3N_4/Ag_3PO_4-5% heterojunction after five consecutive runs. A possible Z-type mechanism is proposed to explain the high activity and stability of the heterojunction.

Keywords: photocatalysis; Ag_3PO_4; g-C_3N_4; heterojunction; degradation

Citation: Yan, X.; Wang, Y.; Kang, B.; Li, Z.; Niu, Y. Preparation and Characterization of Tubelike g-C_3N_4/Ag_3PO_4 Heterojunction with Enhanced Visible-Light Photocatalytic Activity. *Crystals* **2021**, *11*, 1373. https://doi.org/10.3390/cryst11111373

Academic Editors: Aleksej Zarkov, Aivaras Kareiva, Loreta Tamasauskaite-Tamasiunaite and Giuseppe Greco

Received: 23 October 2021
Accepted: 10 November 2021
Published: 11 November 2021

1. Introduction

With the rapid development of modern industry, the problem of water pollution is becoming increasingly serious, as it could harm human health [1–4]. Dye wastewater has large discharge, biological toxicity, complex composition, high content of organic matter, and poor biodegradability, which is a difficult problem in wastewater treatment. Semiconductor photocatalysis technology can transform solar into chemical energy and completely decompose organic matter under mild conditions. It shows great potential and good application prospects in solving environmental pollution and energy shortage [5,6]. The development of highly active photocatalysts is key to the application of photocatalysis technology. Various semiconductor photocatalysts, including TiO_2, ZnO, and SnO_2, were studied [7–10]. However, these conventional photocatalytic materials still have many problems, such as photocatalytic activity only under ultraviolet light, electron-hole recombination, and potential toxicity, which are primary obstacles to further application. Thereby, it is greatly significant to explore photocatalysts with high visible-light activity for the application of semiconductor photocatalysis.

In 2010, Ye et al. [11] reported a novel visible light photocatalyst, Ag_3PO_4, which showed very strong oxidation capacity and high visible-light catalytic activity. In addition, its quantum efficiency reached 90%, far higher than that of other metal oxide semiconductor photocatalysts [12–15]. However, Ag_3PO_4 is prone to photoelectron corrosion and becomes unstable during photocatalytic reaction, which severely limits the application of Ag_3PO_4 in the field of photocatalysis. Therefore, the photocatalytic performance of Ag_3PO_4 was improved by adjusting the morphology [16,17], combining it with carbon materials [18,19], and constructing a heterojunction [20,21]. Among them, the Ag_3PO_4-based heterojunction can effectively separate electrons and holes according to energy band theory, which is regarded as the most effective method to improve the photocatalytic performance of Ag_3PO_4. For example, many semiconductor photocatalysts, including TiO_2, SnO_2, g-C_3N_4, $SrTiO_3$ were used to couple with Ag_3PO_4 to fabricate heterojunction photocatalysts [22–25]. These heterojunctions showed excellent catalytic performance and stability. g-C_3N_4 is of great interest to researchers due to its matching potentials of the conduction and valence bands with Ag_3PO_4. The interfacial contact of heterojunction catalysts greatly affects electron and hole transport. Most of the reported g-C_3N_4/Ag_3PO_4 heterojunction catalysts, however, are bulk g-C_3N_4. The small specific surface area of bulk g-C_3N_4 is not conducive to the formation of heterojunctions with Ag_3PO_4 [26,27].

In this paper, a tubelike g-C_3N_4 was prepared by the molten salt method. A new type of Ag_3PO_4-based heterojunction was formed by chemical precipitation with the tubelike g-C_3N_4. The structure, morphology, and spectral properties of the g-C_3N_4/Ag_3PO_4 heterojunction were characterized by X-ray diffraction patterns, scanning electron microscopy, transmission electron microscopy, UV–vis diffuse reflectance spectra, and photoluminescence spectra. The photocatalytic activity of the photocatalyst was evaluated by the degradation of RhB under visible light irradiation. The effect of g-C_3N_4 content on the photocatalytic performance of the tubelike g-C_3N_4/Ag_3PO_4 heterojunction was investigated. Moreover, the electron transfer mechanism of the tubelike g-C_3N_4/Ag_3PO_4 heterojunction in the degradation of RhB under visible-light irradiation was studied.

2. Materials and Methods

2.1. Synthesis of Tubelike g-C_3N_4

The molten salt method was used to synthesize the tubelike g-C_3N_4 [28,29]. First, 4.5 g of lithium chloride, 5.5 g of potassium chloride, and 1 g of melamine were ground together. The mixture in the crucible was heated in a muffle furnace at 550 °C for 4 h. After the reaction, lithium chloride and potassium chloride were removed by washing with deionized water. Lastly, the tubelike g-C_3N_4 was collected and dried at 80 °C for 24 h. The sample was labeled as TCN.

2.2. Synthesis of Tubelike g-C_3N_4/Ag_3PO_4 Heterojunction

The tubelike g-C_3N_4/ Ag_3PO_4 heterojunction was prepared by chemical precipitation. First, 0.17 g $AgNO_3$ was dissolved in deionized water, and a certain amount of tubelike g-C_3N_4 was dispersed in the above solution. Then, 0.15 g Na_3PO_4 was dissolved in deionized water and slowly dropped into the $AgNO_3$ solution. After magnetic stirring for 30 min, the solution was filtered, washed, and dried at 60 °C for 10 h. The tubelike g-C_3N_4/Ag_3PO_4 heterojunctions with different tubelike g-C_3N_4 mass contents of about 1, 3, 5, and 7% were labeled as TCN/Ag_3PO_4-1%, TCN/Ag_3PO_4-3%, TCN/Ag_3PO_4-5%, and TCN/Ag_3PO_4-7%, respectively. For comparison, the pure Ag_3PO_4 nanoparticle was prepared by the chemical precipitation method without tubelike g-C_3N_4.

2.3. Characterization

The crystal structure was determined by Philips X 'pert MPD (Almelo, The Netherlands). Cu target Kα was used as the radiation source (wavelength, 0.15418 nm), working voltage was 40 kV, working current was 20 mA, scanning range was 10–70°, and scanning speed was 4° min^{-1}. The surface morphology of the sample was observed by scanning electron microscope (S-4800; working voltage, 5 kV, Kyoto, Japan) and transmission electron microscopy (FEI Tecnai G20; working voltage, 200 kV, Hillsboro, OR, USA). X-ray photoelectron spectroscopy (XPS) measurements were performed by using a Kratos Axis Ultra DLD detector (Manchester, UK). The photoluminescence (PL) spectra of the sample at room temperature were measured with a HORIBA Fluorolog-3(Paris, France). The UV–visible diffuse reflectance spectra of the sample were measured with a Shimadzu UV-3600 spectrophotometer (UV-3600, Shimadzu, Kyoto, Japan).

2.4. Photocatalytic Activity Evaluation

We added 0.01 g of the catalyst to a 100 mL RhB solution with an initial concentration of 10 mg/L. In order to achieve adsorption–desorption equilibrium, the RhB solution (with the photocatalyst) was magnetically stirred in the dark for 30 min. Then, the solution was placed under a 300 W xenon lamp (filtering out ultraviolet light) for photocatalytic reaction. The RhB solution was taken every 5 min. After 30 min of centrifugation, the supernatant was removed, and the concentration of RhB solution was measured with a UV-vis spectrophotometer (721 G, Shanghai, China) at 664 nm absorbance value.

3. Results and Discussion

3.1. XRD and XPS Analysis

The XRD patterns of TCN, Ag_3PO_4, and the TCN/Ag_3PO_4 heterojunction are shown in Figure 1. The diffraction peaks of TCN were at 11.7, 20.6, 24.3, 26.6, 29.1, and 32.4°, which reflects different planes from (100), (110), (200), (002), (102), and (210). This is consistent with literature reports [30]. Ag_3PO_4 had obvious diffraction peaks at 20.88, 29.70,33.29, 36.59, 42.49, 47.79, 52.70, 55.02, 57.28, and 61.64°. Compared with the cubic Ag_3PO_4 standard card (PDF 06-0505), the (110), (200), (210), (211), (220), (310), (222), (320), (321), (400) crystal planes corresponded to their crystal phase, respectively. This shows that the pure Ag_3PO_4 phase was successfully prepared by chemical precipitation. The diffraction peak of the TCN/Ag_3PO_4 heterojunction is mainly from Ag_3PO_4, while the characteristic diffraction peak of the tubelike g-C_3N_4 did not appear because of the low content of tubelike g-C_3N_4 in the heterojunction, which is difficult to be detected by X-ray diffraction.

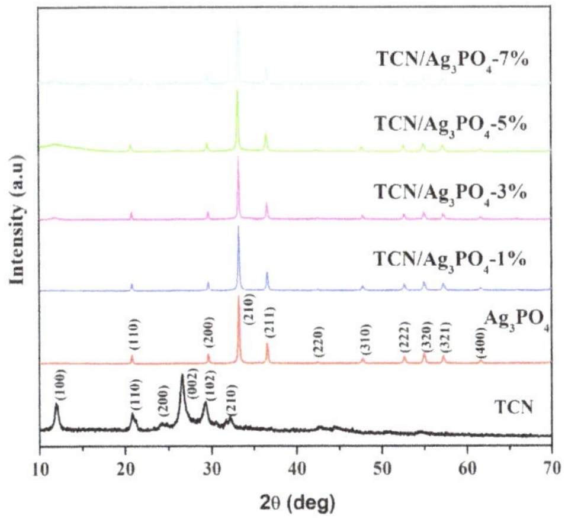

Figure 1. XRD patterns of TCN, Ag_3PO_4, and TCN/Ag_3PO_4 heterojunction.

The chemical composition of the TCN/Ag_3PO_4 heterojunction was characterized by XPS, as shown in Figure 2. Figure 2a is the survey spectrum of the TCN/Ag_3PO_4-5% heterojunction, which shows the presence of Ag, P, O, N, and C. The high-resolution XPS measurements of C 1s, N 1s, Ag 3d, O 1s, and P 2p are shown in Figure 2b–f. There are three peaks in Figure 2b. The peaks with binding energy at 284.6 and 287.5 eV were attributed to the C-C bonds and sp2 carbon in the triazine ring, respectively. The peaks with binding energy at 285.7 eV corresponded to C-O bonds in the hydroxyl state on the sample surface [31]. Results show that there were many hydrophilic hydroxyl groups on the surface of the heterojunction catalyst. These oxygen-containing carbons provided an active surface to form good contact with the Ag_3PO_4 nanoparticle. Figure 2c shows two peaks in the N 1s spectra. The peak with binding energy at 398.2 eV is attributed to sp2 nitrogen in the triazine ring, and another peak with binding energy at 399.7 eV is ascribed to the terminal amino groups [32]. Results further confirmed the existence of g-C_3N_4 in the TCN/Ag_3PO_4 heterojunction. Figure 2d shows the high-resolution spectra of Ag 3d. Peaks with binding energy at 367.8 and 373.8 eV belonged to the 3d5/2 and 3d3/2 characteristic peaks, respectively, with a valence state of Ag^+. Ag 3d is a single peak of splitting, indicating that the Ag element existed only in the form of Ag^+ [33]. Figure 2e shows the high-resolution spectrum of O 1s. The peak of O 1s could be fitted into two peaks, which were mainly the lattice oxygen of Ag_3PO_4 with binding energy at 530.6 eV and the surface hydroxyl oxygen at 532.4 eV [34]. Figure 2f is the high-resolution spectrum of P 2p. The electron binding energy at the center of the characteristic single peak was located at 133.2 eV, suggesting that the P element was in a +5 valence state [35]. Results further confirmed the existence of Ag_3PO_4 in the TCN/Ag_3PO_4 heterojunction.

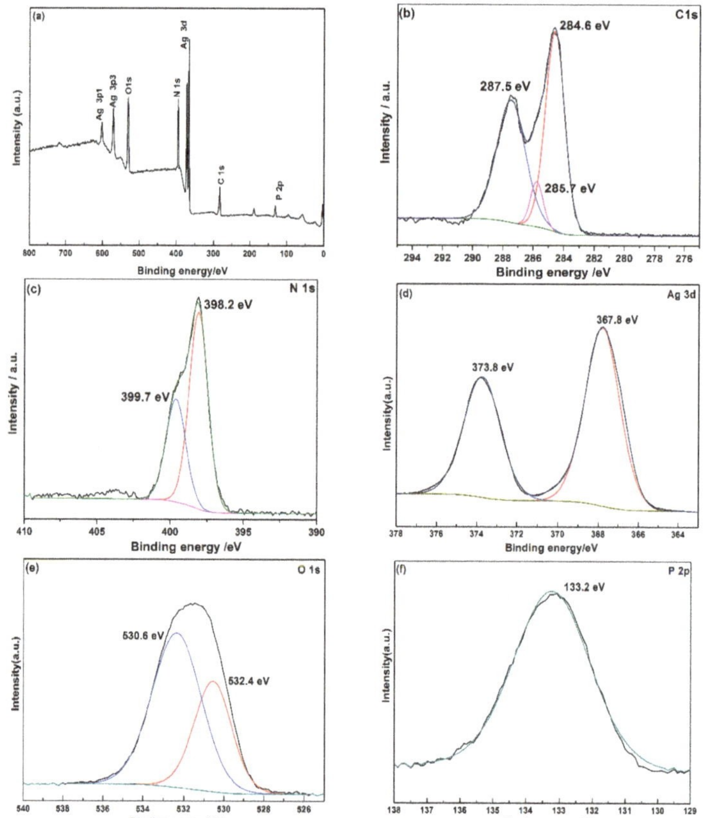

Figure 2. XPS spectra of TCN/ Ag_3PO_4-5% heterojunction: (**a**) XPS survey spectrum; (**b**) C 1s; (**c**) N 1s; (**d**) Ag 3d; (**e**) O 1s; (**f**) P 2p.

3.2. SEM Analysis

Figure 3 shows the SEM images of TCN, Ag_3PO_4, and the TCN/Ag_3PO_4-5% heterojunction, respectively. Figure 3A shows that TCN prepared by the molten salt method had a hollow tubelike shape, a square tube mouth with a smooth surface, and length distribution ranging from several microns to more than 10 microns. Figure 3B shows that Ag_3PO_4 prepared by chemical precipitation was composed of many polycrystalline nanoparticles with an average size of 30–50 nm, and particles presented an obvious agglomeration phenomenon. Figure 3C is the SEM image of the TCN/Ag_3PO_4-5% heterojunction. Results show that the Ag_3PO_4 nanoparticle was uniformly loaded onto the surface of tubelike g-C_3N_4. Figure 3D is the TEM image of TCN/Ag_3PO_4-5% heterojunction. The hollow tubelike g-C_3N_4 structure and Ag_3PO_4 nanoparticle were observed more clearly. Results showed that the tubelike g-C_3N_4 both dispersed the Ag_3PO_4 nanoparticle and formed the heterojunction.

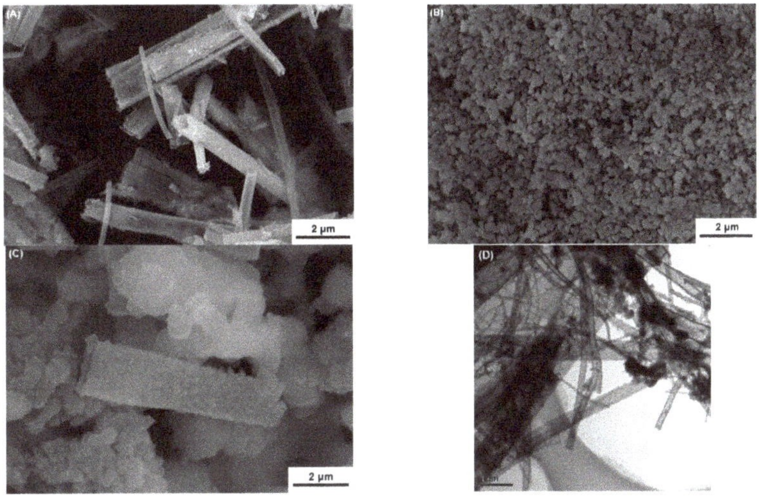

Figure 3. SEM images of (**A**) TCN, (**B**) Ag_3PO_4, (**C**) TCN/Ag_3PO_4-5% heterojunction, and (**D**) TEM image of TCN/Ag_3PO_4-5% heterojunction.

3.3. UV–vis Diffuse Reflectance Spectra and Photoluminescence Spectra Analysis

The UV–vis DRS of TCN, Ag_3PO_4, and the TCN/Ag_3PO_4 heterojunction are shown in Figure 4a. The Ag_3PO_4 showed obvious absorption from the ultraviolet to the visible region, and its band edge absorption was 532 nm [36]. The band edge absorption of TCN was 460 nm, indicating a certain absorption of visible light. With the addition of TCN, the absorption spectra of TCN/Ag_3PO_4 heterojunction had a red shift compared to those of TCN and Ag_3PO_4. The shift is ascribed to the interaction between TCN and Ag_3PO_4. This result implied that more visible light could be harvested by the TCN/Ag_3PO_4 heterojunction.

The separation efficiency of a photoelectron-hole pair greatly affects the performance of the photocatalyst [37]. In Ag_3PO_4 systems, the rapid separation of photogenerated electrons is beneficial to reduce the photocurrent corrosion of Ag_3PO_4. Figure 4b shows the photoluminescence spectra of Ag_3PO_4 and the TCN/Ag_3PO_4 heterojunction, with an excitation wavelength of 365 nm. The emission peak of pure Ag_3PO_4 was obvious at 535 nm, indicating that the electron-hole pair in Ag_3PO_4 had a strong recombination rate. With the addition of TCN, the peak intensity of the TCN/Ag_3PO_4 heterojunction gradually decreased. The peak intensity of the TCN/Ag_3PO_4-5% heterojunction was the lowest, indicating a low recombination rate of charges. Therefore, the formation of the TCN/Ag_3PO_4 heterojunction is beneficial to the separation of electrons and holes, thus effectively improving its catalytic performance.

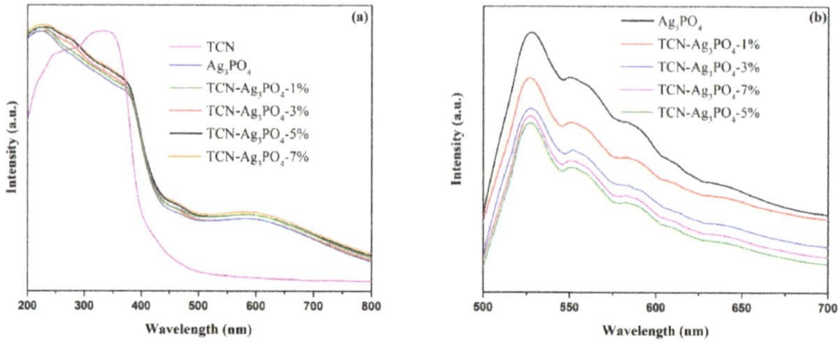

Figure 4. UV–vis diffuse (**a**) reflectance and (**b**) photoluminescence spectra of Ag_3PO_4 and TCN/Ag_3PO_4-5% heterojunction.

3.4. Photocatalytic Activities Analyses

The catalytic activity of different catalysts was investigated by using RhB as a simulated degradation target, as shown in Figure 5a. In the absence of a catalyst, RhB hardly degraded under visible-light irradiation. The TCN and Ag_3PO_4 samples exhibited visible-light photocatalytic activity, and the degradation rate of RhB was 25 and 44%, respectively after 40 min visible light irradiation. Compared with the TCN and Ag_3PO_4, the TCN/Ag_3PO_4 heterojunction exhibited enhanced photocatalytic activity. The TCN/Ag_3PO_4-5% heterojunction resulted in 90% degradation of RhB within 40 min. It is an important factor for the formation of heterojunction with the appropriate amount of TCN. When the amount of TCN is low, Ag3PO4 particles cannot be fully loaded and thus cannot effectively form the heterojunction. On the other hand, excessive TCN absorbs the visible light and forms a masking effect on the heterojunction, thus decreasing its catalytic efficiency. For low-concentration pollutants, the kinetic behaviors for the degradation reaction can be studied with the following equation:

$$\ln(c/c_0) = kt \tag{1}$$

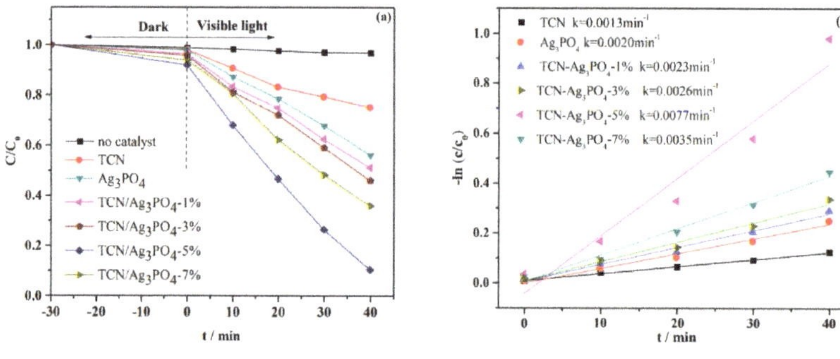

Figure 5. Degradation curves (**a**) and kinetic plot (**b**) of photocatalytic degradation of RhB by TCN, Ag_3PO_4, and TCN/Ag_3PO_4 heterojunction.

The kinetic constants (k) of different samples were calculated and are shown in Figure 5b. The kinetic constant of RhB degradation with TCN/Ag_3PO_4-5% (0.0077 min^{-1}) was about 5.9 and 3.8 times as high as that of TCN (0.0013 min^{-1}) and Ag_3PO_4 (0.0020 min^{-1}), respectively. This result shows that the formation of the TCN/Ag_3PO_4 heterojunction could efficiently enhance the photocatalytic performances of Ag_3PO_4.

To investigate the stability of Ag_3PO_4 and the TCN/Ag_3PO_4 heterojunction, the photocatalyst was recovered and dried for the next cycle of degradation. Degradation conditions remained the same as before, and there were 5 cycles. Experimental results are shown in Figure 6a. In the cyclic experiment, the photocatalytic performance of Ag_3PO_4 obviously decreased, and the degradation rate of RhB for the fifth time was only 15%. The degradation rate of RhB by the TCN/Ag_3PO_4 heterojunction for the fifth time was 85%. Results showed that the TCN/Ag_3PO_4 heterojunction exhibited excellent stability. The samples of Ag_3PO_4 and the TCN/Ag_3PO_4 heterojunction after the cycle experiment were analyzed by X-ray diffraction, and results are shown in Figure 6b. In the XRD of Ag_3PO_4, a new diffraction peak appeared at 38.1°, which is the (111) plane diffraction peak of Ag [38]. Results showed that Ag_3PO_4 was prone to photoelectron corrosion and became unstable during photocatalytic reaction. There was no new diffraction peak in the XRD of the TCN/Ag_3PO_4 heterojunction. Results showed that the TCN/Ag_3PO_4 heterojunction maintained good stability in the cyclic experiment.

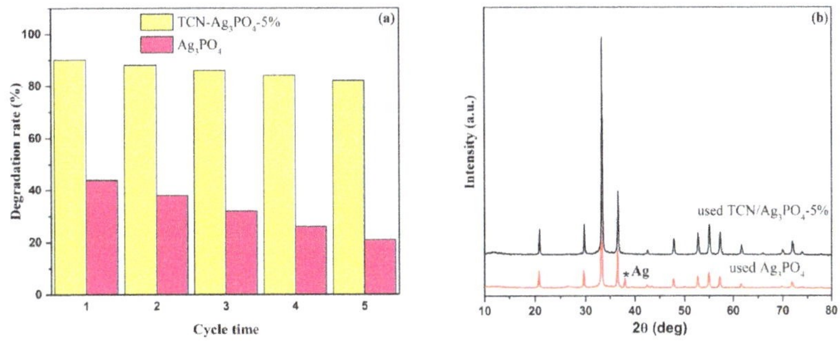

Figure 6. (**a**) RhB degradation rate and (**b**) XRD patterns over Ag_3PO_4 and TCN/Ag_3PO_4 heterojunction in cyclic experiment.

3.5. Photocatalytic Mechanism of TCN/Ag_3PO_4 Heterojunction

To further investigate the photocatalytic reaction mechanism of the TCN/Ag_3PO_4 heterojunction, the TCN/Ag_3PO_4-5% heterojunction was used as photocatalyst, and different scavengers were selected to carry out the photodegradation experiment. Among them, ethylenediaminetetraacetic acid (EDTA) was the electron hole (h^+) scavenger [39], benzoquinone (BQ) was the superoxide radical ($^\bullet O_2^-$) scavenger [40], and isopropanol (IPA) was the hydroxyl radical (•OH) scavenger [41], with a concentration of 1 mmol/L. The effects of various scavengers on the degradation rate of RhB are shown in Figure 7.

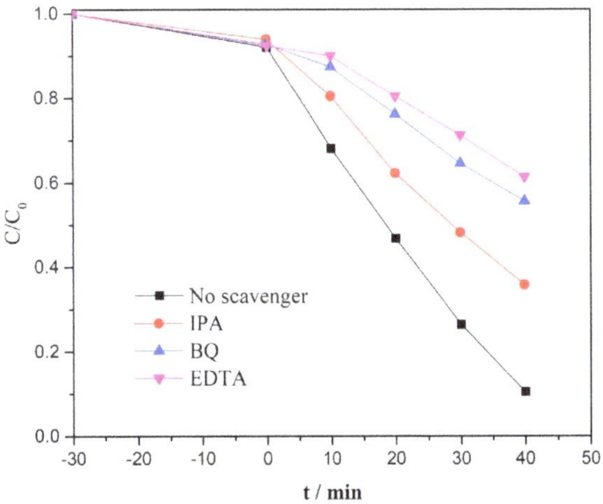

Figure 7. Effect of different scavengers on RhB degradation by TCN/Ag_3PO_4−5% heterojunction under visible−light irradiation.

The photocatalytic efficiency of RhB decreased significantly after the addition of EDTA and BQ, and the degradation rate of RhB was 39% and 45% after 40 min reaction, respectively. Results showed that h + and $•O_2^-$ were captured by EDTA and BQ, respectively, so the number of h + and $•O_2^-$ in the photocatalytic reaction was greatly reduced, resulting in a lower degradation rate of RhB. When IPA was added, the degradation rate of RhB was 80% after 40 min reaction. The IPA mainly captured •OH, thus slightly decreasing the degradation rate of RhB. It is clear that h^+ and $•O_2^-$ were the main active species for RhB degradation, while •OH only played a minor role. According to previous reports [42,43], the valence band (VB) and conduction band (CB) potentials of Ag_3PO_4 were about 2.69 and 0.24 eV, which are both more positive than those of g-C_3N_4 (−1.15 and 1.60 eV), respectively. According to traditional heterojunction theory, the CB of g-C_3N_4 is more negative than that of Ag_3PO_4, and the electrons generated after photoexcitation migrate from the CB of g-C_3N_4 to the CB of Ag_3PO_4. The VB of Ag_3PO_4 is more positive than that of g-C_3N_4, and holes generated after photoexcitation migrate from the VB of Ag_3PO_4 to the VB of g-C_3N_4. $•O_2^-/O_2$ potential (–0.33 eV) is more negative than that CB of Ag_3PO_4, so electrons migrating to the CB of Ag_3PO_4 cannot react with oxygen to produce $•O_2^-$. This is not consistent with the results of photocapture experiments in which $•O_2^-$ is the main active species. A large number of electrons gathering on the crystal surface of Ag_3PO_4 lead to the photocurrent corrosion of Ag_3PO_4. This is inconsistent with the experimental results of the stability of the TCN/Ag_3PO_4 heterojunction. Second, because the potential of the VB of g-C_3N_4 was more negative than the potential of •OH/H_2O (2.27 eV), the holes that migrated to CB of g-C_3N_4 could not oxidize water to produce •OH. This is not consistent with the results of photocapture experiments in which •OH is also an active species. According to the experimental results of the stability and capture of the TCN/Ag_3PO_4 heterojunction, a possible Z-type photocatalytic mechanism is proposed, as shown in Figure 8. The potential position of the VB of g-C_3N_4 was close to the CB of Ag_3PO_4, so it was easy to combine the holes in the VB and the electrons in the CB. Therefore, the electrons in the VB of g-C_3N_4 and the holes in the VB of Ag_3PO_4 were effectively separated. The electrons in the CB of g-C_3N_4 would react with the dissolved O_2 to produce $•O_2^-$, which could directly oxidize RhB. Meanwhile, the holes in the VB of Ag_3PO_4 can also directly oxidize RhB. Its high valence band potential of +2.69 eV could oxidize water to produce •OH, and •OH also had a strong oxidation capacity and can oxidize RhB. The Z-type mechanism is consistent with the experimental results of the photostabilization and photocapture of the heterojunction catalyst.

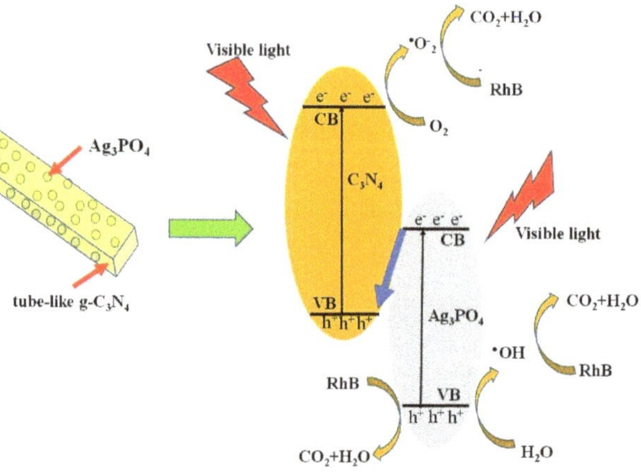

Figure 8. Photocatalytic degradation RhB and charge transfer mechanism in TCN/Ag_3PO_4 heterojunction.

4. Conclusions

A TCN/Ag_3PO_4 heterojunction was successfully prepared by chemical precipitation. The TCN/Ag_3PO_4 heterojunction catalyst exhibited excellent catalytic activity and stability under visible-light irradiation. A possible Z-type mechanism was proposed in which the Ag_3PO_4-excited electrons were directly recombined with holes generated by the tubelike g-C_3N_4, thus rapidly separating photogenerated electrons from the Ag_3PO_4 crystal surface. The TCN/Ag_3PO_4 heterojunction structure both reduced the photoelectronic corrosion of Ag_3PO_4 and improved the photocatalytic activity of the catalyst. As a photocatalyst with high activity and stability, the TCN/Ag_3PO_4 heterojunction has potential for applications in the field of environmental pollution control.

Author Contributions: Data curation, X.Y. and Y.W.; formal analysis, Z.L.; methodology, X.Y. and B.K.; writing—original draft, B.K. and Y.W.; writing—review and editing, X.Y. and Y.N. All authors have read and agreed to the published version of the manuscript.

Funding: This research was funded by the International Project on Scientific and Technological Cooperation in Shaanxi Province of China (no. 2018KW-052), and Research Funds for the Central Universities, CHD (nos. 300102311404 and 300102310301).

Institutional Review Board Statement: Not applicable.

Informed Consent Statement: Not applicable.

Data Availability Statement: Not applicable.

Acknowledgments: The authors thank the International Project on Scientific and Technological Cooperation in Shaanxi Province of China (no. 2018KW-052), and Research Funds for the Central Universities, CHD (nos. 300102311404 and 300102310301) for the financial support.

Conflicts of Interest: The authors declare no conflict of interest.

References

1. Zhou, Q.; Gong, W.; Xie, C.; Yang, D.; Ling, X.; Yuan, X.; Chen, S.; Liu, X. Removal of neutral red from aqueous solution by adsorption on spent cottonseed hull substrate. *J. Hazard. Mater.* **2011**, *185*, 502–506. [CrossRef]
2. Yaseen, D.A.; Scholz, M. Textile dye wastewater characteristics and constituents of synthetic effluents: A critical review. *Int. J. Environ. Sci. Technol.* **2019**, *16*, 1193–1226. [CrossRef]
3. Rueda-Marquez, J.J.; Levchuk, I.; Ibañez, P.F.; Sillanpää, M. A critical review on application of photocatalysis for toxicity reduction of real wastewaters. *J. Clean. Prod.* **2020**, *258*, 120694. [CrossRef]
4. Moradihamedani, P. Recent advances in dye removal from wastewater by membrane technology: A review. *Polym. Bull.* **2021**, *78*, 1–29. [CrossRef]
5. Hoffman, M.R.; Martin, S.T.; Choi, W.; Bahnemann, D.W. Environmental applications of semiconductor photocatalysis. *Chem. Rev.* **1995**, *95*, 69–96. [CrossRef]
6. Chong, M.N.; Jin, B.; Chow, C.W.K.; Saint, C. Recent developments in photocatalytic water treatment technology: A review. *Water. Res.* **2010**, *44*, 2997–3027. [CrossRef]
7. Lu, D.Z.; Zhao, B.; Fang, P.F.; Zhai, S.B.; Li, D.L.; Chen, Z.Q.; Wu, W.H.; Chai, W.Q.; Wu, Y.C.; Qi, N. Facile one-pot fabrication and high photocatalytic performance of vanadium doped TiO_2-based nanosheets for visible-light-driven degradation of RhB or Cr(VI). *Appl. Surf. Sci.* **2015**, *359*, 435–448. [CrossRef]
8. Tian, C.; Zhang, Q.; Wu, A.; Jiang, M.; Liang, Z.; Jiang, B.; Fu, H. Cost-effective large-scale synthesis of ZnO photocatalyst with excellent performance for dye photodegradation. *Chem. Commun.* **2012**, *48*, 2858–2860. [CrossRef] [PubMed]
9. Wang, F.C.; Zhao, J.C.; Wang, X.M.; Mai, B.X.; Sheng, G.Y.; Peng, P.A.; Fu, J.M. Preparation, characterization and photocatalytic activity of nano-sized ZnO/SnO_2 coupled photocatalysts. *Appl. Catal. B Environ.* **2002**, *39*, 269–279.
10. Wang, C.; Li, Y.; Huang, L.; Yang, L.; Wang, H.; Liu, J.; Liu, J.; Song, Z.; Huang, L. Enhanced photocatalytic antibacterial and degradation performance by n-p type 0D/2D SnO_{2-x}/BiOI photocatalyst under LED Light. *Chem. Eng. J.* **2021**, *411*, 128505. [CrossRef]
11. Yi, Z.; Ye, J.; Kikugawa, N.; Kako, T.; Ouyang, S.; Stuart-Williams, H.; Yang, H.; Cao, J.; Luo, W.; Li, Z. An orthophosphate semiconductor with photooxidation properties under visible-light irradiation. *Nat. Mater.* **2010**, *9*, 559–564. [CrossRef]
12. Guan, X.; Shi, J.; Guo, L. Ag_3PO_4 photocatalyst: Hydrothermal preparation and enhanced O_2 evolution under visible-light irradiation. *Inter. J. Hydrogen. Energy.* **2013**, *38*, 11870–11877. [CrossRef]
13. Petala, A.; Spyrou, D.; Frontistis, Z.; Mantzavinos, D.; Kondarides, D.I. Immobilized Ag_3PO_4 photocatalyst for micro-pollutants removal in a continuous flow annular photoreactor. *Catal. Today.* **2019**, *328*, 223–229. [CrossRef]

14. Osman, N.S.; Sulaiman, S.N.; Muhamad, E.N.; Mukhair, H.; Tan, S.T.; Abdullah, A.H. Synthesis of an Ag_3PO_4/Nb_2O_5 photocatalyst for the degradation of dye. *Catalysts* **2021**, *11*, 458. [CrossRef]
15. Li, D.; Wang, W.; Shang, X.; Tang, H.; Zulfiqar, S. Solar-driven photocatalytic water oxidation of Ag_3PO_4/CNTs@$MoSe_2$ ternary composite photocatalyst. *Appl. Surf. Sci.* **2019**, *505*, 144613. [CrossRef]
16. Teng, F.; Liu, Z.; Zhang, A.; Min, L. Photocatalytic performances of Ag_3PO_4 polypods for degradation of dye pollutant under natural indoor weak light irradiation. *Environ. Sci. Technol.* **2015**, *49*, 9489–9494. [CrossRef] [PubMed]
17. Ma, Y.; Li, J.; Jin, Y.; Gao, K.; Cai, H.; Ou, G. The enhancement mechanism of ultra-active Ag_3PO_4 modified by tungsten and the effective degradation towards phenolic pollutants. *Chemosphere* **2021**, *285*, 131440. [CrossRef]
18. Najafabadi, M.N.; Ghanbari, H.; Naghizadeh, R. Graphene/silver-based composites and coating on dead coral for degradation of organic pollution using the z-scheme mechanism. *RSC Adv.* **2021**, *11*, 19890–19901. [CrossRef]
19. Muthukumar, P.; Alex, V.; Pannipara, M.; Al-Sehemi, A.G.; Anthony, S.P. Fabricating highly efficient Ag_3PO_4-Fe_3O_4-GO ternary nanocomposite photocatalyst: Effect of Fe_3O_4-GO preparation methods on photocatalytic activity. *Mater. Res. Bull.* **2021**, *141*, 111337. [CrossRef]
20. Yan, X.; Hui, X.-Y.; Gao, Q.; Yu, G.-J.; Mo, Y.-C.; Ye, Z.-M.; Li, J.-C.; Ma, Z.-Y.; Sun, G.-D. Synthesis and visible light photocatalytic performance of Ag_3PO_4/MoS_2 nanosheets composite photocatalyst. *Chinese J. Inorg. Chem.* **2017**, *33*, 1782–1788.
21. Li, N.; Chen, J.; Chen, X.; Lai, Y.; Yu, C.; Yao, L.; Liang, Y. Novel visible-light-driven $SrCoO_3/Ag_3PO_4$ heterojunction with enhanced photocatalytic performance for tetracycline degradation. *Environ. Sci. Pollut. Res.* **2021**. [CrossRef] [PubMed]
22. Lyu, Y.-H.; Wei, F.; Zhang, T.; Luo, L.; Pan, Y.; Yang, X.; Yu, H.; Zhou, S. Different antibacterial effect of Ag_3PO_4/TiO_2 heterojunctions and the TiO_2 polymorphs. *J. Alloy. Compd.* **2021**, *876*, 160016. [CrossRef]
23. Zhang, L.; Zhang, H.; Huang, H.; Liu, Y.; Kang, Z. Ag_3PO_4/SnO_2 semiconductor nanocomposites with enhanced photocatalytic activity and stability. *New J. Chem.* **2012**, *36*, 1541–1544. [CrossRef]
24. Zhang, F.-J.; Xie, F.-Z.; Zhu, S.-F.; Liu, J.; Zhang, J.; Mei, S.-F.; Zhao, W. A novel photofunctional g-C_3N_4/Ag_3PO_4 bulk heterojunction for decolorization of RhB. *Chem. Eng. J.* **2013**, *228*, 435–441. [CrossRef]
25. Guan, X.; Guo, L. Cocatalytic effect of $SrTiO_3$ on Ag_3PO_4 toward enhanced photocatalytic water oxidation. *ACS Catal.* **2014**, *4*, 3020–3026. [CrossRef]
26. He, Y.M.; Zhang, L.H.; Teng, B.T.; Fan, M.H. New application of Z-Scheme Ag_3PO_4/g-C_3N_4 composite in converting CO_2 to fuel. *Environ. Sci. Technol.* **2015**, *49*, 649–656. [CrossRef]
27. Cui, X.; Tian, L.; Xian, X.; Tang, H.; Yang, X. Solar photocatalytic water oxidation over Ag_3PO_4/g-C_3N_4 composite materials mediated by metallic Ag and graphene. *Appl. Surf. Sci.* **2018**, *430*, 108–115. [CrossRef]
28. Tian, L.; Li, J.; Liang, F.; Wang, J.K.; Li, S.; Zhang, H.J.; Zhang, S.W. Molten salt synthesis of tetragonal carbon nitride hollow tubes and their application for removal of pollutants from wastewater. *Appl. Catal. B Environ.* **2018**, *225*, 307–313. [CrossRef]
29. Schwinghammer, K.; Tuffy, B.; Mesch, M.B.; Wirnhier, C.; Martineau, F.; Taulelle, W.; Lotsch, B.V. Triazine-based carbon nitrides for visible-light-driven hydrogen evolution. *Angew. Chem. Int. Ed.* **2013**, *125*, 2495–2499. [CrossRef]
30. Yan, X.; Ning, G.; Zhao, P. Enhanced visible light photocatalytic reduction of Cr(VI) over a novel square nanotube poly(triazine imide)/TiO_2 heterojunction. *Catalysts* **2019**, *9*, 55. [CrossRef]
31. Du, C.; Song, J.; Tan, S.; Yang, L.; Yu, G.; Chen, H.; Zhou, L.; Zhang, Z.; Zhang, Y.; Su, Y. Facile synthesis of z-scheme ZnO/Ag/Ag_3PO_4 composite photocatalysts with enhanced performance for the degradation of ciprofloxacin. *Mater. Chem. Phys.* **2021**, *260*, 124136. [CrossRef]
32. Ma, S.; Zhan, S.; Jia, Y.; Shi, Q.; Zhou, Q. Enhanced disinfection application of Ag-modified g-C_3N_4 composite under visible light. *Appl. Catal. B Environ.* **2016**, *186*, 77–87. [CrossRef]
33. Xie, Y.; Huang, Z.; Zhang, Z.; Zhang, X.; Wen, R.; Liu, Y.; Fang, M.; Wu, X. Controlled synthesis and photocatalytic properties of rhombic dodecahedral Ag_3PO_4 with high surface energy. *Appl. Surf. Sci.* **2016**, *389*, 56–66. [CrossRef]
34. Zhang, Y.; Xie, C.; Gu, F.L.; Wu, H.; Guo, Q. Significant visible-light photocatalytic enhancement in Rhodamine B degradation of silver orthophosphate via the hybridization of N-doped graphene and poly(3-hexylthiophene). *J. Hazard. Mater.* **2016**, *315*, 23–34. [CrossRef] [PubMed]
35. Miao, X.; Yue, X.; Ji, Z.; Shen, X.; Zhou, H.; Liu, M.; Xu, K.; Zhu, J.; Zhu, G.; Kong, L. Nitrogen-doped carbon dots decorated on g-C_3N_4/Ag_3PO_4 photocatalyst with improved visible light photocatalytic activity and mechanism insight. *Appl. Catal. B Environ.* **2018**, *227*, 459–469. [CrossRef]
36. Liu, J.J.; Fu, X.; Chen, S.F.; Zhu, Y.F. Electronic structure and optical properties of Ag_3PO_4 photocatalyst calculated by hybrid density functional method. *Appl. Phys. Lett.* **2011**, *99*, 191903. [CrossRef]
37. Yan, X.; Qin, J.; Ning, G.; Li, J.; Ai, T.; Su, X.; Wang, Z. A novel poly(triazine imide) hollow tube/ZnO heterojunction for tetracycline hydrochloride degradation under visible light irradiation. *Adv. Powder. Technol.* **2019**, *30*, 359–365. [CrossRef]
38. Chen, Z.; Bing, F.; Liu, Q.; Zhang, Z.G.; Fang, X.M. Novel z-scheme visible-light-driven Ag_3PO_4/Ag/SiC photocatalysts with enhanced photocatalytic activity. *J. Mater. Chem. A* **2015**, *3*, 4652–4658. [CrossRef]
39. Yan, X.; Li, J.; Zhou, H. Molten salts synthesis and visible light photocatalytic activity of crystalline poly(triazine imide) with different morphologies. *J. Mater. Sci. Mater. Electron.* **2019**, *30*, 11706–11713. [CrossRef]
40. Li, J.; Jin, L.; Liu, F.; Liu, X. Synthesis of cocoon-like Ag_3PO_4 and its high-performance in photocatalytic degradation of ciprofloxacin. *Mater. Lett.* **2019**, *242*, 139–142. [CrossRef]

41. Ding, M.; Zhou, J.; Yang, H.; Cao, R.; Zhang, S.; Shao, M.; Xu, X. Synthesis of Z-scheme g-C_3N_4 nanosheets/Ag_3PO_4 photocatalysts with enhanced visible-light photocatalytic performance for the degradation of tetracycline and dye. *Chinese Chem. Lett.* **2020**, *31*, 71–76. [CrossRef]
42. Yan, X.; Gao, Q.; Hui, X.; Yan, C.; Ai, T.; Wang, Z.; Sun, G.; Su, X.; Zhao, P. Fabrication of g-C_3N_4/MoS_2 nanosheet heterojunction by facile ball milling method and its visible light photocatalytic performance. *Rare. Metal. Mat. Eng.* **2018**, *47*, 3015–3020.
43. Sun, M.; Zeng, Q.; Zhao, X.; Shao, Y.; Ji, P.; Wang, C.; Yan, T.; Du, B. Fabrication of novel g-C_3N_4 nanocrystals decorated Ag_3PO_4 hybrids: Enhanced charge separation and excellent visible-light driven photocatalytic activity. *J. Hazard. Mater.* **2017**, *339*, 9–21. [CrossRef] [PubMed]

crystals MDPI

Communication

Synthesis and Characterization of New Guanine Complexes of Pt(IV) and Pd(II) by X-ray Diffraction and Hirshfeld Surface Analysis

Anton Petrovich Novikov, Mikhail Alexandrovich Volkov, Alexey Vladimirovich Safonov, Mikhail Semenovich Grigoriev and Evgeny Vladilenovich Abkhalimov *

Frumkin Institute of Physical Chemistry and Electrochemistry, Russian Academy of Sciences, Leninskii Prospect 31-4, 119071 Moscow, Russia; tony.novickoff@yandex.ru (A.P.N.); mendeleev93@yandex.ru (M.A.V.); alexeysafonof@gmail.com (A.V.S.); mickgrig@mail.ru (M.S.G.)
* Correspondence: abkhalimov@ipc.rssi.ru

Abstract: The aim of the work was to synthesize new perspective compounds of palladium and platinum with nitrogenous bases (guanine), promising for use in biomedicine and catalysis. The article describes the synthesis of new $[PdCl_2(HGua)_2]Cl_2 \cdot H_2O$ and $[PtCl_5(HGua)] \cdot 2H_2O$ compounds using wet chemistry methods. The structure of the obtained single crystals was established by the method of single crystal X-ray diffraction. The complexes have an M-N bond, and the organic ligand is included in the first coordination sphere. The analysis of Hirshfeld surfaces for the obtained complexes and their analogues for the analysis of intermolecular interactions was carried out. In the palladium complex we obtained, π-halogen and π-stacking interactions were found; in analogues, such interactions were not found. π-halogen and halogen interactions were found in structure of platinum complex and its analogues.

Keywords: X-ray diffraction analysis; Hirshfeld surface analysis; guanine; palladium; platinum; metal-nitrogen bond

Citation: Novikov, A.P.; Volkov, M.A.; Safonov, A.V.; Grigoriev, M.S.; Abkhalimov, E.V. Synthesis and Characterization of New Guanine Complexes of Pt(IV) and Pd(II) by X-ray Diffraction and Hirshfeld Surface Analysis. *Crystals* **2021**, *11*, 1417. https://doi.org/10.3390/cryst11111417

Academic Editors: Aivaras Kareiva, Loreta Tamasauskaite-Tamasiunaite and Aleksej Zarkov

Received: 20 October 2021
Accepted: 15 November 2021
Published: 19 November 2021

1. Introduction

The study of Pt and Pd complexes with various organic compounds is of great fundamental and practical interest. One of the areas of research is focused on the synthesis of nanoscale metal clusters [1] and nanoparticles with DNA using Pd and Pt complexes [2–4]. Such works have a wide range of applications, including biosensors, DNA visualization [5], and catalysis [2,6].

First of all, coordination compounds of platinum group metals occupy an important place in anticancer therapy [7]. Cisplatin is most successfully used, although serious side-effects (toxic effect, nonspecificity, development of resistance during treatment) stimulate the search for new coordination compounds of metals (palladium, ruthenium, etc.) [8]. The basis of the operation of cisplatin (cis-diamminedichloroplatinum (II)) lies in the alkylating effect of DNA due to the rapid replacement of chlorine in an aqueous environment. In coincident complexes of cisplatin, when replacing platinum with palladium, the lability of chlorine significantly decreases and for this reason palladium analogues to cisplatin are unusable in anti-cancer therapy. However, there is evidence of the successful use of complexes of paladium II). For example, Pd(II) complexes of dithiocarbamate [9] cis-Diamminedichloropalladium are promising in anticancer therapy, which are characterized by the much lower toxicity in comparison with platinum complexes for tissues [10]. In the work [11], the antitumoral activity of some palladium complexes of compounds against Sarcoma is described. As in the case of platinum, the interaction of palladium with DNA leads to a cytotoxic effect during the formation of inter and intrastrand cross-links due to complexation with purine and pyrimidine bases [12]. Nitrogen bases of DNA and other similar nucleobases are ambidentate (or ambivalent) ligands since they have two or more donor atoms capable of binding metal ions; therefore, Pd(II) complexes with various heterocycles (xanthine, theophylline, theobromine, 3,8-dimethylxanthine, caffeine and 1,3,8-trimethylxanthine), promising for use in medicine were synthetized and described earlier [13].

It is well known that the interaction of platinum and palladium with DNA occurs mainly on guanine-rich DNA fragments [14–18]. In this sense, guanine N(7) is the most commonly accepted atom for primary attachment of platinum compounds to DNA [19–21] because its N(9) position in guanine is unsubstituted. Some studies have demonstrated that N(9) of guanine is also the main site of binding interaction with metals [22]. Guanine is one of the major purine nucleobases present in DNA and RNA. RNA has considerably higher content of guanine. In general, purines are basic components of nucleotides in cell proliferation, because they provide the necessary energy and cofactors to promote cell survival and proliferation, thus purine metabolism is associated with the progression of cancer. High concentrations of purine metabolites have been indicated in tumor cells [23].

Metal complexes of unsubstituted guanine have been investigated by various groups [24–26]. Both purine and pyrimidine DNA derivatives belong to antimetabolites of nucleic metabolism, and their application as active carriers of alkylating agents is based on the biochemical data relating to the accumulation of purines and pyrimidines in a tumour cell. It is for this reason that platinum complexes with purine and pyrimidine bases and their derivatives are considered as potential antitumor agents [27]. In addition to medicine, it was proposed to use guanine palladium complexes in organic catalysis, including in the form of graphene deposited on nanoparticles [28,29].

Some previous works on the synthesis of palladium derivatives of guanine proposed to use guanine derivatives with ethylenediamine and diethylenetriamine complexes of palladium(II) [30]. In their research works, they obtained complexes of palladium and platinum with nucleoside fragments [31,32], but the complexes described in these works have different geometry [33] or X-ray data were not obtained or the structures were not described [34]. Thus, the aim of our work was to obtain Pd(II) and Pt(IV) compounds with unsubstituted guanine as a ligand and describe the crystal structure and intermolecular interactions by the method of Hirshfeld surfaces.

2. Materials and Methods

2.1. Materials

Complexes **1** and **2** were synthesized using hexachloroplatinic (IV) acid hexahydrate ($H_2PtCl_6{\cdot}6H_2O$, ~40% Pt), palladium (II) chloride ($PdCl_2$, ≥99%), guanine ($C_5H_5N_5O$, ≥98%) and hydrochloric acid (HCl, 36.5%), all being ACS Reagent Grade chemicals purchased from Sigma-Aldrich (Darmstadt, Germany) and used without additional purification. All solutions were prepared using distilled water additionally deionized with the Arium setup (Sartorius AG, Goettingen, Germany).

2.2. Synthesis Procedure of $[PdCl_2HGua_2]Cl_2{\cdot}H_2O$ (**1**)

To begin with, we dissolved 15 mg (0.1 mmol) of guanine in 15 mL of 1 mol L^{-1} hydrochloric acid aqueous solution in 25 mL two-necked flask equipped with a thermometer. To prepare H_2PdCl_4 acid, we dissolved 9 mg (0.05 mmol) of $PdCl_2$ in 1 mol L^{-1} hydrochloric acid aqueous solution. Then, both guanine and tetrachloropalladic (II) acid solutions were mixed at room temperature under stirring for 5 min at 23 °C. After that, we raised the temperature to 50 °C and continued stirring for 60 min. After turning off the stirring, the solution was transferred to a vacuum desiccator and left for 24 h over anhydrous $CaCl_2$. To the resulting brown-red crystals, 500 µL of anhydrous methanol were added. The resulting suspension was heated to 40 °C and cooled in air to room temperature; the resulting red crystals were analyzed by XRD.

2.3. Synthesis Procedure of $PtCl_5HGua{\cdot}2H_2O$ (**2**)

We dissolved ~10 mg of guanine in 5 mL of distilled water at room temperature in a 10 mL two-necked flask equipped with a thermometer. To this suspension we added 100 μL of 0.7 mol L^{-1} hexachloroplatinic acid aqueous solution. The mixture was stirred for 5 min at 23 °C, then the temperature was raised to 50 °C and held for 60 min. After turning off the stirring, the flask with the solution was transferred to a vacuum desiccator and left for 24 h over anhydrous $CaCl_2$. To the resulting yellow-green powder was added 500 μL of anhydrous methanol. The resulting suspension was heated to 40 °C and cooled in air to room temperature; the formed thin needles of light green color were analyzed by XRD.

2.4. Single-Crystal XRD Analysis

The crystal structure of all synthesized substances was determined by X-ray structural analysis using an automatic four-circle area-detector diffractometer Bruker KAPPA APEX II with MoKα radiation at a temperature of 100 K (for **1**) and 296 K (for **2**). The cell parameters were refined over the entire data set by using SAINT-Plus software [35]. Absorption corrections were introduced using the SADABS program [36]. The structures were solved using the SHELXT-2018/2 program [37] and refined by full-matrix least squares on F^2 in the anisotropic approximation for all non-hydrogen atoms (SHELXL-2018 [38]). The H atoms were objectively located from the difference Fourier synthesis and refined in an idealized geometry with isotropic temperature factors equal to $1.2U_{eq}$ (N, C) for NH- and CH-groups, and $1.5U_{eq}$ (O) for water molecules. Tables and pictures for structures were generated using Olex2 [39].

Crystal data, data collection, and structure refinement details are summarized in Table 1. All other crystallographic parameters of structures **1** and **2** are indicated in Tables S1–S8. The atomic coordinates were deposited at the Cambridge Crystallographic Data Centre [40], CCDC № 2116634 and 2116635 for **1** and **2**, respectively. The Supplementary crystallographic data can be obtained free of charge from the Cambridge Crystallographic Data Centre via www.ccdc.cam.ac.uk/data_request/cif (accessed on 14 November 2021).

Table 1. Crystal data and structure refinement for structure **1** and **2**.

Identification Code	1	2
Empirical formula	$C_{10}H_{14}Cl_4N_{10}O_3Pd$	$C_5H_{10}Cl_5N_5O_3Pt$
Formula weight	570.51	560.52
Temperature/K	100(2)	296(2)
Crystal system	Triclinic	Monoclinic
Space group	*P*-1	$P2_1/n$
a/Å	7.7566(4)	7.9869(2)
b/Å	10.7651(5)	15.8000(5)
c/Å	11.4193(5)	13.0000(3)
α/°	97.127(3)	90
β/°	99.974(3)	107.173(1)
γ/°	90.128(3)	90
Volume/Å^3	931.58(8)	1567.37(7)
Z	2	4
ρ_{calc}g/cm^3	2.034	2.375
μ/mm^{-1}	1.608	9.812
F(000)	564.0	1048.0
Crystal size/mm^3	0.9 × 0.6 × 0.11	0.4 × 0.1 × 0.08
Radiation	MoKα (λ = 0.71073)	MoKα (λ = 0.71073)
2 Θ range for data collection/	8.23 to 60	8.348 to 60
Index ranges	$-10 \le h \le 10, -15 \le k \le 15, -15 \le l \le 16$	$-10 \le h \le 11, -21 \le k \le 22, -17 \le l \le 18$
Reflections collected	18,831	14,606
Independent reflections	5402 [R_{int} = 0.0711, R_{sigma} = 0.0736]	4556 [R_{int} = 0.0263, R_{sigma} = 0.0286]
Data/restraints/parameters	5402/0/257	4556/3/178

Table 1. *Cont.*

Identification Code	1	2
Empirical formula	$C_{10}H_{14}Cl_4N_{10}O_3Pd$	$C_5H_{10}Cl_5N_5O_3Pt$
Goodness-of-fit on F^2	1.027	1.039
Final R indexes [$I \geq 2\sigma$ (I)]	$R_1 = 0.0477$, $wR_2 = 0.1168$	$R_1 = 0.0206$, $wR_2 = 0.0408$
Final R indexes [all data]	$R_1 = 0.0778$, $wR_2 = 0.1318$	$R_1 = 0.0293$, $wR_2 = 0.0432$
Largest diff. peak/hole/e Å^{-3}	1.30/−1.12	0.86/−0.97

3. Results and Discussions

3.1. Structural Description

Compound **1** crystallizes in the triclinic space group *P*-1, **2** in the monoclinic space group $P2_1/n$. The asymmetric fragments of **1** and **2** are shown in Figure 1. In both structures, the metal atom forms a M-N bond with the nitrogen atom of the five-membered ring of the guanine fragment. Also, in both structures, the nitrogen atom of the six-membered ring, which is in the *para*-position to the oxygen atom of the guanine molecule, is protonated. However, the molecule of complex **1** has a charge of +2, and the molecule of complex **2** is electroneutral. This can be used for selective delivery of the complex in organisms. As counter-ions in structure **1**, there are two chloride anions. In both structures, in addition to the molecules of the complex, there are molecules of crystallization water. There are one and two molecules of water in **1** and **2**, respectively.

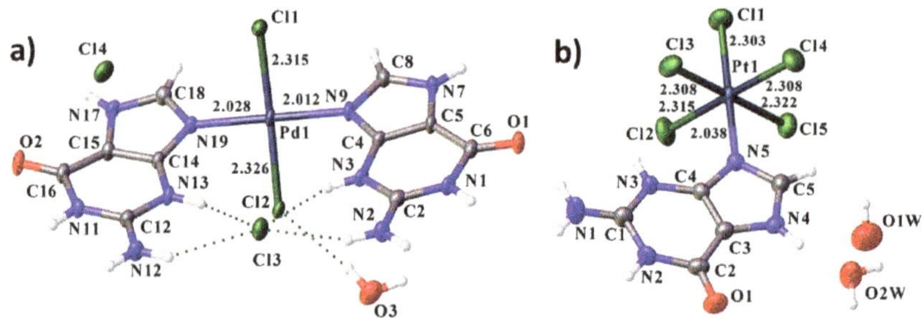

Figure 1. Asymmetric fragments of **1** (**a**) and **2** (**b**) with labeling of atoms and some bond lengths. Displacement ellipsoids are drawn at the 50% probability level.

The guanine fragments in structures **1** and **2** are plane. However, in structure **1** in the second ring, the atoms deviate from the plane more strongly than for the first ring in **2** and for the ring in **1** (maximum displacement of 0.128, 0.047 and 0.003 Å for C15, C2 in 1 and C3 in 2 respectively).

The planes of guanine fragments are significantly rotated relative to the plane perpendicular to the direction determined by the practically linear grouping Cl1-Pd1-Cl2. The torsion angles Cl2-Pd1-N19-C14 of 57.05° and Cl2-Pd1-N9-C4 of −60.32° differ greatly from 90° as in similar compounds [41–43]. However, in this case, the planes of both ligands almost coincide (angle (plane normal-plane normal): 3.000°), which is obviously due to H-bonding with the outer-sphere chloride ion.

The Pd-N distance varies from 2.012 to 2.028 Å. The average distance found with the data from the Cambridge Structural Database (CSD Version 5.42, update of September 2021) is 2.018 Å [40]. The Pt-N distance is slightly longer than Pd-N and is 2.038 Å, which is slightly shorter than the average distance found in the CSD, which is 2.046 Å. The Pd–Cl distances vary from 2.315 to 2.326 Å, which is longer than the average distance found in CSD for the structures with analogous coordination, which is 2.296 Å. The Pt–Cl distances change from 2.303 to 2.322 Å, which is close to 2.310 Å (average distance from CSD). The slight lengthening of bonds may be due to the different participation of chlorine atoms in hydrogen bonds.

In structures **1** and **2**, a complex system of hydrogen bonds is formed. The hydrogen bond parameters are listed in Tables S3 and S7. Crystal water plays an important role in the formation of a system of hydrogen bonds in both structures. In crystal **1**, bifurcate H-bonds are absent, but the chlorine atom Cl3 is linked by four hydrogen bonds to the amino groups of six-membered rings (Figure 2a). In **2**, there is a trifurcate hydrogen bond N3–H3A··· Cl2, N3–H3A··· Cl3, N3–H3A··· O15 with bond lengths 3.377 (2), 3.159 (3), 2.942 (3) Å, respectively. The strongest hydrogen bonds N11–H11··· O11 and N7–H7··· O22 in **1** are formed between nitrogen atoms of five- and six-membered rings and oxygen atoms of guanine fragments. In **2**, the strongest H-bonds are formed between water molecules (O2W··· O1W3 distance is 2.795 (4) Å), nitrogen molecule N4 and water molecule O2W (N4··· O2W distance is 2.746 (3)Å) and amino group and oxygen atom guanine fragment (N4··· O2W distance is 2.734 (4) Å).

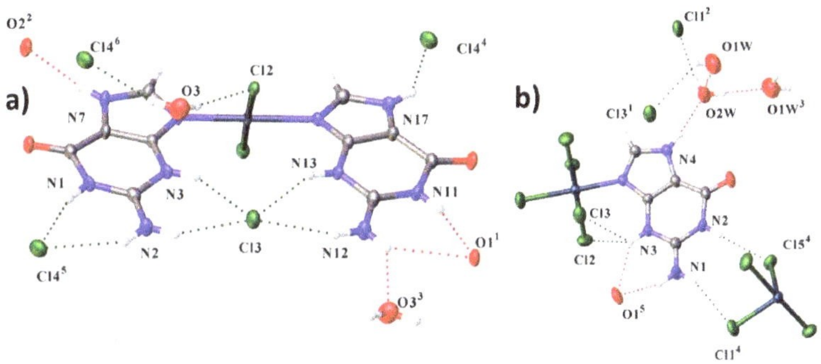

Figure 2. View showing hydrogen bonds in **1** (**a**) and in **2** (**b**). Symmetrical molecules of the complex are not shown in full for clarity.

Crystal packing in **1** can be represented as layered (Figure 3). In **2**, the crystal packing can be represented by columns (Figure 4). In both structures, hydrogen bonds form a three-dimensional framework and connect the layers and columns to each other. The layers are additionally connected by π-halogen interactions in **1**. The complex molecules in **2** in the columns are connected by halogen bonds (XB). In structure **2**, four water molecules form a cluster linked by H-bonds. There is π-staking interaction in **1** of six-membered ring (C14 N19 C18 N17 C15) with five-membered ring (C4 N9 C8 N7 C5) with angle: 3.543°, centroid-centroid distance: 3.855 Å and shift distance 1.851 Å, which additionally stabilizes the structure.

Previously, only a few compounds similar to ours were described. Complexes are described in the literature in which the Pd atom is bound by a Pd-N bond with an organic ligand [41–44]. But in almost all works, electrically neutral complexes are obtained. In our work, the complex has a charge of +2. Besides, the palladium atom binds to two guanine molecules. This fact presupposes the use of our complex as a selective delivery of palladium to various organs and tissues. We have found only one work with structure of the platinum atoms with molecules of guanine or 7-methylguanine [45]. However, in the work of Gupta et al., Pt binds not only to the guanine fragment, but also to the nitrogen atoms of the diethyleneamine fragments. Also, the guanine fragment in compound **2** is not protonated.

3.2. Hirshfeld Surface Analysis

Crystal Explorer 21 was used to calculate the Hirshfeld surfaces (HS) and two-dimensional fingerprint plots [46]. The donor-acceptor groups are visualized using a standard (high) surface resolution and d_{norm} surfaces mapped over a fixed color scale of −0.705 (red) to 1.266 (blue) for **1** and −0.665 to 1.650 a.u. for **2**, as illustrated in

Figures 5a and 6a, respectively. Additionally, the surface of shape-index was constructed, with bright spots on which short interactions are also visible (Figure 7).

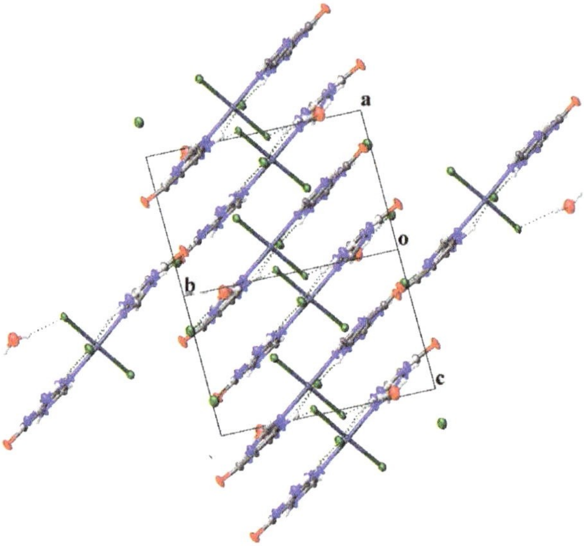

Figure 3. Packing diagram of **1**, view along the [101] direction. 0 indicates the origin of the unit cell and a, b, c—lattice translations.

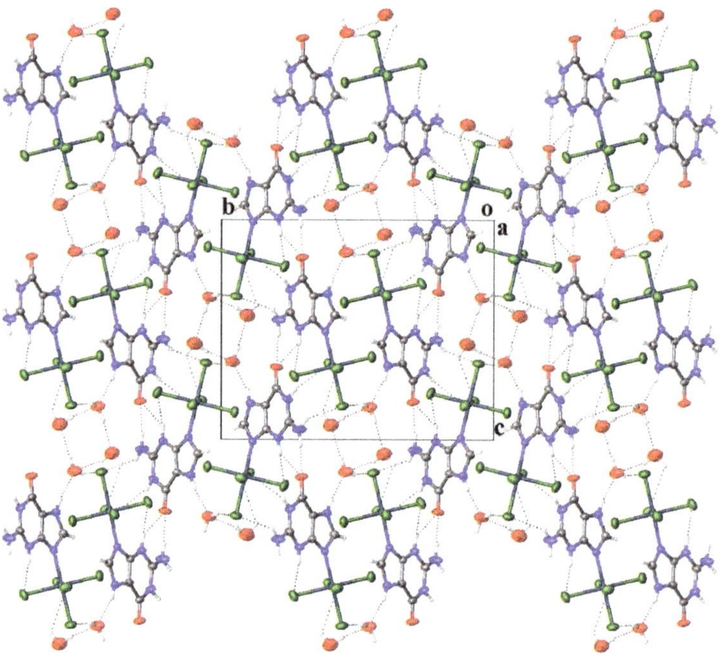

Figure 4. Packing diagram of **2**, view along the [100] direction. 0 indicates the origin of the unit cell and a, b, c—lattice translations.

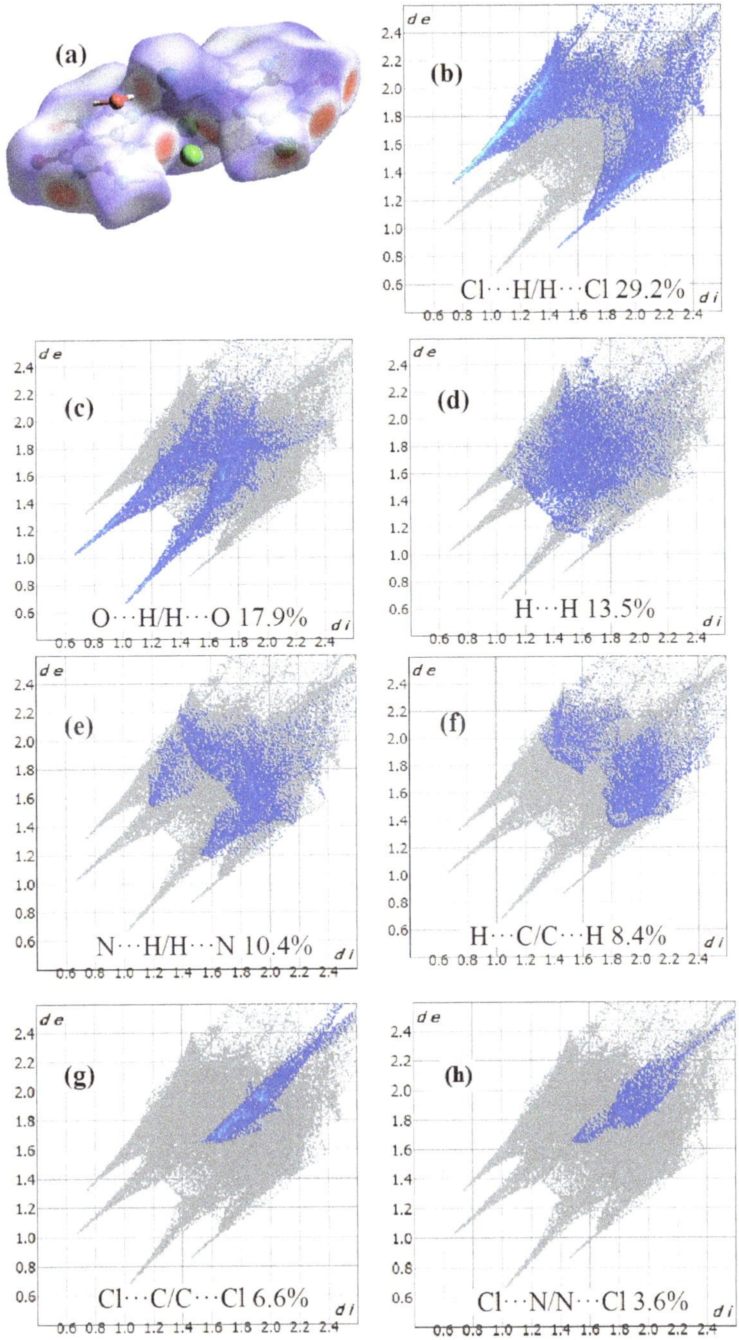

Figure 5. (**a**) Hirshfeld surface d_{norm} for complex **1** and (**b**) two-dimensional fingerprint plot for Cl···H/H···Cl, (**c**) O···H/H···O, (**d**) H···H, (**e**) N···H/H···N, (**f**) H···C/C···H, (**g**) Cl···C/C···Cl, (**h**) Cl···N/N···Cl contacts. Surface color scale: red (distances shorter than sum of vdW radii), through white to blue (distances longer than sum of vdW radii).

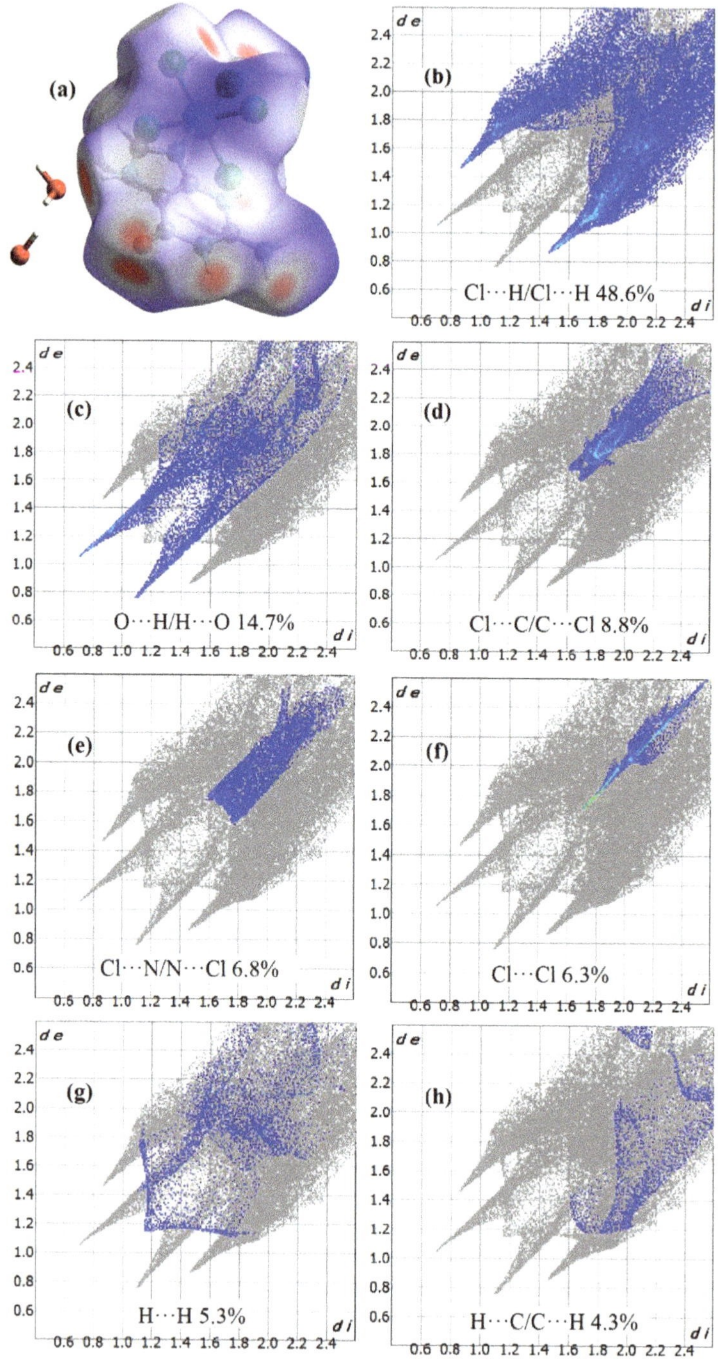

Figure 6. (**a**) Hirshfeld surface d_{norm} for complex **2** and (**b**) two-dimensional fingerprint plot for Cl···H/H···Cl, (**c**) O···H/H···O, (**d**) Cl···C/C···Cl, (**e**) Cl···N/N···Cl, (**f**) Cl···Cl, (**g**) H···H, (**h**) H···C/C···H contacts. Surface color scale: red (distances shorter than sum of vdW radii), through white to blue (distances longer than sum of vdW radii).

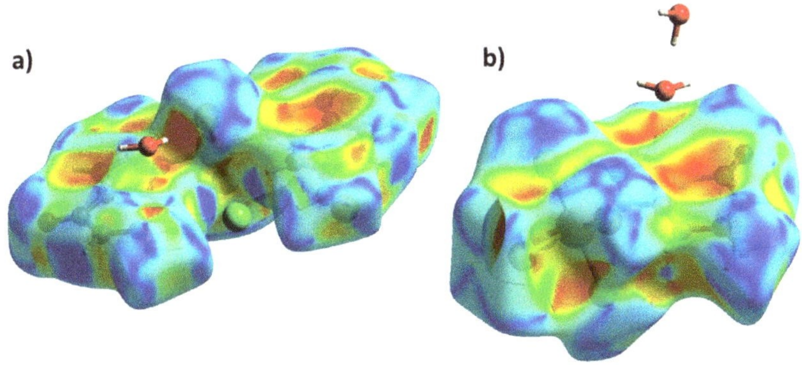

Figure 7. Shape-index of **1** (**a**) and **2** (**b**).

Red spots on the d_{norm} surface for both compounds correspond to hydrogen bonds. In compounds **1** and **2**, the main contribution to intermolecular interactions is made by hydrogen bonds of the Cl···H/H···Cl type (29.2% and 48.6%, respectively) and O···H/H···O type (17.9% and 14.7%, respectively). Such a difference in the contribution of contacts may be associated with a large percentage of chlorine atoms in complex **2** and with a different composition of the complexes. Van der Waals interactions H···H (13.6%), N···H/H···N (10.4%), and H···C/C···H (8.4%) play important role in crystal packing. Cl···C/C···Cl contacts in complexes **1** and **2** (8.8% and 6.6%, respectively) indicate the presence of π-halogen interaction. In the structure with a palladium atom, there are the π-anion interactions between the six-member ring and the chloride anions and chlorine atom of other complex, as described in [47,48]. This is additionally confirmed by short distances (plane N1C2N3C4C5C6 centroid-atom $Cl4^2$ (symmetry code: x, $1 + y$, z): 3.493 Å, plane N11C12N13C14C15C16 centroid-atom Cl4: 3.192 Å and plane N11C12N13C14C15C16 centroid-atom Cl^1 (symmetry code: $1 + x$, y, z): 3.484) (Figure 8). The angles α confirm this fact (67.07, 80.0, and 82.82°, respectively). In structure **2**, π-halogen interaction also occurs, but this time the ring of the guanine fragment with the chlorine atom $Cl4^3$ (plane C1N2C2C3C4N3 centroid-atom $Cl4^3$ (symmetry code: $1-x$, $1-y$, $1-z$) is 3.183 Å and α is 83.97°) [47,49–51] (Figure 9). The short contact Cl···Cl (6.3%) in structure **2** indicates the presence of a halogen bond. The short distance Cl3···Cl5 and the angle Pd-Cl···Cl equal to 163° confirms this fact [52]. Significant contribution to intermolecular interactions in **2** is made by contacts of the type Cl···N/N···Cl, H···H, H···C/C···H.

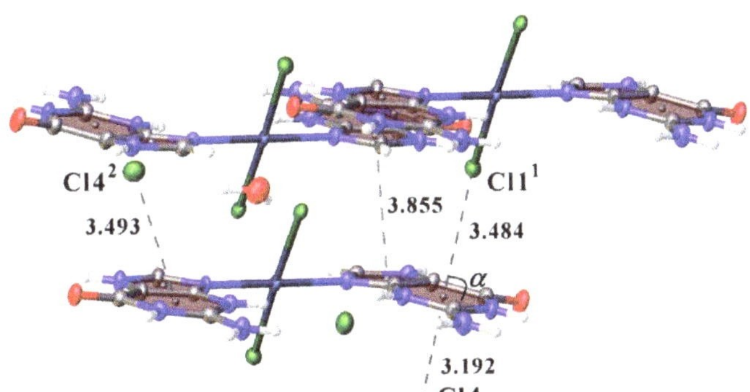

Figure 8. View showing π-stacking and π-halogen interactions with bond lengths between layers in the structure **1**.

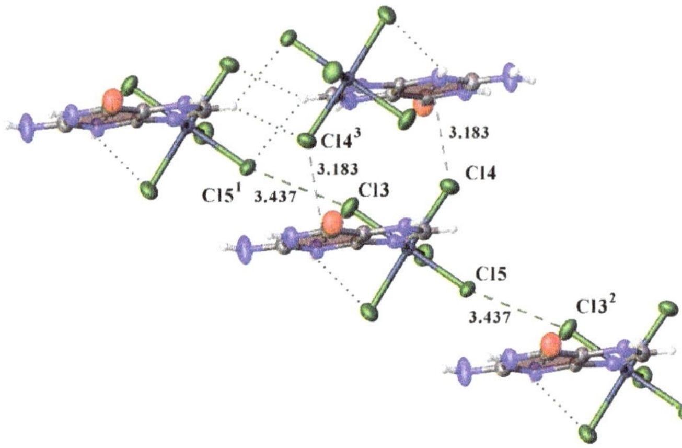

Figure 9. View showing π-halogen interactions and halogen bonds with bond lengths within and between columns respectively in **2**.

We have analyzed the HS for analogues of the title compounds. As an analogue for compound **1**, we found only a few compounds in which there is a Pd-N bond and in which the ligand is represented by five- and six-member rings. We have taken compounds with flat substituents. Moreover, all the analogues we found are electrically neutral. The analysis performed for analogues of the complex **1** showed that in neutral complexes the main contribution to intermolecular interactions is made by van der Waals interactions (Figure 10). Hydrogen bonds contribute less to intermolecular interactions than in **1**. In these compounds, there is no π-halogen interaction, as in our compound. This may be since in **1** the π-halogen interaction occurs predominantly between the chloride anion and the rings of the guanine fragments, while the analogues we have found are electrically neutral. It can be noted that with lengthening of the substituent, the proportion of H···H interactions increases.

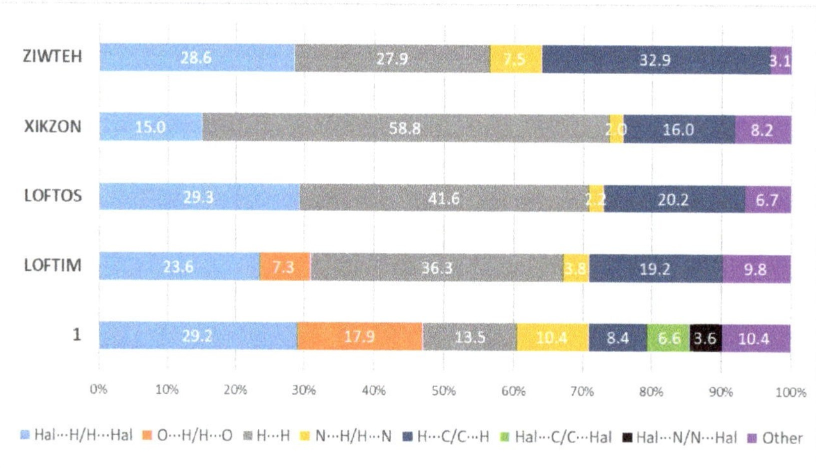

Figure 10. Percentage contributions of contacts to the Hirschfeld surface in structures **1** and taken from CSD. LOFTIM: bis(1H-benzimidazole)-(dibromo)-palladium(II) dimethyl sulfoxide solvate [53]; LOFTOS: bis(iodo)-bis(2-methyl-1H-benzimidazole)-palladium(II) [53]; XIKZON: Dichlorido[bis(2-hexylsulfanyl)-1H-benzimidazole-κN^3]palladium(II) [54]; ZIWTEH: trans-Dichlorido-bis(1H-benzimidazole)-palladium(II) [55].

As an analogue for the HS analysis, we found only two structures with similar geometry (Pt-N bond with the benzoimidazole ligand, Pt(IV) bonded to five halogen atoms) for **2**. Moreover, the CSD contains data only on pentachloroplatinates structures; for other halogens, such structures are absent. The complex in ASITIF is electrically neutral, and in EBECEV it is negatively charged, but in these complexes the charge does not affect strongly the HS. A similar type of packing is formed in all structures, and hydrogen bonds play an important role in the formation of crystal packing (Figure 11). All compounds also contain Cl···C/C···Cl and Cl···N/N···Cl contacts. This shows that, as in our complex, in similar structures, the molecules of the complexes are additionally bound by the π-halogen interaction. Short Cl···Cl contacts show that in similar structures, a halogen bonding is present. It should also be noted that the fraction of van der Waals interactions in similar structures is higher, which may be due to the absence of crystallization water molecules in the structure.

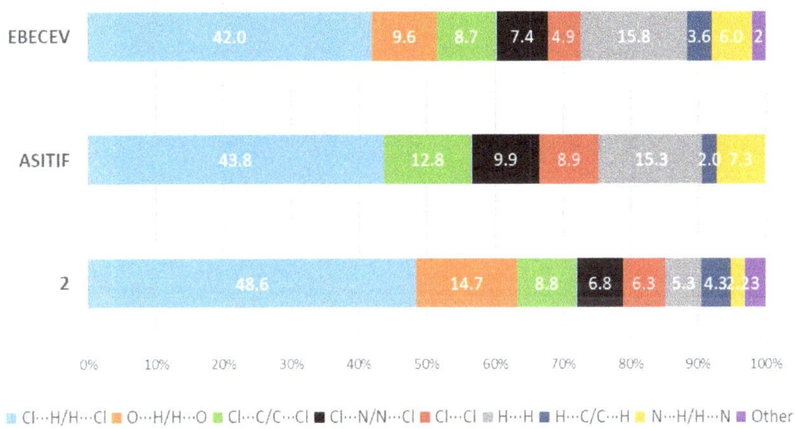

Figure 11. Percentage contributions of contacts to the Hirschfeld surface in structures **2** and taken from CSD. ASITIF: Pentachloro-(9-methyladeninio-N)-platinum(IV) [32]; N-Methylmethanaminium pentachl ro-(5-methyl[1,2,4]triazolo[1,5-a]pyrimidin-7(4H)-one)-platinum(IV) [56].

In this work, complexes of palladium and platinum with guanine were obtained for the first time. When analyzing the literature, we found only a few similar Pt or Pd compounds with M-N bonds with imidazole rings or molecules of nitrogenous bases. In both complexes, a system of hydrogen bonds is formed, which additionally stabilizes the structures in the crystals. The analysis of non-bonded interactions carried out in this work showed their different nature in two complexes under consideration. In structure **1**, the molecules of the complex are linked by both π-stacking interaction and a π-halogen bonding. In **2**, the molecules of the complex are additionally linked to HB by a π-halogen and halogen bonding.

The detailed analysis of non-bonded interactions carried out in this work may be of interest in the context of self-assembly [57]. In the future, when replacing, for example, chlorine atoms with bromine atoms, we can assume that there will be an increase in non-bonded interactions and a better association of complexes in the crystal and in solution. And replacing chlorine with fluorine, on the contrary, can lead to a different type of packing. The analysis of the influence of halogen atoms on the character of interactions was carried out in [58]. It has been shown that the strength of the XB donor increases in the following order as the XB donor capacity increases: F < Cl <Br < I.

Analysis of non-bonded interactions and H-bond systems can also predict the behavior of complexes in solutions. Such an analysis may be of interest from the point of view of predicting the formation of solvates or micelles in the process of creating targeted dosage forms, as well as in modeling receptor activity in silico. At a certain concentration of the solution, these complexes will most likely form aggregates due to non-bonded interactions, but in this case, they will interact with solvent molecules through a system of H-bonds. Platinum and palladium do not participate in intermolecular interactions and are completely screened by their coordination spheres, which suggests that compounds **1** and **2** are resistant to hydrolytic processes.

4. Conclusions

In this work, we obtained the compounds of four-coordinated palladium(II) and six-coordinated platinum(IV) with guanine entering the coordination sphere through the M-N bond for the first time. The compounds were synthesized from acidic solutions, which resulted in additional protonation of the formed complexes at the nitrogen atoms N3. The resulting palladium complex is a doubly charged cation, and the platinum compound is an electrically neutral molecule. The noble metals complexes usage may be interesting for medical purposes. The main intermolecular interactions are shown by the Hirshfeld surface method, and the significant contribution of hydrogen bonds to the crystal packing is proved. π-halogen, π-stacking, and halogen interactions were found in the obtained compounds and compared with their analogues.

Supplementary Materials: The following are available online at https://www.mdpi.com/article/10.3390/cryst11111417/s1, Table S1: Bond Lengths for **1**, Table S2: Bond Angles for **1**, Table S3: Hydrogen Bonds for **1**, Table S4: Torsion Angles for **1**, Table S5: Bond Lengths for **2**, Table S6: Bond Angles for **2**, Table S7: Hydrogen Bonds for **2**, Table S8: Torsion Angles for **2**.

Author Contributions: Conceptualization, E.V.A. and A.V.S.; methodology, M.A.V. and E.V.A.; validation, M.S.G.; investigation, E.V.A., A.P.N., M.S.G. and M.A.V.; writing—original draft preparation, A.P.N., A.V.S. and M.S.G.; writing—review and editing, E.V.A.; visualization, A.P.N.; supervision, M.S.G.; project administration, E.V.A. All authors have read and agreed to the published version of the manuscript.

Funding: The study was supported by the Ministry of Science and Higher Education of the Russian Federation (program no. AAAA-A18-118040590105-4).

Institutional Review Board Statement: Not applicable.

Informed Consent Statement: Not applicable.

Acknowledgments: X-ray diffraction experiments were performed at the Center for Shared Use of Physical Methods of Investigation at the Frumkin Institute of Physical Chemistry and Electrochemistry, RAS.

Conflicts of Interest: The authors declare no conflict of interest.

References

1. Richter, J.; Seidel, R.; Kirsch, R.; Mertig, M.; Pompe, W.; Plaschke, J.; Schackert, H.K. Nanoscale Palladium Metallization of DNA. *Adv. Mater.* **2000**, *12*, 507–510. [CrossRef]
2. Van Nguyen, K.; Minteer, S.D. DNA-functionalized Pt nanoparticles as catalysts for chemically powered micromotors: Toward signal-on motion-based DNA biosensor. *Chem. Commun.* **2015**, *51*, 4782–4784. [CrossRef]
3. Nguyen, K.; Monteverde, M.; Filoramo, A.; Goux-Capes, L.; Lyonnais, S.; Jegou, P.; Viel, P.; Goffman, M.; Bourgoin, J.-P. Synthesis of Thin and Highly Conductive DNA-Based Palladium Nanowires. *Adv. Mater.* **2008**, *20*, 1099–1104. [CrossRef]
4. Ford, W.E.; Harnack, O.; Yasuda, A.; Wessels, J.M. Platinated DNA as Precursors to Templated Chains of Metal Nanoparticles. *Adv. Mater.* **2001**, *13*, 1793–1797. [CrossRef]
5. Pekarik, V.; Peskova, M.; Duben, J.; Remes, M.; Heger, Z. Direct fluorogenic detection of palladium and platinum organometallic complexes with proteins and nucleic acids in polyacrylamide gels. *Sci. Rep.* **2020**, *10*, 12344. [CrossRef]
6. Miller, M.A.; Askevold, B.; Mikula, H.; Kohler, R.H.; Pirovich, D.; Weissleder, R. Nano-palladium is a cellular catalyst for in vivo chemistry. *Nat. Commun.* **2017**, *8*, 15906. [CrossRef]

7. Fong, C.W. Platinum anti-cancer drugs: Free radical mechanism of Pt-DNA adduct formation and anti-neoplastic effect. *Free Radic. Biol. Med.* **2016**, *95*, 216–229. [CrossRef]
8. Clarke, M.J.; Zhu, F.; Frasca, D.R. Non-Platinum Chemotherapeutic Metallopharmaceuticals. *Chem. Rev.* **1999**, *99*, 2511–2533. [CrossRef]
9. Hadizadeh, S.; Najafzadeh, N.; Mazani, M.; Amani, M.; Mansouri-Torshizi, H.; Niapour, A. Cytotoxic Effects of Newly Synthesized Palladium(II) Complexes of Diethyldithiocarbamate on Gastrointestinal Cancer Cell Lines. *Biochem. Res. Int.* **2014**, *2014*, 813457. [CrossRef]
10. Hill, G.A.; Forde, G.; Gorb, L.; Leszczynski, J. cis-Diamminedichloropalladium and its interaction with guanine and guanine-cytosine base pair. *Int. J. Quantum Chem.* **2002**, *90*, 1121–1128. [CrossRef]
11. Butour, J.L.; Wimmer, S.; Wimmer, F.; Castan, P. Palladium(II) compounds with potential antitumour properties and their platinum analogues: A comparative study of the reaction of some orotic acid derivatives with DNA in vitro. *Chem. Biol. Interact.* **1997**, *104*, 165–178. [CrossRef]
12. Tikhomirov, A.G.; Ivanova, N.A.; Erofeeva, O.S.; Gorbacheva, L.B.; Efimenko, I.A. Interaction of Palladium(II) Acido Complexes with DNA. *Russ. J. Coord. Chem. 2003 297* **2003**, *29*, 489–493. [CrossRef]
13. Tewari, B.B. Critical Reviews: Spectroscopic Studies On Palladium (Ii)-Complexes With Xanthine And Its Derivatives At Normal And High External Pressure Part-I: Spectroscopic Studies On Palladium (Ii)-Complexes With Xanthine And Its Derivatives At Normal Pressure. *Boliv. J. Chem.* **2013**, *30*, 122–130.
14. Sigel, H. (Ed.) *Metal Ions in Biological Systems Volume 42: Metal Complexes in Tumor Diagnosis and as Anticancer Agents*, 1st ed.; CRC Press: Boca Raton, FL, USA, 2004; ISBN 9780824754945.
15. Stone, P.J.; Kelman, A.D.; Sinex, F.M. Specific binding of antitumour drug cis-Pt(NH3)2Cl2 to DNA rich in guanine and cytosine. *Nature* **1974**, *251*, 736–737. [CrossRef]
16. Cohen, G.L.; Ledner, J.A.; Bauer, W.R.; Ushay, H.M.; Caravana, C.; Lippard, S.J. Sequence dependent binding of cis-dichlorodiammineplatinum(II) to DNA. *J. Am. Chem. Soc.* **1980**, *102*, 2487–2488. [CrossRef]
17. Tullius, T.D.; Lippard, S.J. cis-Diamminedichloroplatinum(II) binds in a unique manner to oligo(dG).oligo(dC) sequences in DNA—a new assay using exonuclease III. *J. Am. Chem. Soc.* **1981**, *103*, 4620–4622. [CrossRef]
18. Galea, A.M.; Murray, V. The Anti-tumour Agent, Cisplatin, and its Clinically Ineffective Isomer, Transplatin, Produce Unique Gene Expression Profiles in Human Cells. *Cancer Inform.* **2008**, *6*, CIN.S802. [CrossRef]
19. Kistenmacher, T.J.; Orbell, J.D.; Marzilli, L.G. Conformational Properties of Purine and Pyrimidine Complexes of cis-Platinum. In *Platinum, Gold, and Other Metal Chemotherapeutic Agents*; ACS Symposium Series; American Chemical Society: Washington, DC, USA, 1983; Volume 209, pp. 191–207. ISBN 9780841207585.
20. Zeng, W.; Zhang, Y.; Zheng, W.; Luo, Q.; Han, J.; Liu, J.; Zhao, Y.; Jia, F.; Wu, K.; Wang, F. Discovery of Cisplatin Binding to Thymine and Cytosine on a Single-Stranded Oligodeoxynucleotide by High Resolution FT-ICR Mass Spectrometry. *Molecules* **2019**, *24*, 1852. [CrossRef]
21. Baik, M.H.; Friesner, R.A.; Lippard, S.J. Theoretical Study of Cisplatin Binding to Purine Bases: Why Does Cisplatin Prefer Guanine over Adenine? *J. Am. Chem. Soc.* **2003**, *125*, 14082–14092. [CrossRef]
22. Pezzano, H.; Podo, F. Structure of binary complexes of mono- and polynucleotides with metal ions of the first transition group. *Chem. Rev.* **1980**, *80*, 365–401. [CrossRef]
23. Siddiqui, A.; Ceppi, P. A non-proliferative role of pyrimidine metabolism in cancer. *Mol. Metab.* **2020**, *35*. [CrossRef]
24. Cochran, K.; Forde, G.; Hill, G.A.; Gorb, L.; Leszczynski, J. cis-Diamminodichloronickel and Its Interaction with Guanine and Guanine–Cytosine Base Pair. *Struct. Chem.* **2002**, *13*, 133–140. [CrossRef]
25. Mikulski, C.M.; Mattucci, L.; Smith, Y.; Tran, T.B.; Karayannis, N.M. Guanine complexes with first row transition metal perchlorates. *Inorg. Chim. Acta* **1983**, *80*, 127–133. [CrossRef]
26. Savchenkov, A.; Demina, L.; Safonov, A.; Grigoriev, M.; Solovov, R.; Abkhalimov, E. Syntheses and crystal structures of new aurate salts of adenine or guanine nucleobases. *Acta Crystallogr. Sect. C Struct. Chem.* **2020**, *76*, 139–147. [CrossRef]
27. Efimenko, I.A.; Kurbakova, A.P.; Matović, Z.D.; Ponticelli, G. Synthesis and structure of palladium(II) mixed complexes with DNA purine or pyrimidine bases and imidazole derivatives. Part I. *Transit. Met. Chem.* **1994**, *19*, 539–541. [CrossRef]
28. Trost, B.M.; Madsen, R.; Guile, S.D.; Brown, B. Palladium-Catalyzed Enantioselective Synthesis of Carbanucleosides. *J. Am. Chem. Soc.* **2000**, *122*, 5947–5956. [CrossRef]
29. Ng, J.C.; Tan, C.Y.; Ong, B.H.; Matsuda, A.; Basirun, W.J.; Tan, W.K.; Singh, R.; Yap, B.K. Novel palladium-guanine-reduced graphene oxide nanocomposite as efficient electrocatalyst for methanol oxidation reaction. *Mater. Res. Bull.* **2019**, *112*, 213–220. [CrossRef]
30. Uchida, K.; Toyama, A.; Tamura, Y.; Sugimura, M.; Mitsumori, F.; Furukawa, Y.; Takeuchi, H.; Harada, I. Interactions of guanine derivatives with ethylenediamine and diethylenetriamine complexes of palladium(II) in solution: Pd binding sites of the guanine ring and formation of a cyclic adduct, [{Pd(en)(guanine ring)}4]. *Inorg. Chem.* **1989**, *28*, 2067–2073. [CrossRef]
31. Hadjiliadis, N. Pt(II) and Pd(II) interactions with nucleosides-binding sites-new compounds. *Inorg. Chim. Acta* **2016**, *452*, 279–284. [CrossRef]
32. Gaballa, A.; Schmidt, H.; Hempel, G.; Reichert, D.; Wagner, C.; Rusanov, E.; Steinborn, D. Protonated nucleobase ligands: Synthesis, structure and characterization of 9-methyladeninium hexachloroplatinate and pentachloro(9-methyladeninium)platinum(IV). *J. Inorg. Biochem.* **2004**, *98*, 439–446. [CrossRef]

33. Terzis, A.; Mentzafos, D. Trichloro(9-methylguaninium)platinum(II) hydrate and 9-methylguaninium hexachloroplatinate(IV) dihydrate: Synthesis and structure. *Inorg. Chem.* **1983**, *22*, 1140–1143. [CrossRef]
34. Basallote, M.G.; Vilaplana, R. Palladium and platinum guanine complexes. *Transit. Met. Chem.* **1986**, *11*, 232–235. [CrossRef]
35. Bruker AXS Inc. *SAINT-Plus, Version 7.68*; Bruker AXS Inc.: Madison, WI, USA, 2012.
36. Sheldrick, G.M. *SADABS*; Bruker AXS: Madison, WI, USA, 2008.
37. Sheldrick, G.M. SHELXT—Integrated space-group and crystal-structure determination. *Acta Crystallogr. Sect. A Found. Adv.* **2015**, *71*, 3–8. [CrossRef]
38. Sheldrick, G.M. Crystal structure refinement with SHELXL. *Acta Crystallogr. Sect. C Struct. Chem.* **2015**, *71*, 3–8. [CrossRef]
39. Dolomanov, O.V.; Bourhis, L.J.; Gildea, R.J.; Howard, J.A.K.; Puschmann, H. OLEX2: A complete structure solution, refinement and analysis program. *J. Appl. Crystallogr.* **2009**, *42*, 339–341. [CrossRef]
40. Groom, C.R.; Bruno, I.J.; Lightfoot, M.P.; Ward, S.C. The Cambridge Structural Database. *Acta Crystallogr. Sect. B Struct. Sci. Cryst. Eng. Mater.* **2016**, *72*, 171–179. [CrossRef]
41. De la Cruz, R.; Espinet, P.; Gallego, A.M.; Martín-Alvarez, J.M.; Martínez-Ilarduya, J.M. Structural and dynamic studies in solution of anionic dinuclear azolato-bridged palladium(II) complexes. *J. Organomet. Chem.* **2002**, *663*, 108–117. [CrossRef]
42. Navarro-Ranninger, M.C.; Martínez-Carrera, S.; García-Blanco, S. Structure of trans-dichlorobis(1-methylimidazole)palladium(II), [Pd(C4H6N2)2Cl2]. *Acta Crystallogr. Sect. C Cryst. Struct. Commun.* **1983**, *39*, 186–188. [CrossRef]
43. Liebing, P.; Edelmann, F.T. Trifluoromethylated 3-(Pyrazol-1-yl)propanamide (PPA) Ligands. *Helv. Chim. Acta* **2020**, *103*, e2000148. [CrossRef]
44. Qin, Z.; Jennings, M.C.; Puddephatt, R.J. Self-Assembly of Polymer and Sheet Structures from Palladium(II) Complexes by Hydrogen Bonding between Carboxamide Substituents. *Inorg. Chem.* **2001**, *40*, 6220–6228. [CrossRef]
45. Gupta, D.; Nowak, R.; Lippert, B. Pt(ii) complexes of unsubstituted guanine and 7-methylguanine. *Dalt. Trans.* **2010**, *39*, 73–84. [CrossRef]
46. Spackman, P.R.; Turner, M.J.; McKinnon, J.J.; Wolff, S.K.; Grimwood, D.J.; Jayatilaka, D.; Spackman, M.A. CrystalExplorer: A program for Hirshfeld surface analysis, visualization and quantitative analysis of molecular crystals. *J. Appl. Crystallogr.* **2021**, *54*, 1006–1011. [CrossRef]
47. Escobar, C.A.; Artigas, V.; Bacho, M.; Trujillo, A. π-halogen interaction on the crystalline packing of 1,3,5-tris(4-bromophenyl)-1,3,5-triazine-2,4,6-trione·[solvate]. *J. Mol. Struct.* **2022**, *1247*, 131307. [CrossRef]
48. Lucas, X.; Bauzá, A.; Frontera, A.; Quiñonero, D. A thorough anion–π interaction study in biomolecules: On the importance of cooperativity effects. *Chem. Sci.* **2016**, *7*, 1038–1050. [CrossRef] [PubMed]
49. Youn, I.S.; Kim, D.Y.; Cho, W.J.; Madridejos, J.M.L.; Lee, H.M.; Kołaski, M.; Lee, J.; Baig, C.; Shin, S.K.; Filatov, M.; et al. Halogen−π Interactions between Benzene and X 2 /CX 4 (X = Cl, Br): Assessment of Various Density Functionals with Respect to CCSD(T). *J. Phys. Chem. A* **2016**, *120*, 9305–9314. [CrossRef]
50. Zhuo, H.; Li, Q.; Li, W.; Cheng, J. Is π halogen bonding or lone pair···π interaction formed between borazine and some halogenated compounds? *Phys. Chem. Chem. Phys.* **2014**, *16*, 159–165. [CrossRef] [PubMed]
51. Kellett, C.W.; Kennepohl, P.; Berlinguette, C.P. π covalency in the halogen bond. *Nat. Commun.* **2020**, *11*, 3310. [CrossRef] [PubMed]
52. Albright, E.; Cann, J.; Decken, A.; Eisler, S. Halogen···halogen interactions in diiodo-xylenes. *CrystEngComm* **2017**, *19*, 1024–1027. [CrossRef]
53. Sadaf, H.; Imtiaz-ud-Din; Fettouhi, M.; Fazal, A.; Ahmad, S.; Farooqi, B.A.; Nadeem, S.; Ihsan-ul-Haq; Ahmad, W. Synthesis, crystal structures and biological activities of palladium(II) complexes of benzimidazole and 2-methylbenzimidazole. *Polyhedron* **2019**, *170*, 537–543. [CrossRef]
54. Huang, Z.-J.; Du, L.; Xie, M.-J.; Chen, J. Dichlorido[bis(2-hexylsulfanyl)-1 H -benzimidazole-κ N 3]palladium(II). *Acta Crystallogr. Sect. E Struct. Reports Online* **2007**, *63*, m2474. [CrossRef]
55. Sadaf, H.; Imtiaz-ud-Din; Zahra, S.S.; Ihsan-ul-Haq; Nadeem, S.; Tahir, M.N.; Ahmad, S.; Andleeb, S. Synthesis, X-ray structures and biological properties of palladium(II) complexes of 1,2-dimethylimidazole and benzimidazole. *Polyhedron* **2019**, *160*, 101–107. [CrossRef]
56. Łakomska, I.; Fandzloch, M.; Wojtczak, A.; Szłyk, E. Platinum(IV) coordination compounds containing 5-methyl-1,2,4-triazolo[1,5-a]pyrimidin-7(4H)-one as nonleaving ligand. Molecular and cytotoxicity in vitro characterization. *Spectrochim. Acta Part A Mol. Biomol. Spectrosc.* **2011**, *79*, 497–501. [CrossRef] [PubMed]
57. Chongboriboon, N.; Samakun, K.; Dungkaew, W.; Kielar, F.; Sukwattanasinitt, M.; Chainok, K. Halogen-Bonding-Driven Self-Assembly of Solvates of Tetrabromoterephthalic Acid. *Crystals* **2021**, *11*, 198. [CrossRef]
58. Politzer, P.; Murray, J.S.; Clark, T. Halogen bonding: An electrostatically-driven highly directional noncovalent interaction. *Phys. Chem. Chem. Phys.* **2010**, *12*, 7748. [CrossRef]

crystals

MDPI

Article

Effect of Synthesis Method on Properties of Layered Double Hydroxides Containing Ni(III)

Olga V. Nestroinaia, Irina G. Ryltsova and Olga E. Lebedeva *

Institute of Pharmacy, Chemistry and Biology, Belgorod State National Research University, 308015 Belgorod, Russia; nestroynaya@bsu.edu.ru (O.V.N.); ryltsova@bsu.edu.ru (I.G.R.)
* Correspondence: olebedeva@bsu.edu.ru

Abstract: Unstable oxidation state +3 of nickel can be stabilized in the structure of layered double hydroxides, the resulting crystallinity and properties being dependent on the synthesis method. Three different wet chemical methods (co-precipitation at variable pH, co-precipitation followed by hydrothermal treatment, co-precipitation with microwave treatment) were used to synthesize Mg/Ni–Al layered double hydroxides containing triply charged nickel cations. Lattice parameters of the samples synthesized by various methods were found to differ from each other by about 1.5%. The most crystallized sample was obtained by hydrothermal synthesis. The oxidation state of nickel in the LDH samples was confirmed by XPS. TEM mapping gave evidence of the uniform distribution of nickel in all the samples. The LDHs' reduction with hydrogen and thermal transformations of the phase composition and morphology of LDHs were studied in detail. The properties of the samples synthesized by the different methods were shown to be quite similar.

Keywords: layered double hydroxides; nickel(III); co-precipitation; hydrothermal treatment; microwave treatment; XRD; XPS; TPR-H_2

Citation: Nestroinaia, O.V.; Ryltsova, I.G.; Lebedeva, O.E. Effect of Synthesis Method on Properties of Layered Double Hydroxides Containing Ni(III). *Crystals* **2021**, *11*, 1429. https://doi.org/10.3390/cryst11111429

Academic Editor: Aivaras Kareiva

Received: 23 October 2021
Accepted: 18 November 2021
Published: 21 November 2021

1. Introduction

Layered double hydroxides (LDHs) or hydrotalcite-like compounds are layered basic salts. The general formula for this class of compounds is $[M^{2+}_{1-x}M^{3+}_x(OH)_2]^{x+}[(A^{n-})_{x/n} \cdot mH_2O$, where M^{2+} and M^{3+} are metal ions located in octahedral positions of brucite-like layers, and A^{n-} stands for inorganic or organic anions that compensate the positive charge of brucite-like layers [1,2]. The structural features of hydrotalcite-like compounds give them a number of specific ion-exchange [3–5], sorption [6–8], electrical [9–11], catalytic [12–14], and magnetic properties [15–17].

The properties of LDHs are known to be significantly affected by cation–anion composition. Due to the easily reconstructed structure, it is possible to introduce cations of various metals into brucite-like layers, which makes it possible to regulate the properties of layered double hydroxides [18–21]. From the other point of view, LDHs are able to influence the properties of cations, changing them or, on the contrary, preserving them. The typical example is the stabilization of the unstable oxidation state of metals in a matrix of layered double hydroxides. Previously, our scientific group synthesized and characterized layered double hydroxides with unstable Ce(III) [22], Sn(II) [23], or Ni(III) [24] incorporated into brucite-like layers of LDHs.

Compounds containing trivalent nickel, however exotic they seem to be, are considered to be promising catalysts for the carbonation of polypropylene [25], the hydrocarboxylation of acetylene [26], the photocatalytic generation of hydrogen from water [27], and the electrocatalytic production of biodiesel [28].

The current study was aimed at the optimization of the synthesis method for obtaining nickel(III)-containing LDHs and the characterization of the effect of the synthesis method on the properties of the materials, in particular the degree of crystallinity, morphology, uniformity of element distribution, etc.

2. Materials and Methods

2.1. Materials and Preparation

The synthesis of Mg/AlNi layered double hydroxides was carried out by three different wet methods: co-precipitation at variable pH, co-precipitation with hydrothermal treatment, co-precipitation with microwave treatment. The preset degree of substitution of nickel for aluminum was about 25 at.%.

The following salts were used as sources of metal cations: $Mg(NO_3)_2 \cdot 6H_2O$, $Ni(NO_3)_2 \cdot 6H_2O$, $Al(NO_3)_3 \cdot 9H_2O$. All reagents were of analytical grade.

Aqueous solution of a mixture of NaOH and Na_2CO_3 with a molar ratio of reagents of 8:1 was used as a precipitant solution, and sodium hypochlorite NaOCl was applied as an oxidizer for nickel. The sodium hypochlorite solution was taken in a 1.5-fold molar excess with respect to the nickel content set during the synthesis.

Co-precipitation at variable pH was carried out by adding dropwise a precipitant solution to a solution containing the required amounts of salts. The total concentration of ions in the initial reaction mixture was 1 M, and the molar ratio of cations M^{2+}:M^{3+} = 3:1. During the synthesis, the pH of the mixture was kept approximately equal to 9–10. The precipitate was aged for 48 h under mother liquor. The sample obtained by this method is designated as Mg/AlNi25-c.

Hydrothermal synthesis of LDH was carried out in an Autoclave Engineers Parker autoclave at a temperature of 120 °C and excessive pressure of about 1 atm for 8 h. The volume of the autoclave was 50 mL, and 90% of it was filled with the reagents. The process was performed under stirring at 350 rpm. The ratio of the components in the reaction mixture and the pH value were identical to those described above for the co-precipitation synthesis procedure. The sample obtained by this method is designated as Mg/AlNi25-ht.

Co-precipitation with microwave treatment was carried out in a MARS-6 reactor for 10 min at a power of 700 W. The composition of the reaction mixture was similar to that used in the co-precipitation method. The sample obtained by this method is designated as Mg/AlNi25-mw.

In all cases, the precipitated products obtained by various methods were separated from the mother liquor, washed with distilled water, and dried at 120 °C.

To obtain the oxide forms, the samples were calcined at a temperature of 600 °C for 2 h (Mg/AlNi25-ox).

2.2. Apparatus and Procedures

2.2.1. X-ray Diffraction (XRD)

The phase composition was identified by powder X-ray diffraction (XRD). XRD was performed on a Rigaku Ultima IV diffractometer (CuKα—radiation) (Rigaku Co., Tokyo, Japan) in measurement range 2θ 5–80° with a scanning step of 0.02°, at a speed of 3°/min. XRD patterns were registered using Bragg–Brentano reflection geometry. Before the measurements, the powders of LDHs were ground in agate mortar for 30 min. Phase identification was performed using PXRD base PDF-2 (2014) with PDXL software (Rigaku Corporation). The cell parameters were calculated by Rietveld method using PDXL software (version 1.8.0.3, Rigaku Corporation) as well.

2.2.2. Transmission Electron Microscopy (TEM)

The morphology of the synthesized LDHs was studied by transmission electron microscopy on a JEM-2100 microscope (JEOL, Tokyo, Japan) at an operating voltage of 200 kV. Sample preparation was carried out by grinding the samples to a powder state with further dispersion in acetone under the action of ultrasound and deposition on a carbon film.

2.2.3. Scanning Electron Microscopy (SEM) Combined with EDXA

The elemental composition of the synthesized samples was determined using high-resolution scanning electron microscope QUANTA 200 3D FEI (FEI Company,

Mahwah, NJ, USA), equipped with an energy dispersion analyzer of chemical composition at an operating voltage of 20 kV.

2.2.4. X-ray Photoelectron Spectroscopy (XPS)

The chemical state of atoms on the surface of synthesized layered double hydroxides was studied by X-ray photoelectron spectroscopy (XPS). The purpose of this analysis was to confirm oxidation of nickel(II) to nickel(III). The XPS measurements were carried out on a Thermo Fisher Scientific ESCALAB 220iXL spectrometer (VG Scientific, West Sussex, UK) with an energy range of 0–5000 eV and a monochromatic Al Kα X-ray source (E = 1486.6 eV). The spot size on the sample was adjusted from 200 to 900 μm.

2.2.5. Textural Characteristics

Textural characteristics were studied by low-temperature nitrogen adsorption on a Microtrac BELSORP-mini (Microtrac Bel Co., Osaka, Japan) specific surface area and porosity analyzer II at a temperature of 77 K. In order to remove physically sorbed water and carbon dioxide, the samples were heated at a temperature of 100–110 °C. The values of the specific surface area of the samples were calculated using BET equation.

2.2.6. Temperature-Programmed Reduction (TPR-H_2)

The reducibility of LDHs and the products of their thermal destruction were studied by temperature-programmed reduction (TPR).

Temperature-programmed reduction with hydrogen was carried out in an 8-channel stationary reactor system with a continuous flow, operating in parallel. Each quartz reactor (i.e., 6 mm) was filled with 50 mg of the corresponding sample with a particle size of 250–315 μm, which had been pretreated at different temperatures: LDH at 200 °C and thermal degradation products at 500 °C in Ar flow (80 mL/min) at a heating rate of 10 °C/min. The reduction was carried out in an H_2/Ar flow (5:95) with a temperature interval of 10 °C/min. Hydrogen consumption and water formation were monitored using OmniStar quadrupole mass spectrometer (Pfeiffer Vacuum, Asslar, Germany).

2.2.7. Thermogravimetric and Differential Thermal Analysis

Thermogravimetric and differential thermal analysis were performed on the NETZSCH STA 449 F3 Jupiter synchronous thermal analysis device (Netzsch, Selb, Germany). Measurements of changes in mass and thermal effects were carried out in the temperature range between 30 °C and 950 °C. The samples were heated in nitrogen atmosphere.

3. Results

The structure and phase composition of the synthesized samples were determined using X-ray diffraction. Powder XRD patterns of LDHs synthesized by various methods are shown in Figure 1. For all samples, the presence of five reflexes corresponding to basal reflections (003), (006), (009/012), (015), and (018) was recorded, and also the presence of two reflexes (110) and (113) forming a doublet at 2θ of 60° was detected. Based on the results obtained, one can conclude that all three methods give rise to single-phase samples with a well-crystallized layered structure.

The most intense and narrow reflexes are observed in the X-ray patterns of the Mg/AlNi 25-ht sample synthesized under hydrothermal conditions. This indicates that the sample has the highest degree of crystallinity. A well-split doublet at 2θ = 60° is also a confirmation of this fact.

For a sample obtained by co-precipitation followed by microwave treatment, the intensity of reflexes is significantly lower.

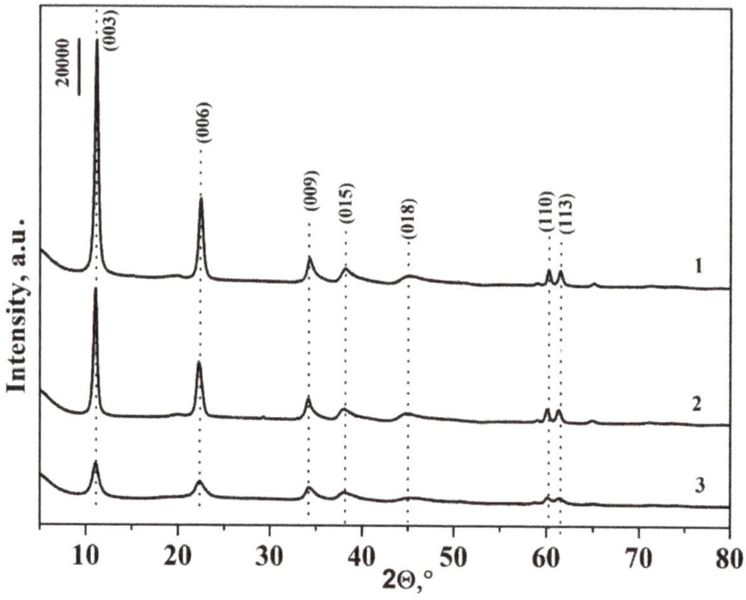

Figure 1. Powder XRD patterns: 1—Mg/AlNi 25-ht, 2—Mg/AlNi 25-c, 3—Mg/AlNi 25-mw.

The elemental composition of the obtained samples was determined using energy dispersion analysis. Figure 2 shows typical EDAX spectra of synthesized materials. Only those elements that were introduced during the synthesis were detected in the spectra. There are no signals of extra elements.

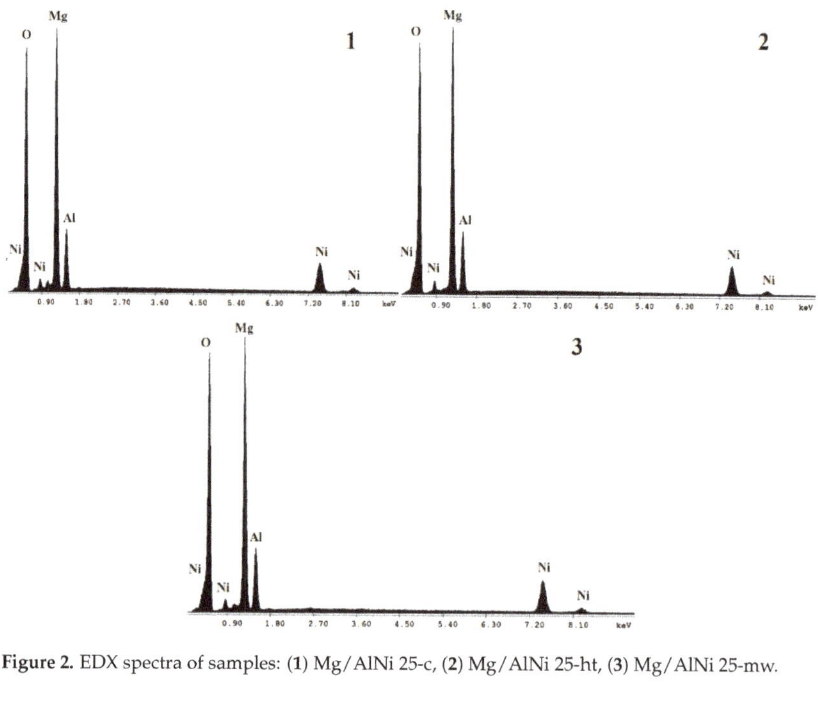

Figure 2. EDX spectra of samples: (**1**) Mg/AlNi 25-c, (**2**) Mg/AlNi 25-ht, (**3**) Mg/AlNi 25-mw.

Table 1 shows the atomic fractions of metal cations calculated from EDAX. According to the data obtained, the molar ratio M^{2+}/M^{3+} and the degree of substitution of aluminum cations for Ni(III) correspond to those set during the synthesis.

Table 1. Elemental composition, crystal lattice parameters and textural characteristics of LDHs synthesized by different methods.

Sample	Metal Ion Content, Atomic Fraction			M^{2+}/M^{3+}	c, Å	a, Å	S, m^2/g
	Mg^{2+}	Al^{3+}	Ni^{3+}				
Mg/Al Ni 25-c	1.224	0.325	0.124	2.72	24.01 ± 0.05	3.06 ± 0.01	61
Mg/Al Ni 25-ht	1.235	0.333	0.121	2.72	23.77 ± 0.05	3.08 ± 0.01	72
Mg/Al Ni 25-mw	1.201	0.329	0.134	2.61	24.06 ± 0.05	3.06 ± 0.01	70

For all the samples, mapping analysis was conducted to confirm the uniformity of element distribution (Figure 3). From the data presented, it can be seen that atoms of all metals are present in the sample and are evenly distributed throughout its volume. For samples obtained by different methods, the pictures are similar.

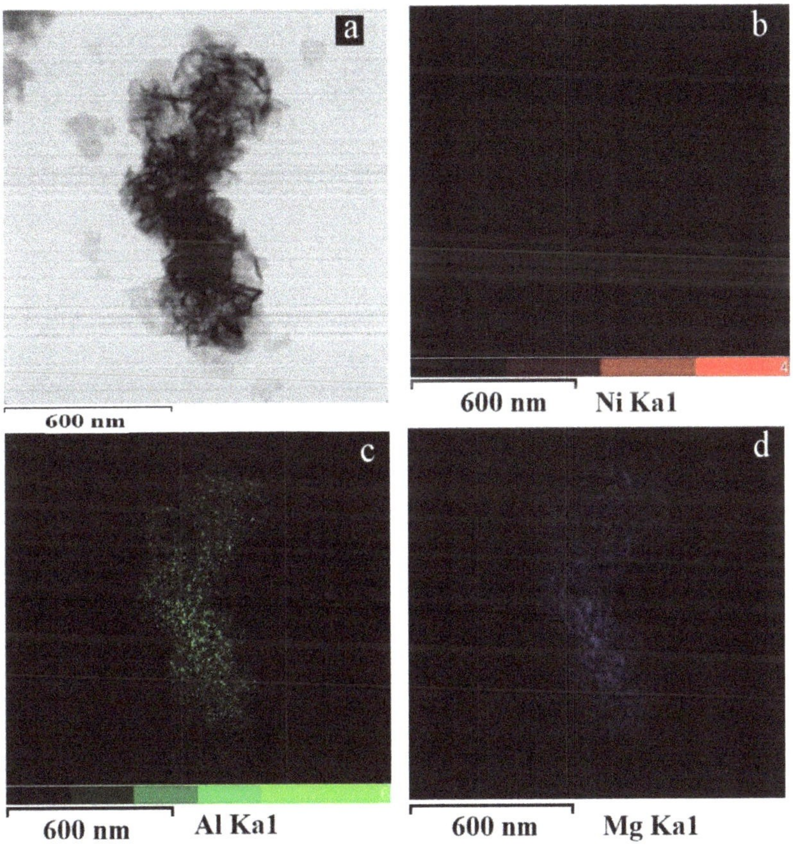

Figure 3. TEM mapping of Mg, Al and Ni for Mg/AlNi 25-ht sample: (**a**) TEM image, (**b**) nickel space distribution, (**c**) aluminum distribution, (**d**) magnesium distribution.

The experimental XRD in R-3m symmetry cell parameters is also presented in Table 1. The parameters c and a of the crystal lattice correlate with the size of the interlayer space and the distance between neighboring cations in the brucite-like layer, respectively. The difference in the parameters for various samples is not more than 1.5%.

The specific surface area of the layered double hydroxides is also included in Table 1. The values are comparable for the three samples.

The morphology of the obtained samples was studied by transmission electron microscopy (TEM) (Figure 4). For the Mg/AlNi25-c and Mg/AlNi25-ht samples, well-crystallized aggregates of lamellar hexagonal particles dominate. For the Mg/AlNi25-mw sample, crystals of a different shape are observed along with hexagonal particles.

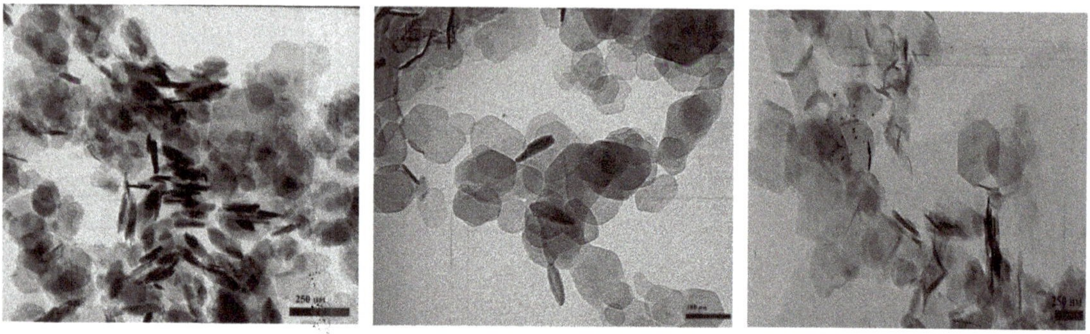

(a) (b) (c)

Figure 4. TEM micrographs of: (**a**) Mg/AlNi 25-c, (**b**) Mg/AlNi 25-ht, (**c**) Mg/AlNi 25-mw.

The thermal stability of the samples was studied by TG-DSC analysis. Figure 5 shows the TG-DTA curves for the Mg/AlNi25-c sample. There are two main stages of mass loss on the thermogravimetric curve. Two endothermic effects were recorded on the DTA curves, which were consistent with the TG data.

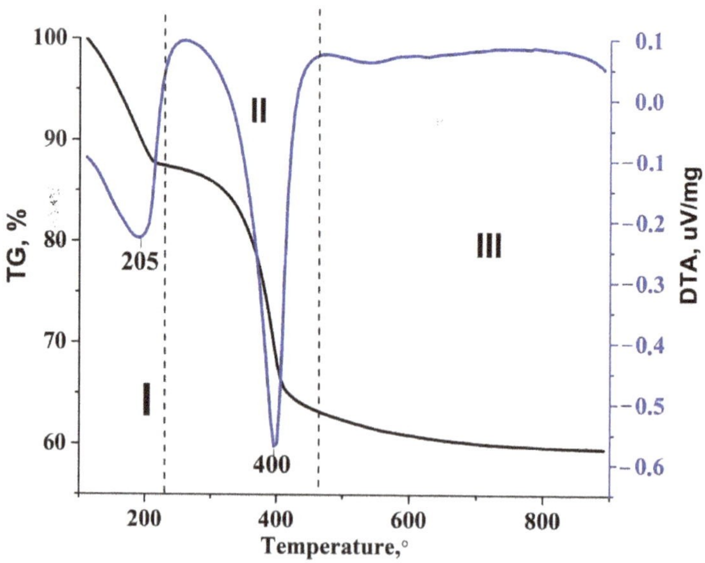

Figure 5. TG-DTA curves of Mg/AlNi-25-c.

Typically for LDHs, the first stage of mass loss is caused by the loss of both sorbed (20–130 °C) and crystallized interlayer (130–240 °C) water [29]. However, the additional effect for Ni(III) is hidden in this range. Nickel(III) is known to transform easily into a more stable divalent state at a temperature exceeding 138 °C [30]. This process is accompanied by the release of molecular oxygen and a change in color from black to green. It can be assumed that in the range of 130–240 °C, a loss of mass partially results from the release of oxygen during the reduction of Ni(III) to Ni(II).

The second stage of mass loss is associated with the dehydroxylation of brucite-like layers and the removal of carbonates from the interlayer space.

For all samples synthesized by different methods, the TG-DTA curves have the same appearance. However, for sample Mg/AlNi-25-mw, a 5–6-degree shift of the minima of endothermic effects towards lower temperatures was recorded.

For the sample Mg/AlNi-25-c, the thermal behavior and phase transitions were studied in detail. Figure 6 presents the XRD patterns of the sample calcined at different temperatures. It can be seen that up to the temperature of 300 °C, the structure of hydrotalcite is preserved though a decrease in the intensity of recorded peak starts, which is likely to originate from the water removal from the interlayer space. At 350 °C, diffractograms contain only two reflections belonging to the class of hydrotalcites.

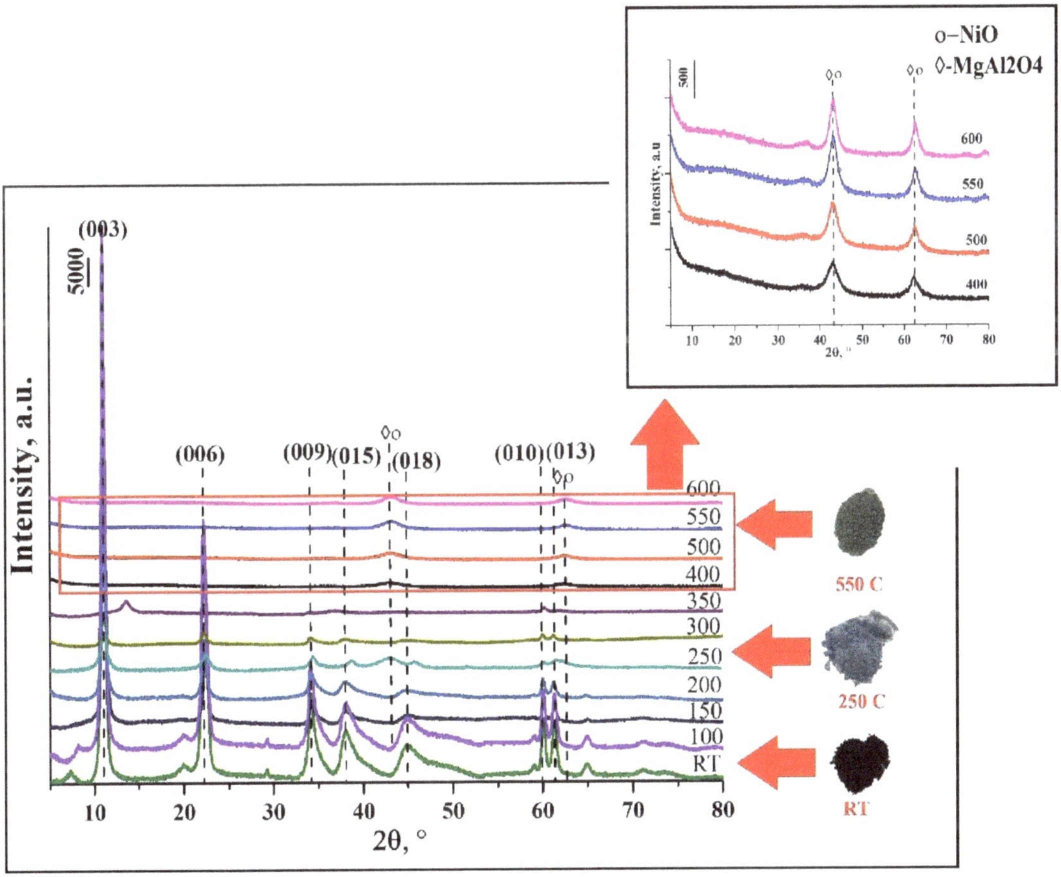

Figure 6. Powder XRD patterns of Mg/AlNi-25-c calcined at different temperatures.

An increase in temperature above 400 degrees leads to the complete destruction of the LDH structure and the formation of a mixture of oxides, which at a temperature of 500–600 °C are converted into spinel phases.

Figure 6 also shows the change in the color of the LDH depending on the heating temperature. At room temperature, the sample is black; further heating to 250 °C leads to a change in the color of the sample to gray-blue, which is most likely associated with the onset of the reduction of nickel. Further heating leads to the complete transformation of Ni(III) into Ni(II).

The thermal transformations for two other LDH samples were demonstrated to be essentially the same.

The change in the morphology of LDHs upon heating was studied using transmission electron microscopy. In Figure 7, it is clearly seen that with heating the hexagonal flakes gradually cease to be distinguishable. After calcination at 200 °C, they are still clearly visible in micrographs, and it is almost impossible to distinguish them after calcination at 400 °C. Oxides are formed at 400 °C, and at 600 °C oxides are agglomerated into spinel phases.

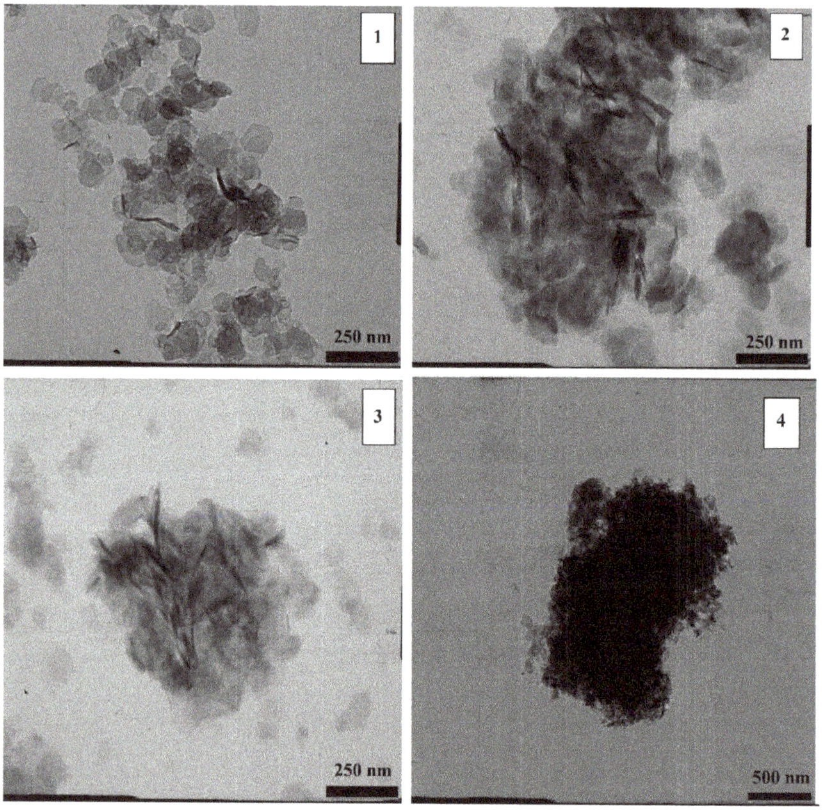

Figure 7. TEM micrographs of Mg/AlNi 25-ht heated at different temperatures: (**1**) RT, (**2**) 200 °C, (**3**) 400 °C, (**4**) 600 °C.

The specific surface area of the Mg/AlNi 25-c sample calcined at 600 °C is 150 m^2/g.

In the current research, the determination of the oxidation state of nickel in the LDH samples was the most challenging. XPS permitted us to obtain information about the nickel oxidation state. Binding energies for Ni(III) and Ni(II) have close values, and this fact causes difficulties in the identification of nickel species. In the publication of Haoyan Liang et al., it was shown that the peak at 854.7 eV corresponds to Ni(III), and the peak at 856.1 eV corresponds to Ni(II) [31]. These values were taken as references. The binding energy for nickel from NiO obtained by calcining LDHs acted for a comparison.

Figure 8a shows overview X-ray photoelectron spectra of the samples of Mg/AlNi-25 synthesized under different conditions, as well as the spectrum of the sample Mg/AlNi-25-ox after calcination at 600 °C. The survey XPS spectra for all LDH samples have an identical form, and the signals of all elements have comparable intensities with the exception of the signal corresponding to the energy O 1s.

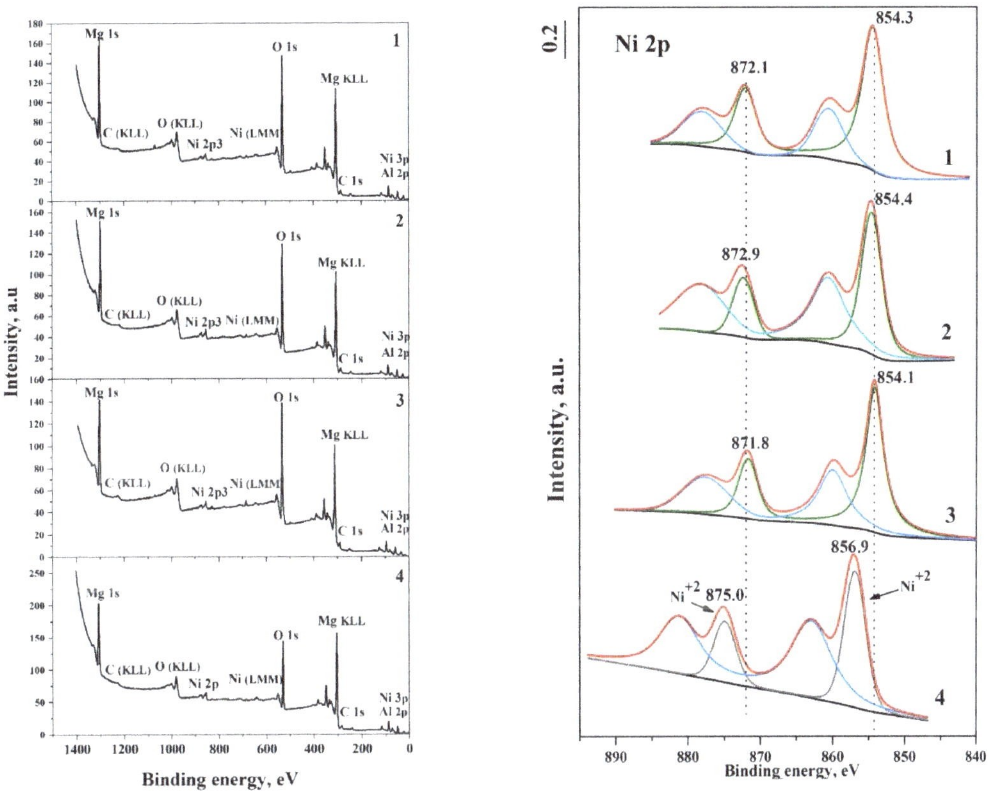

(a) (b)

Figure 8. (**a**) XPS survey spectra and (**b**) spectra of the levels of Ni 2p atoms of the Mg/AlNi-25 sample, synthesized under different conditions: 1—Mg/AlNi-25-c, 2—Mg/AlNi-25-ht, 3—Mg/AlNi-25-mw, 4—Mg/AlNi25-ox.

Figure 8b shows the XPS spectra of the Ni2p in LDH samples and mixed oxides. The presence of two main lines (Ni2p1/2 and Ni2p3/2) and two satellites was recorded for all samples. The difference between binding energies of nickel in LDHs and in the calcined sample gives evidence of the different oxidation states of nickel in the LDHs and in calcined samples. Ni(III) is definitely present in all the LDH samples.

Unfortunately, the content of nickel(III) in the materials could not be determined quantitatively because of the presence of aluminum cations and the overlapping Al and Ni lines in XPS spectra.

Oxidative properties of Ni(III) in LDH were studied by hydrogen TPR. Figure 9a shows the TPR profiles of Mg/AlNi-25-c, 2—Mg/AlNi25-ox. The TPR profile of LDH contains weak signals of hydrogen absorption at low temperatures in the range of 100–400 °C, a signal at 400–550 °C and the beginning of a very broad signal in the high-temperature region (above 750 °C).

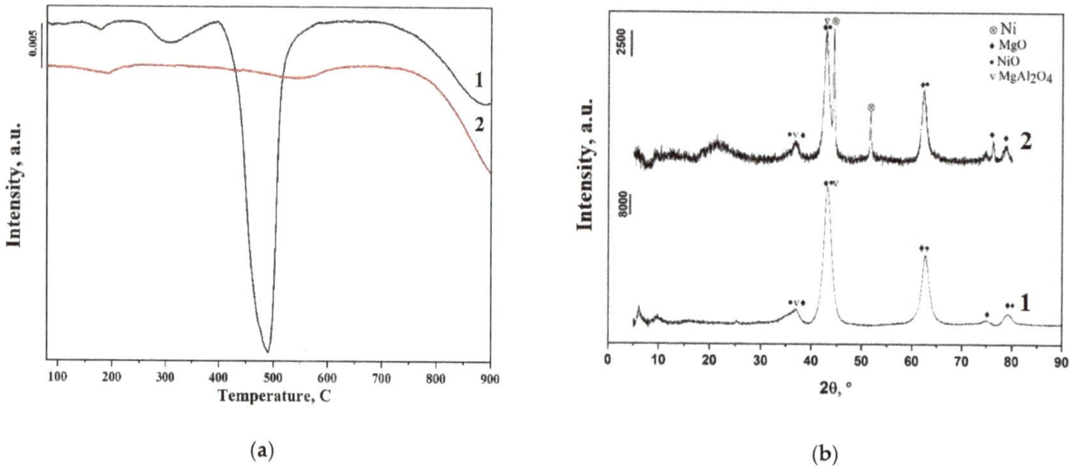

Figure 9. (**a**) H_2-TPR profiles of 1—Mg/AlNi-25-c, 2—Mg/AlNi25-ox; (**b**) X-ray diffraction patterns for Mg/AlNi-25-c: 1—original, 2—reduced.

It was already mentioned that nickel(III) oxide, when heated above 138 °C, is reduced to NiO. However, nickel(III) in layered double hydroxides seems to be stabilized. The low-temperature signals are so weak that they can be attributed to the reduction of nickel ions located on the surface. The peak located in the temperature range of 400–550 °C is likely to refer directly to the reduction of the majority of Ni(III) located inside brucite-like layers.

The appearance of a signal in the region of high temperatures is probably due to the reduction of Ni(II) to Ni^0.

Figure 9b shows X-ray diffraction patterns for the Mg/AlNi-25-c sample after thermal destruction and temperature-programmed reduction. According to the XRD results for the Mg /AlNi-25 sample after TPR-H_2, the presence of peaks corresponding to both nickel oxide NiO and metallic nickel Ni^0 is observed, which signifies the incomplete reduction of nickel.

4. Discussion

Nickel(III) is known to be quite unstable. Only a few publications concerning compounds of trivalent nickel appear each decade. The successful synthesis of the single-phase hydrotalcite-like layered double hydroxide with Ni(III) via co-precipitation and the primary characteristics of this material were described in our previous publication [24]. However, it was not obvious that nickel could be incorporated in brucite-like layers of LDHs under conditions other than slow co-precipitation with subsequent long-term ageing. The efficiency of an application of a short-term microwave procedure was especially doubtful. Nevertheless, the XRD data described above demonstrate the formation of the LDH structure as a result of three different synthesis methods—co-precipitation, hydrothermal synthesis and microwave-assistant synthesis. Hence, one can conclude that the location of Ni(III) in the brucite-like layers stabilizes this non-typical oxidation state of nickel regardless of the route of synthesis. This may be owing to either steric protection inside the layer or the higher stability of the state of the ion located in the layer.

Nickel(III) stabilization in the structure of the brucite-like layer of the LDH is unambiguously confirmed by X-ray photoelectron spectroscopy (XPS). According to the XPS data presented above, nickel cations are in a trivalent state in all LDH samples synthesized by different methods.

The main difference between the three samples was observed in their crystallinity. Obviously, the sample synthesized by the microwave-assistant method has lower and broader reflexes in the XRD pattern (Figure 1) and more obscure morphology of particles in TEM images (Figure 4). In addition, TG-DSC results correlate with the other data since the sample in question starts thermal transformation at slightly lower temperatures than others. This might be caused by the relatively disordered lattice of the sample. However, no extraneous phases were detected in this sample. Moreover, the results of TEM mapping (Figure 3) confirm the uniform distribution of nickel in the samples. It is worth noting that the microwave-assistant method of LDH preparation has significant advantages due to its short time and relatively low energy consumption, which in some cases may play a more important role than the high crystallinity of LDHs.

The parameters of the crystal lattice for the three samples have a slight distinction, while the chemical composition is essentially the same (Table 1). It is necessary to take into account that the parameters of the LDH lattice may depend on the amount of water in the interlayer space and on the order of the brucite-like layer. Most likely, the lowest *c* value for the sample synthesized under hydrothermal conditions is due to its highest crystallinity. The obtained results on parameter values, in general, indicate a successful isomorphic substitution of aluminum in the hydrotalcite structure for Ni(III) ions with a larger radius compared to Al(III). The evidence of the incorporation of nickel into the hydrotalcite structure is an increase in the parameter *c*, which for all synthesized samples exceeds the value of 22.9 Å typical for aluminum magnesium hydrotalcite.

The synthesis of layered double hydroxides is known to result in materials replete with stacking faults [32,33]. This kind of disorder is due to LDH polytypism and the stacking faults are an intergrowth of $3R_1$ and $2H_1$ polytypes [32]. The fault formation can be detected by the analysis of XRD patterns. In particular, a comparison of the diffractograms of our samples with the examples given in [32,33] permits us to suppose that the XRD patterns correspond to the $3R_1$ polytype with $2H_1$ stacking faults. It is noteworthy that all three samples synthesized by different methods demonstrate quite similar XRD patterns with comparable shape and asymmetry of peaks. Apparently, this leads to the conclusion that the synthesis method used does not greatly affect the order–disorder in the LDHs.

Thermal transformations of LDHs containing Ni(III) are quite usual for this class of compounds. The only unexpected fact was the hindered reducibility of the samples by hydrogen. This fact could be considered as an additional example of Ni(III) stabilization in the LDH lattice. Perhaps it is an argument in favor of the assumption concerning hindered access of the reagents towards the Ni(III) located in the structure of the metal hydroxide layer.

Thus, the efficient methods of LDH synthesis are not restricted to classical co-precipitation. The application of different approaches was demonstrated not to significantly affect the structure and properties of LDHs.

Author Contributions: Conceptualization, O.E.L. and I.G.R.; methodology, O.E.L.; investigation, O.V.N. and I.G.R.; writing—original draft preparation, O.V.N.; writing—review and editing, O.E.L.; visualization, O.V.N.; supervision, O.E.L. All authors have read and agreed to the published version of the manuscript.

Funding: This research received no external funding.

Acknowledgments: The authors thank the staff and administration of the Center of Common Use of Equipment of NRU "BelSU" for the technical support of the research.

Conflicts of Interest: The authors declare no conflict of interest.

References

1. Cavani, F.; Trifiro, F.; Vaccari, A. Hydrotalcite-type anionic clays: Preparation, properties and applications. *Catal. Today* **1991**, *11*, 173–301. [CrossRef]
2. Forano, C.; Hibino, T.; Leroux, F.; Taviot-Gueho, C. *Handbook of Clay Science*; Bergaya, F., Theng, B.K.G., Lagaly, G., Eds.; Elsevier: Amsterdam, The Netherlands, 2006; Volume 1, pp. 1021–1095.
3. Haraketi, M.; Hosni, K.; Srasra, E. Intercalation behavior of salicylic acid into calcined Cu–Al-layered double hydroxides for a controlled release formulation. *Surf. Eng. Appl. Electrochem.* **2017**, *53*, 360–370. [CrossRef]
4. Dwiasi, D.W.; Mudasir, M.; Roto, R. Ion Exchange of Benzoate in Ni-Al-Benzoate Layered Double Hydroxide by Amoxicillin. *Open Chem.* **2019**, *17*, 1043–1049. [CrossRef]
5. Rives, V.; Arco, M.; Martín, C. Intercalation of drugs in layered double hydroxides and their controlled release: A review. *Appl. Clay Sci.* **2014**, *88–89*, 239–269. [CrossRef]
6. Lei, C.; Zhu, X.; Zhu, B.; Jiang, C.; Le, Y.; Yu, J. Superb adsorption capacity of hierarchical calcined Ni/Mg/Al layered double hydroxides for Congo red and Cr(VI) ions. *J. Hazard. Mater.* **2017**, *321*, 801–811. [CrossRef] [PubMed]
7. Zhang, D.; Liu, X.Y.; Zhao, H.T.; Yang, L.; Lü, T.; Jin, M.Q. Application of hydrotalcite in soil immobilization of iodate (IO^{3-}). *RSC Adv.* **2018**, *38*, 21084–21091. [CrossRef]
8. Johnston, A.L.; Lester, E.; Williams, O.; Gomesa, R.L. Understanding Layered Double Hydroxide properties as sorbent materials for removing organic pollutants from environmental water. *J. Environ. Chem. Eng.* **2021**, *9*, 105197. [CrossRef]
9. Dipali, S.; Sachin, A.P.; Seong, H.; Jae, C. CoFe layered double hydroxide for enhanced electrochemical performance. *J. Electroanal. Chem.* **2020**, *862*, 11–40.
10. Wang, W.; Zhang, N.; Shi, Z.; Ye, Z.; Gao, Q.; Zhi, M.; Hong, Z. Preparation of Ni–Al layered double hydroxide hollow microspheres for supercapacitor electrode. *Chem. Eng. J.* **2018**, *338*, 55–61. [CrossRef]
11. Iguchi Sh Kikkawa, S.; Teramura, K.; Hosokawa, S.; Tanaka, T. Investigation of the electrochemical and photoelectrochemical properties of Ni–Al LDH photocatalysts. *Phys. Chem. Chem. Phys.* **2016**, *18*, 13811–13819. [CrossRef]
12. Meng, Y.; Luo, W.; Xia Sh Ni, Z.H. Preparation of Salen–Metal Complexes (Metal = Co or Ni) Intercalated ZnCr-LDHs and Their Photocatalytic Degradation of Rhodamine B. *Catalysts* **2017**, *7*, 143. [CrossRef]
13. Mantovani, K.M.; Westrup KC, M.; da Silva Junior, R.M.; Jaerger, S.; Wypych, F.; Nakagaki, S. Oxidation catalyst obtained by the immobilization of layered double hydroxide/Mn(iii) porphyrin on monodispersed silica spheres. *Dalton Trans.* **2018**, *47*, 3068–3073. [CrossRef]
14. Mac Leod, T.; Kopylovich, M.M.; Guedes da Silva, F.; Mahmudov, K.; Pombeiro, A. Copper(II) complexes of arylhydrazones of β-diketones immobilized on Zn–Al layered double hydroxides as effective recyclable catalysts for peroxidative oxidation of alkanes. *Appl. Catal. A* **2012**, *439*, 15–23. [CrossRef]
15. Nestroinaia, O.V.; Ryl'tsova, I.G.; Yapryntsev, M.N.; Lebedeva, O.E. Effect of the synthesis method on the phase composition and magnetism of layered double hydroxides. *Inorg. Mater.* **2020**, *7*, 747–753. [CrossRef]
16. Abellen, G.; Carrasco, J.; Coronado, E. Room temperature magnetism in layered double hydroxides due to magnetic nanoparticles. *Inorg. Chem.* **2013**, *52*, 7828–7830. [CrossRef] [PubMed]
17. Li, Q.; Xing, L.; Lu, X.; Li, N.; Mingxiang, X. Magnetic properties of Mg/Co(II)–Al/Fe(III) layered double hydroxides. *Inorg. Chem. Commun.* **2015**, *52*, 46–49. [CrossRef]
18. Chowdhury, P.R.; Bhattacharyya, K.G. Ni/Ti layered double hydroxide: Synthesis, characterization and application as a photocatalyst for visible light degradation of aqueous methylene blue. *Dalton. Trans.* **2015**, *44*, 6809–6824. [CrossRef] [PubMed]
19. Li, S.S.; Wang, L.; Li, Y.D.; Zhang, L.H.; Wang, A.X.; Xiao, N.; Gao, Y.Q.; Li, N.; Song, W.Y.; Ge, L.; et al. Novel photocatalyst incorporating Ni-Co layered double hydroxides with P-doped CdS for enhancing photocatalytic activity towards hydrogen evolution. *Appl. Catal. B* **2019**, *254*, 145–155. [CrossRef]
20. Abellan, G.; Carrasco, J.A.; Coronado, E.; Romero, J.; Varela, M. Alkoxide intercalated CoFe-layered double hydroxides as precursors of colloidal nanosheet suspensions: Structural, magnetic and electrochemical properties. *J. Mater. Chem. C* **2014**, *2*, 3723. [CrossRef]
21. Ma, K.; Cheng, J.P.; Zhang, J.; Li, M.; Liu, F.; Zhang, X. Dependence of Co/Fe ratios in Co-Fe layered double hydroxides on the structure and capacitive properties. *Electrochim. Acta* **2016**, *198*, 231–240. [CrossRef]
22. Golovin, S.N.; Yapryntsev, M.N.; Ryltsova, I.G.; Veligzhanin, A.A.; Lebedeva, O.E. Novel cerium-containing layered double hydroxide. *Chem. Pap.* **2020**, *74*, 367–370. [CrossRef]
23. Damindarova, V.N.; Ryl'tsova, I.G.; Tarasenko, E.A.; Wang, X.; Lebedeva, O.E. Tin-Containing Layered Double Hydroxides. *Pet. Chem.* **2020**, *60*, 440–450. [CrossRef]
24. Ryltsova, I.G.; Nestroinaya, O.V.; Lebedeva, O.E.; Schroeter, F.; Roessner, F. Synthesis and characterization of layered double hydroxides containing Nickel in unstable oxidation state +3 in cationic sites. *J. Solid State Chem.* **2018**, *265*, 332–338. [CrossRef]
25. Gong, J.; Liu, J.; Wan, D.; Chen, X. Catalytic carbonization of polypropylene by the combined catalysis of activated carbon with Ni_2O_3 into carbon nanotubes and its mechanism. *Appl. Catal. A* **2012**, *449*, 112–120. [CrossRef]
26. Lin, T.J.; Meng, X.; Shi, L. Catalytic hydrocarboxylation of acetylene to acrylic acid using Ni_2O_3 and cupric bromide as combined catalysts. *J. Mol. Catal. A Chem.* **2015**, *396*, 77–83. [CrossRef]
27. Sayed, M.E.; Abukhadra, M.R.; Salam, M.A.; Yakout, S.M.; Aziz, I.M. Photocatalytic hydrogen generation from raw water using zeolite/polyaniline@ Ni_2O_3 nanocomposite as a novel photo-electrode. *Energy* **2019**, *187*, 115943. [CrossRef]

28. Sarno, M.; Ponticorvo, E.; Scarpa, D. Novel Pt-Ni/NiO/Ni_2O_3 based electrodes for electrocatalytic biodiesel production from waste palm oil. *Mater. Today Proc.* **2020**, *20*, 69–73. [CrossRef]
29. Tongamp, W.; Zhang, Q.; Saito, F. Preparation of meixnerite (Mg–Al–OH) type layered double hydroxide by a mechanochemical route. *J. Mater. Sci.* **2007**, *42*, 9210–9215. [CrossRef]
30. Su, Q.; Gu, L.; Yao, Y.; Zhao, J.; Ji, W.; Ding, W.; Au, C.-T. Layered double hydroxides derived Nix(MgyAlzOn) catalysts: Enhanced ammonia decomposition by hydrogen spillover effect. *Appl. Catal. B* **2017**, *201*, 451–460. [CrossRef]
31. Liang, H.; Lin, J.; Jia, H.; Chen, S.; Qi, J.; Cao, J.; Feng, J. Hierarchical NiCo-LDH/NiCoP@NiMn-LDH hybrid electrodes on carbon cloth for excellent supercapacitors. *J. Mater. Chem. A* **2018**, *6*, 15040–15046. [CrossRef]
32. Shivaramaiah, R.; Navrotsky, A. Energetics of Order−Disorder in Layered Magnesium Aluminum Double Hydroxides with Interlayer Carbonate. *Inorg. Chem.* **2015**, *54*, 3253–3259. [CrossRef] [PubMed]
33. Sławiński, W.A.; Sjåstad, A.O.; Fjellvåg, H. Stacking Faults and Polytypes for Layered Double Hydroxides: What Can We Learn from Simulated and Experimental X-ray Powder Diffraction Data? *Inorg. Chem.* **2016**, *55*, 12881–12889. [CrossRef] [PubMed]

crystals

MDPI

Article

Solvothermal Synthesis of Calcium-Deficient Hydroxyapatite via Hydrolysis of α-Tricalcium Phosphate in Different Aqueous-Organic Media

Rasa Karalkeviciene [1], Eva Raudonyte-Svirbutaviciene [1,2], Justina Gaidukevic [1], Aleksej Zarkov [1] and Aivaras Kareiva [1,*]

1 Institute of Chemistry, Faculty of Chemistry and Geosciences, Vilnius University, Naugarduko St. 24, LT-03225 Vilnius, Lithuania; rasa.karalkeviciene@nvspl.lt (R.K.); eva.raudonyte-svirbutaviciene@chgf.vu.lt (E.R.-S.); justina.gaidukevic@chf.vu.lt (J.G.); aleksej.zarkov@chf.vu.lt (A.Z.)

2 Institute of Geology and Geography, Nature Research Centre, Akademijos Str. 2, LT-08412 Vilnius, Lithuania

* Correspondence: aivaras.kareiva@chgf.vu.lt

Citation: Karalkeviciene, R.; Raudonyte-Svirbutaviciene, E.; Gaidukevic, J.; Zarkov, A.; Kareiva, A. Solvothermal Synthesis of Calcium-Deficient Hydroxyapatite via Hydrolysis of α-Tricalcium Phosphate in Different Aqueous-Organic Media. *Crystals* **2022**, *12*, 253. https://doi.org/10.3390/cryst12020253

Academic Editor: Carlos Rodriguez-Navarro

Received: 31 January 2022
Accepted: 11 February 2022
Published: 13 February 2022

Abstract: In the present work, the effects of various organic solvents (solvent nature and fraction within the solution) and solvothermal conditions on the formation of calcium-deficient hydroxyapatite (CDHA) via hydrolysis of α-tricalcium phosphate (α-TCP) are investigated. The wet precipitation method is applied for α-TCP synthesis, and the hydrolysis reaction is performed in solutions with different water-to-organic solvent ratios under solvothermal conditions at 120 °C for 3 h and at 200 °C for 5 h. Ethyl alcohol, isopropyl alcohol, and butyl alcohol did not inhibit the hydrolysis of α-TCP, while methyl alcohol and ethylene glycol have a more prominent inhibitory effect on the hydrolysis, hence the formation of single-phased CDHA. From all the solvents analysed, ethylene glycol has the highest impact on the sample morphology. Under certain water to ethylene glycol ratios and solvothermal conditions, samples containing a significant fraction of rods are obtained. However, samples prepared with ethylene glycol are characterised by a particularly low BET surface area.

Keywords: calcium hydroxyapatite; α-tricalcium phosphate; water-organic solvent system; solvothermal synthesis

1. Introduction

Hydroxyapatite ($Ca_{10}(PO_4)_6(OH)_2$, HAp) is a major inorganic component in human hard tissue and is one of the most investigated calcium phosphates (CaPs) [1–3]. Due to its bone-like chemical composition and crystalline structure, HAp is extensively applied as bioceramic material for bone grafting [4,5]. In addition to that, HAp has found applications in drug delivery [6], chromatography [7], and is a very promising material for the treatment of air, water, and soil pollution [8–11]. Since HAp is considered to be an environmentally benign functional material, and due to its remarkable adsorption capacity, HAp could be extremely useful in the field of environmental management [8]. Calcium-deficient hydroxyapatite (CDHA, $Ca_{10-x}(HPO_4)_x(PO_4)_{6-x}(OH)_{2-x}$) is HAp with a Ca/P ratio from 1.50–1.67 [12]. Previous studies have reported a larger specific surface area and superior incorporating efficacy of CDHA when compared to other CaPs [13,14]. The chemical composition of HAp can be modified from the stoichiometric form to the Ca-deficient form by selecting an appropriate Ca/P molar ratio [15].

Solubility, specific surface area, surface wettability, and hence the adsorption characteristics of HAp crystals, depend greatly on their morphology and crystallinity [5,11]. Moreover, HAp contains the following two types of crystal planes: a (b)-plane, rich in positively charged Ca^{2+} ions, and a c-plane, exposing negatively charged phosphate and hydroxyl groups [10]. Hence, if the crystal growth along a specific direction is induced, HAp could gain new desirable properties and be applied, for instance, as a selective adsorbent of negatively or positively charged compounds [9].

HAp could be obtained via different synthesis methods, including solid-state reactions, sol-gel technology, chemical precipitation, hydrolysis, hydrothermal, or solvothermal methods [2,16–18]. Among them, solvothermal synthesis is probably the most popular method, providing the possibility to obtain well-crystallized, single-phase HAp [1,19,20]. In addition to that, many groups have already reported the solvothermal synthesis of specifically shaped (plate-like and rod-like) HAp crystals using various additives in the reaction solution [1,20–22]. Several groups suggested that various alcohols could be successfully applied as HAp morphology-controlling agents during the solvothermal process. For instance, Goto et al. [16] has reported the synthesis of needle-like HAp crystals using ethyl alcohol-water solutions. Guo et al. [23] has observed that isopropyl alcohol has affected the crystallite size and crystallinity degree of the HAp crystals but had little effect on the sample morphology. However, as far as we know, no comprehensive study has ever been performed to compare the effects of different organic solvents on the hydrolysis of α-tricalcium phosphate (α-TCP) under solvothermal conditions. In this study, we aim to fill in this gap. To this end, solvothermal reactions were performed with different proportions of water-organic solvent. The organic solvents used were as follows: ethylene glycol, methyl alcohol, ethyl alcohol, isopropyl alcohol, and butyl alcohol. The effects of the solvothermal conditions, nature of the solvent, and organic solvent fraction in the solution on the phase purity and morphological features of HA were investigated.

2. Materials and Methods

2.1. Synthesis

First, metastable α-TCP was synthesized by wet precipitation method as a precursor for the further conversion to CDHA. An appropriate amount (3.42 g) of calcium nitrate tetrahydrate ($Ca(NO_3)_2 \cdot 4H_2O$, ≥99%, Roth, Karlsruhe, Germany) was dissolved in 20 ml of deionized water. A portion (1.27 g) of diammonium hydrogen phosphate ($(NH_4)_2HPO_4$, ≥98%, Roth, Karlsruhe, Germany) was dissolved in 15 ml of deionized water in a separate beaker. After dissolution, concentrated ammonium hydroxide (NH_4OH, 25%, Roth, Karlsruhe, Germany) was added to the latter solution until pH of the solution reached 10. After stirring for one minute, an aqueous solution of $Ca(NO_3)_2 \cdot 4H_2O$ was added rapidly. A white precipitate formed, which was stirred for 10 minutes at 400 rpm. The obtained precipitate was subsequently vacuum filtered and washed with an appropriate volume of deionized water and isopropyl alcohol [24]. The synthesis product was dried overnight in an oven at 50 °C. The dried powders were ground in agate mortar and annealed in a furnace at 700 °C for 5 h at a heating rate of 5 °C/min.

Solvothermal reactions were performed with different proportions of water and ethylene glycol (EG, >99%, Roth, Karlsruhe, Germany), water-methyl alcohol (MeOH, >99.9%, Roth, Karlsruhe, Germany), water-ethyl alcohol (EtOH, >96%, Roth, Karlsruhe, Germany), water-isopropyl alcohol (PrOH, >99.5%, Roth, Karlsruhe, Germany), and water-butyl alcohol (BuOH, >99.5%, Roth, Karlsruhe, Germany). The water to alcohol v/v ratios of 0:100, 20:80, 40:60, 60:40, and 80:20 were applied. For the synthesis, 0.3 g of α-TCP powder was placed into 90 ml polytetrafluoroethylene-lined stainless-steel pressure vessels and diluted with 20 ml of water-organic solvent mixture. Solvothermal treatment was performed at 120 °C for 3 h and at 200 °C for 5 h. Finally, the resulting powders were filtered, washed with EtOH, and dried at 50 °C overnight.

The sample notations and treatment conditions are given in Table 1. Water to alcohol v/v ratios of 0:100, 20:80, 40:60, 60:40, and 80:20; 100:0 were applied to all the solvents under both conditions of solvothermal treatment.

Table 1. Sample codes, solvents applied, and solvothermal treatment conditions.

Notation	Solvothermal Conditions	Organic Solvent Applied	Water-to-Organic Solvent (W:O)
120-W-EG	120 °C, 3 h	ethylene glycol	
120-W-MeOH	120 °C, 3 h	methyl alcohol	
120-W-EtOH	120 °C, 3 h	ethyl alcohol	0:100
120-W-PrOH	120 °C, 3 h	isopropyl alcohol	20:80
120-W-BuOH	120 °C, 3 h	butyl alcohol	40:60
200-W-EG	200 °C, 5 h	ethylene glycol	60:40
200-W-MeOH	200 °C, 5 h	methyl alcohol	80:20
200-W-EtOH	200 °C, 5 h	ethyl alcohol	0:100
200-W-PrOH	200 °C, 5 h	isopropyl alcohol	
200-W-BuOH	200 °C, 5 h	butyl alcohol	

2.2. *Characterization*

Powder X-ray diffraction data were collected on a Rigaku MiniFlex II diffractometer (Rigaku, The Woodlands, TX, USA) operating in Bragg–Brentano ($\theta/2\theta$) geometry, using Ni-filtered Cu $K\alpha$ radiation. The data were collected within a 2θ angle range from 10 to 60° at a step width of 0.01° and speed of 5°/min. Infrared (FTIR) spectra were recorded in the range of 4000–400 cm^{-1} employing Bruker ALPHA ATR spectrometer (Bruker, Billerica, Ma, USA). In order to study the morphology of the samples, a field-emission scanning electron microscope (FE-SEM) Hitachi SU-70 (FE-SEM, Hitachi, Tokyo, Japan) was used.

Textural properties of the prepared samples were estimated from N_2 adsorption/desorption isotherms at −196 °C using a Micromeritics TriStar 3020 analyser (Micromeritics, Norcross, GA, USA). Before the measurements, all the samples were outgassed in the N_2 atmosphere at 100 °C. The total surface area (S_{BET}) was estimated using the Brunauer–Emmet–Teller (BET) equation, while Barrett–Joyner–Halenda (BJH) equation was used to calculate pore size distribution of the samples [25].

3. Results and Discussion

The characteristics of the α-TCP precursor are presented in Figure 1. As it could be seen from the XRD diffraction pattern (Figure 1a), all the peaks match the standard XRD data of monoclinic $Ca_3(PO_4)_2$ (ICDD #00-070-0364) very well. The starting powders consisted of agglomerates of nanodimensional, mostly uniform elongated particles of irregular shape (Figure 1c). The sample exhibited type IV isotherms and displayed an H3 hysteresis loop (Figure 1b). Based on the pore size distribution results, illustrated in the inset image of Figure 1b, the sample was mainly characterised by pores smaller than 10 nm, albeit larger pores up to 55 nm were also present. The BET surface area (S_{BET}) of the precursor was 10.22 m^2 g^{-1}.

Under the reaction with water, α-TCP hydrolyses and converts to CDHA as described by the following equation [16]:

$$3\alpha\text{-}Ca_3(PO_4)_2 + H_2O \rightarrow Ca_9(HPO_4)(PO_4)_5(OH) \quad (1)$$

A sufficient amount of water is required for the first stage to occur. The phase crystallinity and purity of synthesized CDHA powders were investigated by XRD analysis, which revealed some differences among the obtained products. The samples treated with organic solvent only (0:100) showed no evidence of CDHA formation. Due to the absence of water, no hydrolysis reaction occurred, and the phase of such samples remained α-TCP (ICDD 00-070-0364). This was true for all the organic solvents used under different solvothermal treatments.

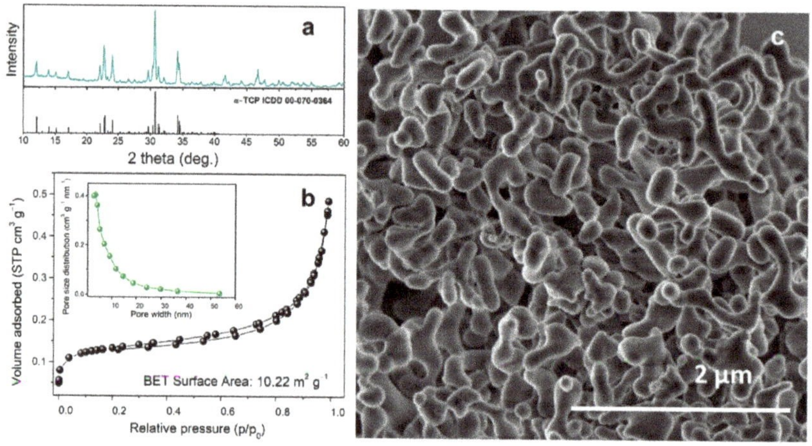

Figure 1. Characteristics of the obtained α-TCP precursor: XRD pattern (**a**), nitrogen adsorption-desorption isotherms and the corresponding BJH pore-size distribution (**b**), and SEM image (**c**).

In the case of EtOH, PrOH, and BuOH, the introduction of even a small fraction (20:80) of water resulted in the formation of single-phased CDHA (ICDD 00-76-0694), while an increasing water content provided the same results. This was observed under various applied solvothermal conditions. Under harsher solvothermal conditions (200 °C for 5 h), the formation of monetite was observed in the presence of EtOH, PrOH, and BuOH. This was especially notable in the case of BuOH. For comparison between the solvents, XRD patterns of the samples prepared under different solvothermal conditions using a W:O ratio of 40:60 are given in Figure 2 as representative.

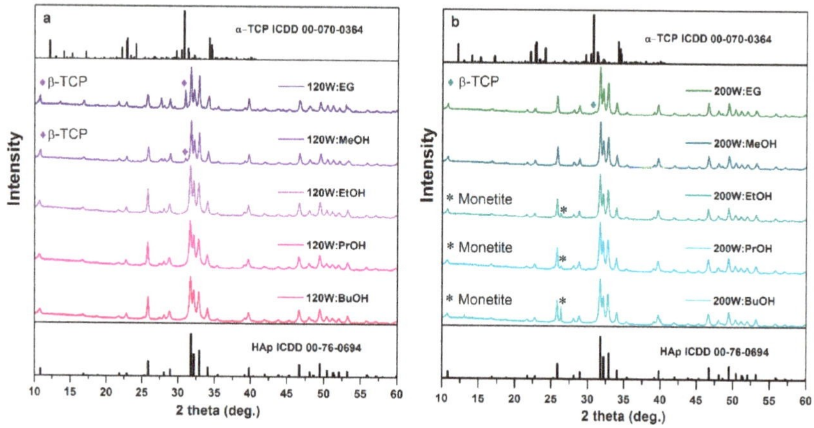

Figure 2. XRD patterns of the samples prepared using water-to-organic solvent ratio of 40:60 after a solvothermal treatment at 120 °C for 3 h (**a**) and at 200 °C for 5 h (**b**).

In contrast, MeOH and EG had a stronger inhibitory effect on α-TCP hydrolysis. These effects were especially notable for EG under the milder solvothermal conditions. Figure 3 shows the powder XRD patterns of the samples prepared under different solvothermal treatments (120 °C for 3 h and 200 °C for 5 h) using varying water to MeOH and water to EG ratios. After a treatment at 120 °C for 3 h, the sample with a water to EG ratio of 20:80 remained a single phase α-TCP (ICDD 00-070-0364, Figure 3b). Increasing water content induced the formation of CDHA, but a strong peak attributed to β-tricalcium phosphate (β-TCP, ICDD 00-070-2065) was visible in the sample 120-W-EG-40:60, while only a trace of β-TCP could be observed in the XRD pattern of 120-W-EG-60:40 (Figure 3b). MeOH has also inhibited the formation of CDHA, albeit to a lesser extent. The sample 120-W-MeOH-20:80 contained large fractions of CDHA, β-TCP and α-TCP. Traces of β-TCP were detected in the sample 120-W-MeOH-40:60, while the samples with a larger amount of water consisted of single-phase CDHA (Figure 3a).

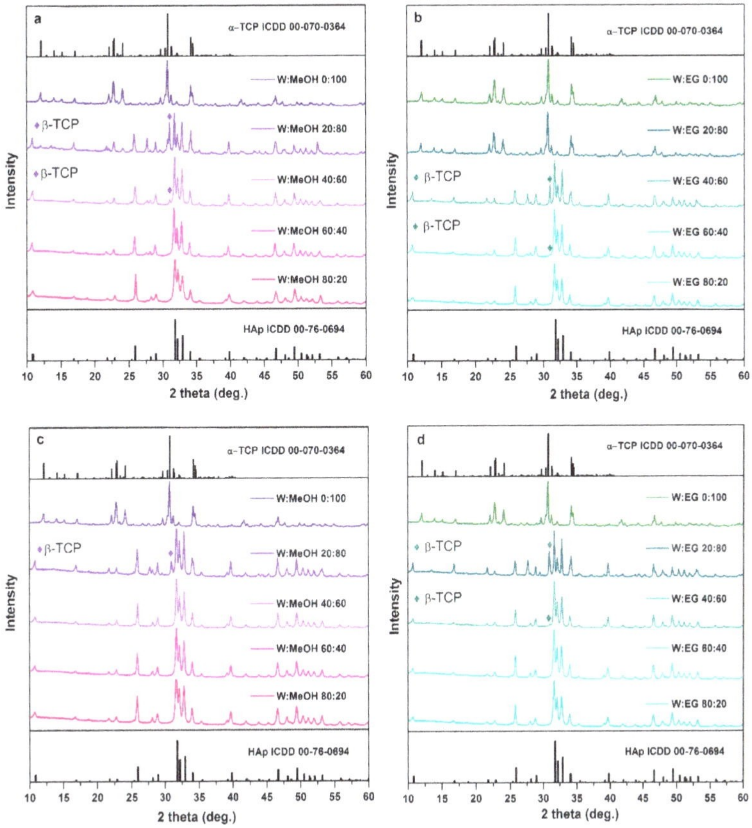

Figure 3. XRD patterns of the samples prepared using different water to MeOH (**a**,**c**) and water to EG (**b**,**d**) ratios under the following different solvothermal conditions: 120 °C for 3 h (**a**,**b**) and 200 °C for 5 h (**c**,**d**).

The increased temperature and prolonged time of the solvothermal synthesis caused the following shift in the inhibitory effect: single-phase CDHA was observed for the sample 200-W-EG-60:40, only a trace of β-TCP was detected in the XRD pattern of 200-W-EG-40:60, whereas the formation of CDHA was also obvious in 200-W-EG-20:80, even though a significant fraction of β-TCP was still present in the latter sample (Figure 3d). A higher temperature and longer reaction time have also resulted in a decreased β-TCP fraction in the sample 200-W-MeOH-20:80 (Figure 3c).

The FTIR spectra of the products prepared by the solvothermal treatment are presented in Figure 4. The FTIR range of 1500–400 cm^{-1} was chosen as representative since the main bands attributed to HAp and TCP polymorphs could be observed in this region, and the differences between the samples were hardly distinguishable in the full range spectra. The stretching modes of the hydroxyl group usually observed at 3572 cm^{-1}

were hardly visible in our case, and their intensity was similar in all the samples [26]. Several absorption bands at around 1095–960 and 636–550 cm^{-1} were observed in all the samples. The bands centred at 561–556 and 603–599 cm^{-1} are assigned to ν_4 O–P–O bending mode of CDHA [3]. Bands centred at 1020–1017 and 1090–1084 cm^{-1} correspond to ν_3 asymmetric P–O stretching vibrations, while the peak centred at 961–960 cm^{-1} corresponds to symmetric P–O stretching vibrations (ν_1) of CDHA [3]. The peak centred at 633–625 cm^{-1} corresponds to the bending vibrational mode of the hydroxyl (–OH) group [3,16]. An absorption band centred at 871–868 cm^{-1} is assigned to the P–O(H) stretching mode of the HPO_4^{2-} group, which is present in the structure of calcium-deficient CDHA [12]. The aforementioned bands were visible in the FTIR spectra of all the samples.

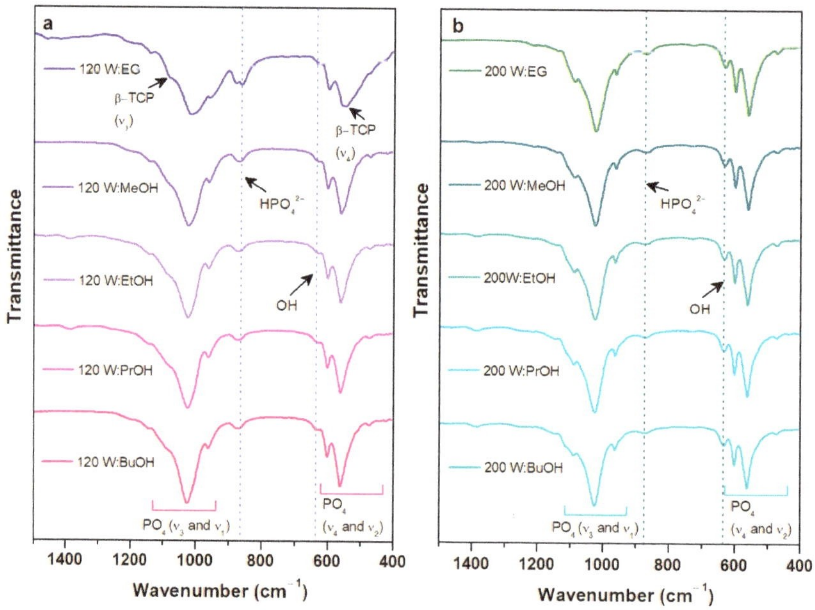

Figure 4. FTIR spectra of the samples prepared using water-to-organic solvent ratio of 40:60 after solvothermal treatment at 120 °C for 3 h (**a**) and at 200 °C for 5 h (**b**).

The additional bands attributed to β-TCP phase were only visible in the FTIR spectra of the samples prepared using larger fractions of EG (120-W-EG-20:80; 120-W-EG-40:60; 200-W-EG-20:80). As it could be seen from Figure 4a, bands at 544 and 1083 cm^{-1} of the sample 120-W-EG-40:60 could be attributed to β-TCP phase (ν_4 and ν_3, respectively) [27]. Such results are in agreement with the XRD data.

The morphology of the obtained samples varied from plate-shaped to rod-shaped. Samples fabricated without organic solvents consisted of plate-shaped crystals arranged into flower-like structures. In this study, only slight effects on morphology were observed due to the introduction of EtOH and PrOH. Under the milder solvothermal conditions (120 °C for 3 h), large proportion of EtOH and PrOH (120-W-EtOH-20:80; 120-W-PrOH-20:80) caused the formation of large plates with no prominent self-assembly (Figures S2 and S3). With an increasing proportion of water (W:O 40:60; 60:40; and 80:20), the formation of narrower plates and some rods was observed; moreover, the crystals were arranged in flower-like structures. When the reaction time and temperature were increased, higher proportions of EtOH, PrOH, and BuOH resulted in the formation of rods. Moreover, more rods have formed at the same W:O ratio under harsher conditions. However, plate-like crystals were still prevalent in all of the samples. SEM images of the samples prepared under different solvothermal conditions using a W:O ratio of 40:60 are presented in Figure 5.

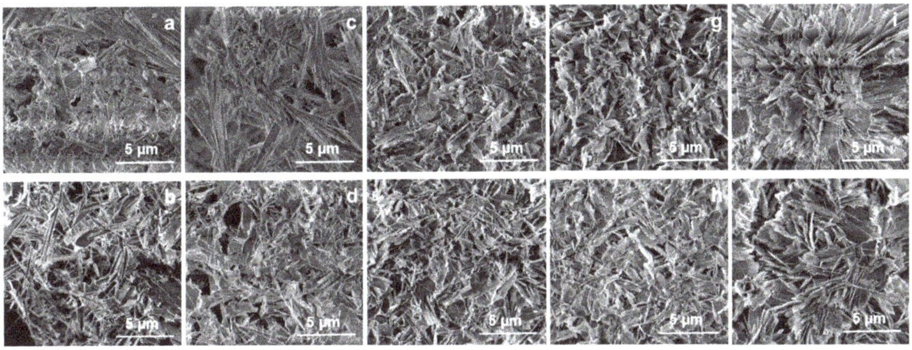

Figure 5. SEM images of the samples after solvothermal treatment with W:O ratio 40:60: (**a**) 120-W-EG-40:60; (**b**) 200-W-EG-40:60; (**c**) 120-W-MeOH-40:60; (**d**) 200-W-MeOH-40:60; (**e**) 120-W-EtOH-40:60; (**f**) 200-W-EtOH-40:60; (**g**) 120-W-PrOH-40:60; (**h**) 200-W-PrOH-40:60; (**i**) 120-W-BuOH-40:60; (**j**) 200-W-BuOH-40:60.

The use of MeOH (Figures 5c,d and S1) and BuOH (Figure 5i,j) had a more prominent effect on the sample morphology than the previously described solvents (Figures S3 and S4). In this case, the samples were also dominated by plate-like crystals, but a trend of long and narrow plate formation was observed. Moreover, more rods were present in the MeOH and BuOH treated samples when compared to the samples prepared in W-EtOH and W-PrOH solutions.

From all the solvents analysed, EG had the highest impact on the sample morphology. Under the milder solvothermal conditions (120 °C for 3 h), the formation of HAp was completely suspended in the sample 120-W-EG-20:80 (Figure 6c). Sample 120-W-EG-40:60 consisted of large plates, some rods, and some particles of different shapes, which could probably be attributed to β-TCP phase (Figure 6e). Sample 120-W-EG-60:40 was characterised by a large number of rods in addition to the plates (Figure 6g). In contrast, no rods were observed in a sample prepared with a minimal amount of EG (120-W-EG-80:20; Figure 6i). Prolonged reaction time and increased temperature resulted in a rod-dominated morphology of the CDHA samples. The sample 200-W-EG-20:80 was characterised by larger and smaller rods, as well as some minor particles of different shapes, which should be attributed to β-TCP phase (Figure 6d). A slightly lower proportion of EG (samples 200-W-EG-40:60 and 200-W-EG-60:40; Figure 6f,h) resulted in the formation of both plate-shaped and a large number of rod-shaped crystals. No rods were observed in a sample prepared with a minimal amount of EG (200-W-EG-80:20; Figure 6j).

It is assumed that the solvothermally assisted formation of rod-like crystals comprises the following two main stages: the nucleation step (reaction of ions), when small crystalline nuclei are formed in a supersaturated matrix, and the growth step, during which nuclei grow into their final shape and size [28]. In our case, the changes in crystal morphology might be related to the decreased supply of water when more organic solvents are introduced to the system. This would limit the hydrolysis reaction of α-TCP [16]. Previous studies stated that the increasing amount of alcohol in the aqueous reaction solution reduces the solubility of α-TCP and hence limits the supply of Ca^{2+} and PO_4^{3-} ions [16,29,30]. Such an effect is related to the changes in dielectric constant ($\varepsilon_r(\omega)$) of the solution: with a decreasing dielectric constant of the solvent, solubility decreases due to the decreased solvation energy [31,32]. Dielectric constant of pure water is 78.5 at 25 °C, whereas the dielectric constants of alcohols are significantly lower. The dielectric constants at 25 °C of the organic solvents used in this study are as follows: $\varepsilon_r(\omega)_{EG}$ = 38.5; $\varepsilon_r(\omega)_{MeOH}$ = 32.70; $\varepsilon_r(\omega)_{EtOH}$ = 24.3; $\varepsilon_r(\omega)_{PrOH}$ = 19.92; $\varepsilon_r(\omega)_{BuOH}$ = 17.5 [32]. Variations in $\varepsilon_r(\omega)$ of water-organic solvent mixtures depend on the composition of the solution, but in general, $\varepsilon_r(\omega)$ values decrease with the increasing fraction of organic solvent [33]. Solvents with different physicochemical properties influence solubility, crystal nucleation, and growth rate, which in turn has an effect on the crystallinity and morphology of the final products [32]. When the hydrolysis reaction is not suspended and the solution is supersaturated with Ca^{2+} and $PO_4{}^{3-}$ ions, nucleation takes dominance over the crystal growth and smaller crystals are formed. On the contrary, when the supply of Ca^{2+} and $PO_4{}^{3-}$ ions is limited, fewer nuclei form and larger crystals tend to grow [28,32]. Other properties of different solvents might have also affected the processes of α-TCP hydrolysis and HAp crystallisation. For instance, the presence of EG in the reaction mixture would significantly change the viscosity of the suspension [34,35]. Subsequently, ion mobility and diffusion rates would be reduced, which would in turn inhibit the hydrolysis reaction and retard the nucleation process [35]. The viscosity of the reaction media decreases with the increasing temperature, and thus the inhibitory effects of EG are less significant when the solvothermal synthesis is performed at higher temperatures (Figure 3b,d).

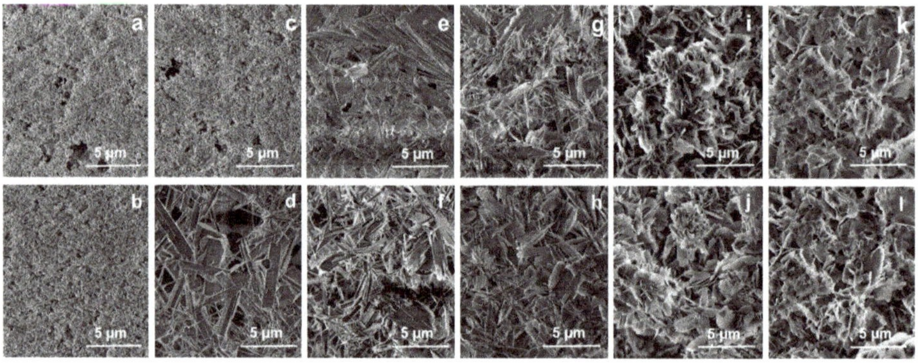

Figure 6. SEM images of the samples prepared with different ratios of water and ethylene glycol: (**a**) 120-W-EG-0:100; (**b**) 200-W-EG-0:100; (**c**) 120-W-EG-20:80; (**d**) 200-W-EG-20:80; (**e**) 120-W-EG-40:60; (**f**) 200-W-EG-40:60; (**g**) 120-W-EG-60:40; (**h**) 200-W-EG-60:40; (**i**) 120-W-EG-80:20; (**j**) 200-W-EG-80:20; (**k**) 120-W-EG-100:0; (**l**) 200-W-EG-100:0.

It is worth noting that the results obtained in this study differ from those reported by Goto et al. [16], who managed to prepare needle-like CDHA crystals arranged into flower-like structures by using water-ethyl alcohol solutions. We assume such discrepancies could originate due to the different starting materials used and slightly different solvothermal conditions applied. In their study, Goto et al. applied commercial α-TCP (Taihei Chemical Industrial Co., Ltd., Osaka, Japan) synthesized at a high temperature and consisting of large particles, while in our study we used low-temperature synthesized metastable α-TCP.

Figure 7 shows the nitrogen adsorption-desorption isotherms and the corresponding BJH pore-size distribution for CDHA particles. According to the new classification by the IUPAC, all the samples exhibited type IV isotherms and displayed H3 hysteresis loops. This type of isotherm indicates the existence of mesopores in the structure of all the samples. The hysteresis loop type H3 is associated with the existence of aggregated plate-like particles [36]. Such results are in agreement with SEM data (Figures 5 and 6). Based on the pore size distribution results illustrated in the inset image of Figure 7a,b, there were no significant differences between the samples. All the samples were characterised by a wide pore size distribution. Multi-scale pores ranged from 2.6 to 128 nm, suggesting that both mesopores and macropores were present in the CDHA structure. All the samples except those prepared with EG contained both mesopores of smaller sizes (from 2.6 nm up to 9.0 nm) and a small number of larger mesopores (from 9.0 nm up to 128 nm). On the contrary, in the sample 120-W-EG-40:60, mainly mesopores of larger width (from 23.0 nm up to 50.2 nm) were found. We assume the pores of smaller sizes presented in the structure of this sample were probably partially blocked by the viscous ethylene glycol. As it can be seen from Figure 7c, the mesoporous 120-W-PrOH-40:60 had the highest BET surface area (S_{BET}) of 24.3 m^2g^{-1}. An extremely low surface area was observed for the samples prepared using EG: S_{BET} of 2.1 m^2g^{-1} and 11.3 m^2g^{-1} were obtained for the samples 120-W-EG-40:60 and 200-W-EG-40:60, respectively. One reasonable explanation for this decrease is that residues of EG may have increased blockage of the nitrogen gas penetration. Moreover, this decrease in S_{BET} could be the result of the structural changes occurring during the synthesis of CDHA. The remaining samples were characterised by similar S_{BET} values ranging from 9.9 to 22.4 m^2g^{-1}.

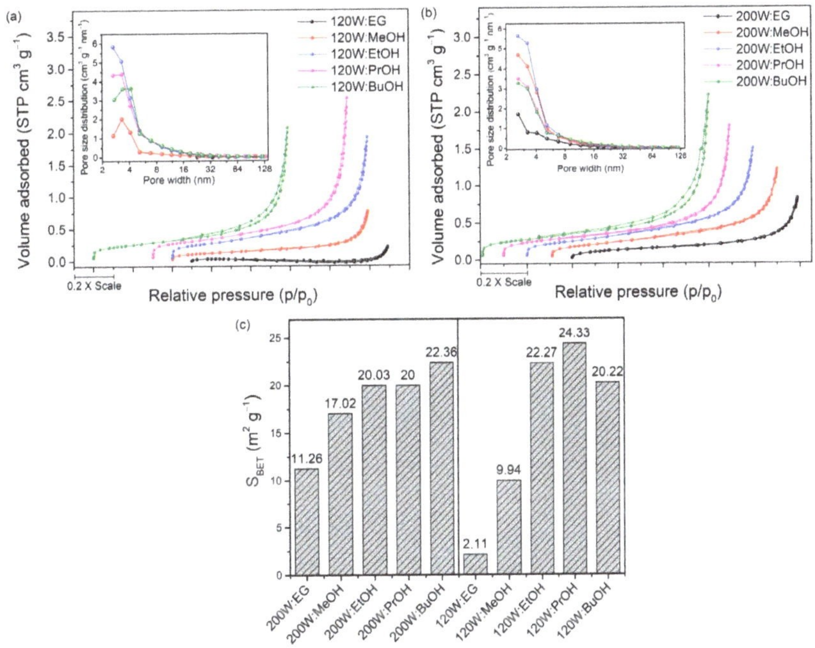

Figure 7. Nitrogen adsorption-desorption isotherms and the corresponding BJH pore-size distribution for the samples after solvothermal treatment with a W:O ratio of 40:60: (**a**) 120 °C; (**b**) 200 °C; and (**c**) BET surface area for the samples after solvothermal treatment with a W:O ratio of 40:60 at 120 °C and 200 °C.

4. Conclusions

A comprehensive study was performed to compare the effects of different organic solvents on the hydrolysis of α-TCP and its conversion to CDHA under different solvothermal conditions. Methyl alcohol and ethylene glycol had a stronger inhibitory effect on α-TCP hydrolysis than ethyl alcohol, isopropyl alcohol, and butyl alcohol. This was especially notable under milder solvothermal conditions. The morphology of the obtained samples varied from plate-shaped to rod-shaped. Samples containing some rods were obtained by applying certain ethyl alcohol and isopropyl alcohol proportions, albeit plate-like structures were still prevailing. The use of water-methyl alcohol and water-butyl alcohol mixtures leads to the formation of more rods in addition to the long and narrow plates. From all the solvents analysed, ethylene glycol had the highest impact on the sample morphology. Under certain water to ethylene glycol ratios and solvothermal conditions, samples containing a significant fraction of rods were obtained.

Supplementary Materials: The following supporting information can be downloaded at: https://www.mdpi.com/article/10.3390/cryst12020253/s1, Figure S1: images of the samples after solvothermal treatment with W-MeOH; Figure S2: SEM images of the samples after solvothermal treatment with W-EtOH; Figure S3: SEM images of the samples after solvothermal treatment with W-PrOH; Figure S4: SEM images of the samples after solvothermal treatment with W-BuOH.

Author Contributions: Conceptualization, R.K., E.R.-S., J.G., A.Z. and A.K.; methodology, R.K., E.R.-S. and A.Z., formal analysis, R.K.; investigation, R.K. and J.G., resources, E.R.-S., A.Z. and A.K., data curation, R.K., E.R.-S., J.G., A.Z. and A.K., writing—original draft preparation, E.R.-S., writing—review and editing, R.K., E.R.-S., J.G., A.Z. and A.K., supervision, A.Z. and A.K.; funding acquisition, E.R.-S., A.Z. and A.K. All authors have read and agreed to the published version of the manuscript.

Funding: This research has received funding by the European Social Fund under the No 09.3.3-LMT-K-712 "Development of Competences of Scientists, other Researchers and Students through Practical Research Activities" measure (project No 09.3.3-LMT-K-712-23-0070).

Data Availability Statement: Data is contained within the article.

Conflicts of Interest: The authors declare no conflict of interest.

References

1. In, Y.; Amornkitbamrung, U.; Hong, M.H.; Shin, H. On the Crystallization of Hydroxyapatite under Hydrothermal Conditions: Role of Sebacic Acid as an Additive. *ACS Omega* **2020**, *5*, 27204–27210. [CrossRef] [PubMed]
2. Cüneyt Tas, A. Synthesis of biomimetic Ca-hydroxyapatite powders at 37 °C in synthetic body fluids. *Biomaterials* **2000**, *21*, 1429–1438. [CrossRef]
3. Gopi, D.; Shinyjoy, E.; Karthika, A.; Nithiya, S.; Kavitha, L.; Rajeswari, D.; Tang, T. Single walled carbon nanotubes reinforced mineralized hydroxyapatite composite coatings on titanium for improved biocompatible implant applications. *RSC Adv.* **2015**, *5*, 36766–36778. [CrossRef]
4. Šupová, M. Problem of hydroxyapatite dispersion in polymer matrices: A review. *J. Mater. Sci. Mater. Med.* **2009**, *20*, 1201–1213. [CrossRef] [PubMed]
5. Zhuang, Z.; Fujimi, T.J.; Nakamura, M.; Konishi, T.; Yoshimura, H.; Aizawa, M. Development of a,b-plane-oriented hydroxyapatite ceramics as models for living bones and their cell adhesion behavior. *Acta Biomater.* **2013**, *9*, 6732–6740. [CrossRef] [PubMed]
6. Melde, B.J.; Stein, A. Periodic Macroporous Hydroxyapatite-Containing Calcium Phosphates. *Chem. Mater.* **2002**, *14*, 3326–3331. [CrossRef]
7. Bernardi, G. Chromatography of Nucleic Acids on Hydroxyapatite. *Nature* **1965**, *206*, 779–783. [CrossRef]
8. Zhao, J.; Hu, Q.; Lei, Y.; Gao, C.; Zhang, P.; Zhou, B.; Zhang, G.; Song, W.; Lou, X.; Zhou, X. Facile synthesis of ultralong hydroxyapatite nanowires using wormlike micelles as soft templates. *CrystEngComm* **2021**, *23*, 5498–5503. [CrossRef]
9. Foroutan, R.; Peighambardoust, S.J.; Aghdasinia, H.; Mohammadi, R.; Ramavandi, B. Modification of bio-hydroxyapatite generated from waste poultry bone with MgO for purifying methyl violet-laden liquids. *Environ. Sci. Pollut. Res.* **2020**, *27*, 44218–44229. [CrossRef]
10. Ibrahim, M.; Labaki, M.; Giraudon, J.M.; Lamonier, J.F. Hydroxyapatite, a multifunctional material for air, water and soil pollution control: A review. *J. Hazard. Mater.* **2020**, *383*, 121139. [CrossRef]
11. Hao, L.; Lv, Y.; Song, H. The morphological evolution of hydroxyapatite on high-efficiency Pb2+ removal and antibacterial activity. *Microchem. J.* **2017**, *135*, 16–25. [CrossRef]
12. Sinusaite, L.; Popov, A.; Raudonyte-Svirbutaviciene, E.; Yang, J.-C.; Kareiva, A.; Zarkov, A. Effect of Mn doping on hydrolysis of low-temperature synthesized metastable alpha-tricalcium phosphate. *Ceram. Int.* **2021**, *47*, 12078–12083. [CrossRef]
13. Deng, Y.; Liu, M.; Chen, X.; Wang, M.; Li, X.; Xiao, Y.; Zhang, X. Enhanced osteoinductivity of porous biphasic calcium phosphate ceramic beads with high content of strontium-incorporated calcium-deficient hydroxyapatite. *J. Mater. Chem. B* **2018**, *6*, 6572–6584. [CrossRef] [PubMed]
14. Ravi, N.D.; Balu, R.; Sampath Kumar, T.S. Strontium-Substituted Calcium Deficient Hydroxyapatite Nanoparticles: Synthesis, Characterization, and Antibacterial Properties. *J. Am. Ceram. Soc.* **2012**, *95*, 2700–2708. [CrossRef]
15. Sekine, Y.; Motokawa, R.; Kozai, N.; Ohnuki, T.; Matsumura, D.; Tsuji, T.; Kawasaki, R.; Akiyoshi, K. Calcium-deficient Hydroxyapatite as a Potential Sorbent for Strontium. *Sci. Rep.* **2017**, *7*, 2064. [CrossRef]
16. Goto, T.; Kim, I.Y.; Kikuta, K.; Ohtsuki, C. Hydroxyapatite formation by solvothermal treatment of α-tricalcium phosphate with water–ethanol solution. *Ceram. Int.* **2012**, *38*, 1003–1010. [CrossRef]
17. Kaviyarasu, K.; Mariappan, A.; Neyvasagam, K.; Ayeshamariam, A.; Pandi, P.; Palanichamy, R.R.; Gopinathan, C.; Mola, G.T.; Maaza, M. Photocatalytic performance and antimicrobial activities of HAp-TiO_2 nanocomposite thin films by sol-gel method. *Surf. Interfaces* **2017**, *6*, 247–255. [CrossRef]

18. Teshima, K.; Lee, S.; Sakurai, M.; Kameno, Y.; Yubuta, K.; Suzuki, T.; Shishido, T.; Endo, M.; Oishi, S. Well-Formed One-Dimensional Hydroxyapatite Crystals Grown by an Environmentally Friendly Flux Method. *Cryst. Growth Des.* **2009**, *9*, 2937–2940. [CrossRef]
19. Suchanek, K.; Bartkowiak, A.; Perzanowski, M.; Marszałek, M. From monetite plate to hydroxyapatite nanofibers by monoethanolamine assisted hydrothermal approach. *Sci. Rep.* **2018**, *8*, 15408. [CrossRef]
20. Tomozawa, M.; Hiromoto, S. Microstructure of hydroxyapatite- and octacalcium phosphate-coatings formed on magnesium by a hydrothermal treatment at various pH values. *Acta Mater.* **2011**, *59*, 355–363. [CrossRef]
21. Wang, Y.-c.; Wang, J.-n.; Xiao, G.-y.; Huang, S.-y.; Xu, W.-l.; Yan, W.-x.; Lu, Y.-p. Investigation of various fatty acid surfactants on the microstructure of flexible hydroxyapatite nanofibers. *CrystEngComm* **2021**, *23*, 7049–7055. [CrossRef]
22. Horiuchi, N.; Shibata, K.; Saito, H.; Iwabuchi, Y.; Wada, N.; Nozaki, K.; Hashimoto, K.; Tanaka, Y.; Nagai, A.; Yamashita, K. Size Control Synthesis of Hydroxyapatite Plates and Their Application in the Preparation of Highly Oriented Films. *Cryst. Growth Des.* **2018**, *18*, 5038–5044. [CrossRef]
23. Guo, X.; Xiao, P. Effects of solvents on properties of nanocrystalline hydroxyapatite produced from hydrothermal process. *J. Eur. Ceram. Soc.* **2006**, *26*, 3383–3391. [CrossRef]
24. Sinusaite, L.; Grigoraviciute-Puroniene, I.; Popov, A.; Ishikawa, K.; Kareiva, A.; Zarkov, A. Controllable synthesis of tricalcium phosphate (TCP) polymorphs by wet precipitation: Effect of washing procedure. *Ceram. Int.* **2019**, *45*, 12423–12428. [CrossRef]
25. Brunauer, S.; Emmett, P.H.; Teller, E. Adsorption of Gases in Multimolecular Layers. *J. Am. Chem. Soc.* **1938**, *60*, 309–319. [CrossRef]
26. Bogdanoviciene, I.; Beganskiene, A.; Tõnsuaadu, K.; Glaser, J.; Meyer, H.J.; Kareiva, A. Calcium hydroxyapatite, $Ca_{10}(PO_4)_6(OH)_2$ ceramics prepared by aqueous sol–gel processing. *Mater. Res. Bull.* **2006**, *41*, 1754–1762. [CrossRef]
27. Carrodeguas, R.G.; De Aza, S. α-Tricalcium phosphate: Synthesis, properties and biomedical applications. *Acta Biomater.* **2011**, *7*, 3536–3546. [CrossRef]
28. Sadat-Shojai, M.; Khorasani, M.-T.; Dinpanah-Khoshdargi, E.; Jamshidi, A. Synthesis methods for nanosized hydroxyapatite with diverse structures. *Acta Biomater.* **2013**, *9*, 7591–7621. [CrossRef]
29. Larsen, M.J.; Thorsen, A.; Jensen, S.J. Ethanol-induced formation of solid calcium phosphates. *Calcif. Tissue Int.* **1985**, *37*, 189–193. [CrossRef]
30. Lerner, E.; Azoury, R.; Sarig, S. Rapid precipitation of apatite from ethanol-water solution. *J. Cryst. Growth* **1989**, *97*, 725–730. [CrossRef]
31. Termine, J.D.; Peckauskas, R.A.; Posner, A.S. Calcium phosphate formation in vitro: II. Effects of environment on amorphous-crystalline transformation. *Arch. Biochem. Biophys.* **1970**, *140*, 318–325. [CrossRef]
32. Wu, J.; Lü, X.; Zhang, L.; Huang, F.; Xu, F. Dielectric Constant Controlled Solvothermal Synthesis of a TiO_2 Photocatalyst with Tunable Crystallinity: A Strategy for Solvent Selection. *Eur. J. Inorg. Chem.* **2009**, *2009*, 2789–2795. [CrossRef]
33. Wyman, J. The dielectric constant of mixtures of ethyl alcohol and water from -5 TO 40°. *J. Am. Chem. Soc.* **1931**, *53*, 3292–3301. [CrossRef]
34. Bohne, D.; Fischer, S.; Obermeier, E. Thermal, Conductivity, Density, Viscosity, and Prandtl-Numbers of Ethylene Glycol-Water Mixtures. *Ber. Der Bunsenges. Für Phys. Chem.* **1984**, *88*, 739–742. [CrossRef]
35. Ma, M.-G.; Zhu, Y.-J.; Chang, J. Monetite Formed in Mixed Solvents of Water and Ethylene Glycol and Its Transformation to Hydroxyapatite. *J. Phys. Chem. B* **2006**, *110*, 14226–14230. [CrossRef]
36. Thommes, M.; Kaneko, K.; Neimark, A.V.; Olivier, J.P.; Rodriguez-Reinoso, F.; Rouquerol, J.; Sing, K.S.W. Physisorption of gases, with special reference to the evaluation of surface area and pore size distribution (IUPAC Technical Report). *Pure Appl. Chem.* **2015**, *87*, 1051–1069. [CrossRef]

crystals

MDPI

Article

Synthesis, Crystal Structure, and Hirshfeld Surface Analysis of Hexachloroplatinate and Tetraclorouranylate of 3-Carboxypyridinium—Halogen Bonds and π-Interactions vs. Hydrogen Bonds

Anton Petrovich Novikov *, Mikhail Alexandrovich Volkov, Alexey Vladimirovich Safonov and Mikhail Semenovich Grigoriev

Frumkin Institute of Physical Chemistry and Electrochemistry, Russian Academy of Sciences, Leninskii prospect 31-4, 119071 Moscow, Russia; mendeleev93@yandex.ru (M.A.V.); alexeysafonof@gmail.com (A.V.S.); mickgrig@mail.ru (M.S.G.)

* Correspondence: tony.novickoff@yandex.ru

Abstract: This work aimed to synthesize new platinum and uranium compounds with nicotinic acid. In this article we describe the synthesis of two new anionic complexes $(HNic)_2[PtCl_6]$ and $(HNic)_2[UO_2Cl_4]$ using wet chemistry methods. The structure of the obtained single crystals was established by single-crystal X-ray diffraction. The Hirshfeld surface analysis of the obtained complexes and their analogue $(HNic)_2[SiF_6]$ was carried out for the analysis of intermolecular interactions. Hydrogen bonds (H···Hal/Hal···H and O···H/H···O) make the main contribution to intermolecular interactions in all compounds. Other important contacts in cations in all compounds are H···H, C···H/H···C and C···Hal/Hal···C; in anions H···Hal/Hal···H. The Pt-containing complex has a halogen-π interaction and halogen bonds, but Si-containing complex has a π–π staking interaction; these types of interactions are not observed in the U-containing compound.

Keywords: X-ray diffraction analysis; Hirshfeld surface analysis; nicotinic acid; platinum; uranium; π-interactions; halogen bonds; hydrogen bonds

Citation: Novikov, A.P.; Volkov, M.A.; Safonov, A.V.; Grigoriev, M.S. Synthesis, Crystal Structure, and Hirshfeld Surface Analysis of Hexachloroplatinate and Tetraclorouranylate of 3-Carboxypyridinium—Halogen Bonds and π-Interactions vs. Hydrogen Bonds. *Crystals* **2022**, *12*, 271. https://doi.org/10.3390/cryst12020271

Academic Editors: Aleksej Zarkov, Aivaras Kareiva and Loreta Tamasauskaite-Tamasiunaite

Received: 30 January 2022
Accepted: 15 February 2022
Published: 17 February 2022

1. Introduction

The synthesis and investigation of structural features of biomolecular *f*-element compounds are of great fundamental and practical interest in various fields including medicine, radioecology, and toxicology. In our work, we pay attention to niacin, also known as nicotinic acid or vitamin B_3, which plays a significant role in cellular energy metabolism [1]. Since nicotinic acid contains not only a carboxyl group but also a nitrogenous heterocycle, it has great potential to form compounds with metals using various mechanisms. One of the mechanisms is realized through non-covalent interactions between the metal atom and the nitrogen atom or π-electron density of the ring [2]. Weak non-covalent interactions are important for the design, operation, and efficiency of molecular sensors and switches [3,4]. These contacts include anion–π interaction (including the usual anion–π, anion–π–cation and anion–π–π interactions) as well as π-stacking, or cation–π-interactions. Since interactions of this type provide rapid association-dissociation between molecules, they are of great importance in the formation of biological self-assembling systems and pharmaceuticals [5].

Previous works have noted the role of intermolecular hydrogen bonds (HB) and π–π interactions in cocrystallization [6] and self-association [7,8] for nicotinic acid and its derivatives. The use of nicotinic acid in organic frameworks with various metals is also based on various non-valent interactions [9,10]. Tetranuclear and hexanuclear platinum-containing metallacycles based on non-covalent interactions are known [11]. Nicotinic acid metal complexes show various biological activities (antimicrobial, anti-inflammatory, analgesic, anti-tubercular, anticancer, etc.) [12].

Organic platinum compounds (dichlorodiammineplatinum, diammineplatinum, etc.) are used as cytostatics in the treatment of various forms of cancer [13,14]. Currently, a search is underway for new platinum compounds for use in medicine, including the use of nicotinic acid [15,16].

The coordination chemistry of *f*-elements, especially their halogen derivatives, makes it possible to form many intermolecular interactions. This opens up a large range of functionalities for such compounds. In [4], the platinum complexes [$(tBu_2bpy)Pt(C{\equiv}CAr)_2$] ($tBu_2bpy$ = 4,4′-bis-tert-butyl-2,2′-bipyridine, Ar = 4-pyridyl, 1; 3-pyridyl, 2; 2-pyridyl, 3; 4-ethynylpyridyl, 4; 2-thienyl, 5; pentafluorophenyl, 6) were synthesized and the following types of interactions have been described: C-H···π (C≡C), C-H···N(py), Cl···Cl, and C-H···F-C.

The study of uranium compounds with biomolecules is important in toxicology for detecting their toxic effects in cells [17]. In radioecology, the study of various uranium compounds with biologically active molecules is important for assessing possible mechanisms of uranium immobilization in the environment in plants, algae, fungi, and microorganisms. Previously, we have shown the effect of uranium immobilization due to carbonyl groups of o-polysaccharide bacteria living in uranium-contaminated groundwater [18]. In [19], uranium complexes with nicotinic acid in solutions were observed, and their formation constants were determined. The formation of uranium complexes with nicotinic acid has also been shown in [20].

Numerous works have noted the ability of uranium-containing compounds to participate in intermolecular interactions [21–24]. Thus, the complexes of uranyl cation with 4-halopyridinium ions (Hal = Cl, Br, I) can be divided into two categories based on the different modes of hydrogen bonding and halogen–halogen interactions present in the crystal structures [25].

Besides its involvement in nuclear energy, organic complexes of uranium are used in photoluminescence [26], magnetism [27], and catalytic [28,29].

In the major part of known Pt and U compounds with nicotinic acid, this acid enters into the coordination sphere of central atoms to form M-N or M-O coordination bonds. Thus, we aimed to obtain compounds of Pt(IV) and U(VI) with nicotinic acid in cationic form and to describe the crystal structure and intermolecular interactions.

2. Materials and Methods

2.1. Materials

Complexes 1 and 2 were synthesized using hexachloroplatinic (IV) acid hexahydrate ($H_2PtCl_6 \times 6H_2O$, ~40% Pt), hydrochloric acid (HCl, 12.5 M), and nicotinic acid ($C_6H_5NO_2$ > 98%) all being ACS reagent grade chemicals purchased from Sigma-Aldrich (St. Louis, Missouri, United States) and used without additional purification. Uranyl acetate dihydrate ($UO_2(CH_3COO)_2 \times 2H_2O$ > 99%) was purchased from PA "IZOTOP". (Moscow, Russia) To prepare the solutions, twice-distilled water was used, whose electric resistance was not less than 18 MOhm cm.

2.2. Synthesis of Complexes bis(3-pyridinecarboxylic Acid) Hexachloroplatinate, $(HNic)_2[PtCl_6]$ (I)

In a 10 mL two-neck flask equipped with a thermometer and a reflux condenser, 10 mg of nicotinic acid was dissolved in 5 mL of 0.1 M HCl at room temperature. After dissolving all the nicotinic acid, 0.1 mL of an aqueous solution of 0.7 mol/L platinic acid was added to the flask. The mixture was stirred for 5 min at room temperature, then the temperature was raised to 70 °C and held for 15 min. The resulting solution was cooled to room temperature without stopping stirring the flask, then removed the return condenser and thermometer; the flask with the solution was transferred to a vacuum desiccator and left for 12 h over anhydrous $Mg(ClO_4)_2$. The resulting yellow needles were washed with two 5 mL portions of cold methanol and dried in an oil pump vacuum at room temperature. The yield was 81% for nicotinic acid. (Mr = 654 g/mol).

2.3. Synthesis of Complexes bis(3-pyridinecarboxylic Acid) Uranile Tetrachloride, $(HNic)_2[UO_2Cl_4]$ (II)

In a 10 mL two-necked flask equipped with a thermometer and a reflux condenser, 10 mg of nicotinic acid was dissolved in 2 mL of 0.1 M HCl. After dissolving all the nicotinic acid, 3 ml of a 0.01 M solution of uranyl acetate in 0.1 M HCl was added. The solution was stirred at room temperature for 5 min, then the mixture's temperature was gradually raised to 70 °C and held for 10 min. The resulting solution was cooled to room temperature without stopping stirring the flask, then removed the return condenser and thermometer; the flask with the solution was transferred to a vacuum desiccator and left for 12 h over anhydrous $Mg(ClO_4)_2$. The resulting yellow fluorescent crystals were washed with 5 mL of cold methanol and dried in vacuo at room temperature. The yield was 97% for nicotinic acid.

The precipitate **I** obtained was characterized by X-ray powder diffraction analysis; the diffraction pattern (Figure S1) was in a good agreement with the theoretical one and does not contain peaks corresponding to the starting substances. For compound **II**, the amount of the precipitate was insufficient for X-ray powder analysis. Some good crystals were selected for single-crystal X-ray diffraction studies.

2.4. Powder XRD Analysis

X-ray phase analysis was performed on a PANalytical AERIS diffractometer (Malvern, Worcestershire, United Kingdom; Almelo, Netherlands) with a Cu anode (40 kV). The analysis was performed over a 28 angle range from 3 to 70°.

2.5. Single-Crystal XRD Analysis

The crystal structure of all synthesized substances was determined by X-ray structural analysis using an automatic four-circle area-detector diffractometer Bruker KAPPA APEX II (Billerica, MA, USA) with MoKα radiation at a temperature of 100 and 296 K (for **I**) and 100 K (for **II**). The cell parameters were refined over the entire data set and data reduction by using SAINT-Plus software [30]. Absorption corrections were introduced using the SADABS program [31]. The structures were solved directly with the *SHELXS97* [32] and refined by SHELXL-2018/3 [33]. The N- and C-bound hydrogen atoms were placed at calculated positions. The O-bound H atoms were located from different Fourier maps and refined with fixed isotropic displacement parameters [$U_{iso}(H) = 1.2\ U_{eq}(O)$]. The hydrogen atoms of the OH groups in **I** were disordered over two positions with occupancies close to 0.5 (in final refinement, the occupancies were fixed at 0.5 at both temperatures). Tables and pictures for structures were generated by Olex2 [34].

Crystal data, data collection, and structure refinement details are summarized in Table 1. All other crystallographic parameters of structures **I** and **II** are indicated in Tables S1–S10. The atomic coordinates were deposited at the Cambridge Crystallographic Data Centre [35], CCDC № 2145387-2145389 for **I** at 296 K, at 100 K, and **II**, respectively. The Supplementary crystallographic data can be obtained free of charge from the Cambridge Crystallographic Data Centre via www.ccdc.cam.ac.uk/data_request/cif (accessed on 30 January 2022).

Table 1. Crystal data and structure refinement for structures **I** and **II**.

	I		II
Empirical formula	$C_{12}H_{12}N_2O_4Cl_6Pt$	$C_{12}H_{12}N_2O_4Cl_6Pt$	$C_{12}H_{12}N_2O_6Cl_4U$
Formula weight	656.03	656.03	660.07
Temperature/K	296(2)	100(2)	100(2)
Crystal system	monoclinic	monoclinic	monoclinic
Space group	$P2_1/n$	$P2_1/n$	$P2_1/n$
a/Å	9.0534(3)	8.9552(14)	6.6653(2)
b/Å	9.4852(3)	9.4270(16)	18.4856(5)
c/Å	11.9423(3)	11.8579(19)	7.4266(2)
α/°	90	90	90
β/°	110.521(1)	110.400(8)	95.0330(10)
γ/°	90	90	90
Volume/Å^3	960.45(5)	938.3(3)	911.52(4)
Z	2	2	2
Z′	0.5	0.5	0.5
ρ_{calc}g/cm^3	2.268	2.322	2.405
μ/mm^{-1}	8.159	8.352	9.521
F(000)	620.0	620.0	612.0
Crystal size/mm^3	0.2 × 0.18 × 0.1	0.32 × 0.12 × 0.1	0.16 × 0.14 × 0.06
Radiation	MoKα (λ = 0.71073)	MoKα (λ = 0.71073)	MoKα (λ = 0.71073)
2Θ range for data collection/°	8.194 to 59.998	8.44 to 59.988	8.182 to 70
Index ranges	$-11 \le h \le 12, -12 \le k \le 13, -16 \le l \le 16$	$-12 \le h \le 12, -13 \le k \le 13, -16 \le l \le 16$	$-8 \le h \le 10, -29 \le k \le 29, -11 \le l \le 11$
Reflections collected	10,006	14,377	18,010
Independent reflections	2794 [R_{int} = 0.0205, R_{sigma} = 0.0201]	2735 [R_{int} = 0.0559, R_{sigma} = 0.0416]	4002 [R_{int} = 0.0283, R_{sigma} = 0.0253]
Data/restraints/parameters	2794/2/122	2735/2/121	4002/0/115
Goodness-of-fit on F^2	1.048	1.030	1.284
Final R indexes [I >= 2σ (I)]	R_1 = 0.0160, wR_2 = 0.0372	R_1 = 0.0235, wR_2 = 0.0529	R_1 = 0.0310, wR_2 = 0.0546
Final R indexes [all data]	R_1 = 0.0234, wR_2 = 0.0404	R_1 = 0.0323, wR_2 = 0.0568	R_1 = 0.0398, wR_2 = 0.0562
Largest diff. peak/hole/e Å^{-3}	0.78/−0.69	0.82/−1.21	1.47/−2.32

3. Results and Discussion

3.1. Structural Description

Both compounds crystallize in the space group $P2_1/n$ with two formula units in the unit cell. The asymmetric units are illustrated in Figure 1. Platinum and uranium atoms are in special positions two *a* and 2 *b*, respectively. In both compounds, the platinum and uranium atoms are octahedrally coordinated. Pt—Cl distances are changed from 2.3090(8) Å to 2.3180(8) Å in **I** at 100 K, and the average distance is 2.313 Å, which can be seen from Table 2. With the increase in temperature to 296 K, this distance slightly increases and amounts to 2.317 Å. At both temperatures, the Cl—Pt—Cl angles in I are close to 90° (Table 2).

Table 2. Selected geometric parameters (Å, °) for **I**.

100 K				296 K	
Pt1—Cl1	2.3180(8)	Cl2—Pt1—Cl3	90.89(3)	Pt1—Cl1	2.3220(6)
Pt1—Cl2	2.3090(8)	Cl2—Pt1—Cl1	89.06(3)	Pt1—Cl3	2.3139(5)
Pt1—Cl3	2.3107(8)	Cl3—Pt1—Cl1	90.43(3)	Pt1—Cl2	2.3144(5)

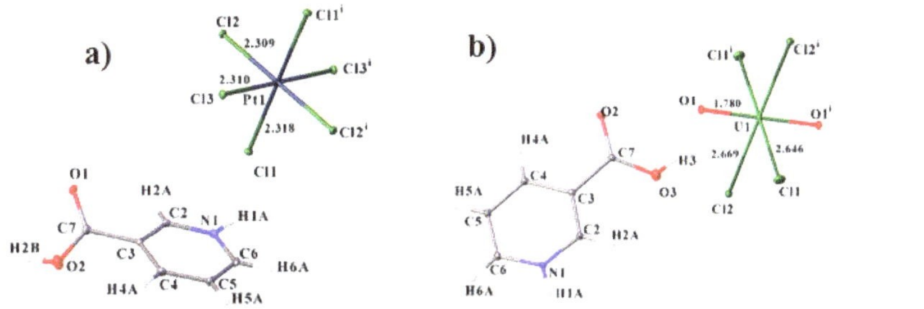

Figure 1. Molecular structure of **I** (**a**) and **II** (**b**) at 100 K, including atom labeling. Only one position of the disordered H-atom of the carboxyl group is shown in **I**. Displacement ellipsoids are drawn at the 50% probability level.

Uranyl tetrachloride forms a bipyramid. The average U–Cl distance in **II** is 2.658 Å, which is slightly shorter than in previous works in the CSD, where it was 2.67 Å. The U—O distance is 1.780(2) Å, which is longer than the average distance at 100 K in CSD, where it was 1.772 Å. This may be due to the participation of the oxygen atom of the anion in H-bonds. The angles Cl—U—Cl and O—U—Cl are close to 90°, as shown in Table 3. The position of the protonated oxygen atom in **I** and **II** is different relative to the nitrogen atom of the six-membered ring (torsion angles C4-C3-C7-O2(3) are −28.85° in **I** and −162.50° in **II**).

Table 3. Selected geometric parameters (Å, °) for **II**.

U1—O1	**1.780(2)**	**O1—U1—Cl1**	**89.81(8)**
U1—Cl1	2.6461(8)	O1—U1—Cl2	89.86(8)
U1—Cl2	2.6694(7)	O1—U1—Cl1	89.81(8)

The hydrogen bond system of **I** is shown in Figure 2a. In **I**, one bifurcate hydrogen bond O2–H2B···Cl1^{i} and O2–H2B···Cl3^{i} [symmetry code: (i) $-x+1$, $-y+1$, $-z+2$] is formed (Table 4). The strongest hydrogen bond is formed by N1–H1A with O1^{iii} of another cation with an N···O distance of 2.728(4) Å [symmetry code: (iii) $x-1/2$, $-y+3/2$, $z-1/2$]. In **II**, one bifurcate hydrogen bond is also formed (Figure 2b). N1 – H1A forms a bifurcate H-bond with Cl2^{i} and O2^{ii} [symmetry codes: (i) $x+1$, y, $z+1$; (ii) x, y, $z+1$]. In both compounds, the CH groups of the six-membered rings form weak H-bonds of the C—H···Cl type, which provide additional bonding of cations and anions in the compounds, such as in [36,37]. In **I**, only one C—H···Cl bond is formed, in **II** three (Table 5), but two of them in **II** are weaker than in **I**.

Table 4. Hydrogen-bond geometry for **I** at 100 K (Å, °).

D—H···A	D—H	H···A	D···A	D—H···A
O2—H2B···Cl1 ^{i}	0.85(2)	2.70(5)	3.486(3)	155(10)
O2—H2B···Cl3 ^{i}	0.85(2)	2.79(8)	3.401(3)	131(9)
O2—H2C···Cl3 ^{ii}	0.86(2)	2.61(4)	3.413(3)	155(8)
N1—H1A···O1 ^{iii}	0.88	1.87	2.728(4)	163.0
C2—H2A···Cl1 ^{iv}	0.95	2.66	3.500(3)	148.0

Symmetry codes: (i) −x+1, −y+1, −z+2; (ii) x−1/2, −y+1/2, z+1/2; (iii) x−1/2, −y+3/2, z−1/2; (iv) −x+1/2, y+1/2, −z+3/2.

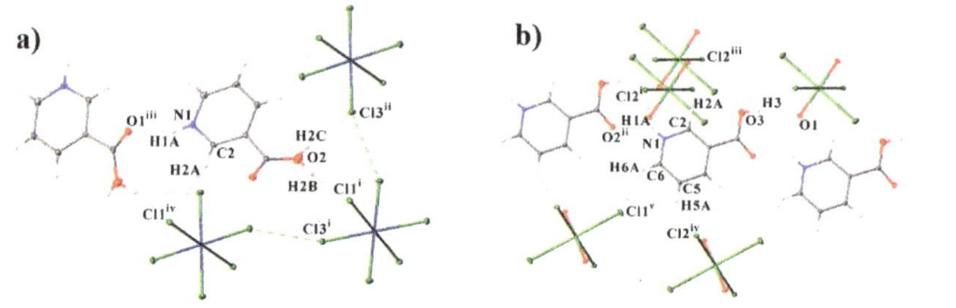

Figure 2. View showing the hydrogen and halogen bonds in **I** (**a**) [Symmetry codes: (i) $-x+1$, $-y+1$, $-z+2$; (ii) $x-1/2$, $-y+1/2$, $z+1/2$; (iii) $x-1/2$, $-y+3/2$, $z-1/2$; (iv) 3/2−x, 1/2+y, 3/2−z.] and hydrogen bonds in **II** (**b**) [Symmetry codes: (i) $x+1$, y, $z+1$; (ii) x, y, $z+1$]. Both positions of the disordered hydrogen atom of carboxyl group in **I** are shown.

Table 5. Hydrogen-bond geometry for **II** (Å, °).

D—H···A	D—H	H···A	D···A	D—H···A
O3—H3···O1	0.857(3)	2.287(2)	3.125(3)	165.7(2)
N1—H1A···Cl2 ^{i}	0.88	2.81	3.380(3)	123.9
N1—H1A···O2 ^{ii}	0.88	2.00	2.721(3)	138.3
C2—H2A···Cl2 ^{iii}	0.95	2.69	3.611(3)	163.4
C5—H5A···Cl2 ^{iv}	0.95	2.82	3.543(3)	133.8
C6—H6A···Cl1 ^{v}	0.95	2.67	3.411(3)	135.4

Symmetry codes: (i) x+1, y, z+1; (ii) x, y, z+1; (iii) −x+1, −y+1, −z+1; (iv) x+1/2, −y+3/2, z+1/2; (v) −x+3/2, y+1/2, −z+3/2.

The crystal packing in **I** can be represented as cationic and anionic layers parallel to the plane (-101) (Figure 3a). The cations in the layers are linked by hydrogen bonds of the N–H···O type and are connected to the anionic layers by hydrogen bonds of the O–H···Cl type. Anions in **I** are interconnected by halogen bonds Cl2···Cl3^{v} with a distance of 3.2074(12) Å [symmetry code: (v) 3/2−x, 1/2+y, 3/2−z], as in [38]. The crystal packing in **II** can also consist of cationic and anionic layers parallel to the plane (010). In this case, cations interact with anions and other cations, but anions do not interact, unlike in **I**.

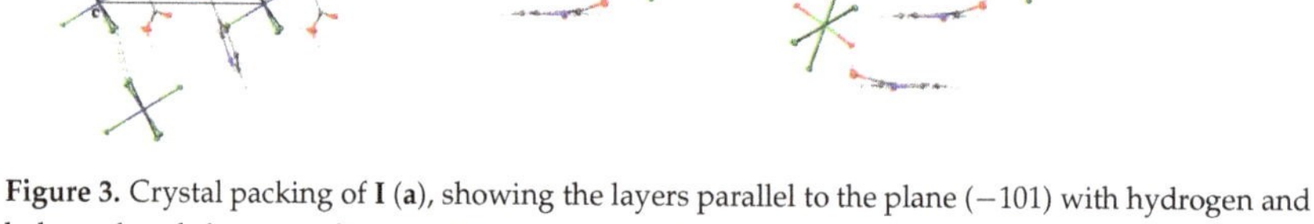

Figure 3. Crystal packing of **I** (**a**), showing the layers parallel to the plane (−101) with hydrogen and halogen bonds between them, and **II** (**b**), showing the layers parallel to plane (010) with hydrogen bonds between them.

In the earlier research works only one compound of singly protonated nicotinic acid with an octahedral anion was found, bis(3-carboxypyridinium) hexafluorosilicate (**III**) [39]. In [39], hexafluorosilicate was also added to nicotinic acid by hydrogen bonds. Compound **III**, like ours, crystallizes in $P2_1/n$ sp. gr. However, other layers are formed in **III**, in which the anions do not interact with each other and in which there are no halogen bonds. However, in **III,** layers are stabilized due to the presence of π-stacking interaction between the rings, which is absent in **I** and **II**.

It has been established from the literature data that, upon interaction with nicotinic acid, platinum usually participates in Pt–N bonds and is attached directly to the nitrogen atom of the six-membered ring, as in [40,41]. Only one uranium-containing compound with nicotinic acid is known [42].

3.2. Hirshfeld Surface Analysis

Hirshfeld surface (HS) analysis was proposed as a new means of separating space in molecular crystals. The Hirshfeld surface covers the molecule and determines the volume of space in which the electron density of the promolecule exceeds the density of all neighboring molecules [43]. Fingerprint plots are a convenient way of summarizing the intermolecular contacts present in crystals, decomposing this fingerprint plot into features to identify specific interactions. This method can be used to analyze π-π stacking interactions [44], halogen and hydrogen bonds [45,46], and other weak non-covalent interactions [47,48].

The Crystal Explorer 21 [49] program was used to analyze the crystal and Hirshfeld surface analysis interactions. The donor-acceptor groups are visualized using a standard (high) surface resolution and d_{norm} surfaces, as illustrated in Figure 4. Red spots on the surface of the d_{norm} plot indicate intermolecular contacts involving the hydrogen and halogen bonds. The brightest red spots correspond to the strongest hydrogen bond N—H···O in both compounds and O–H···O in **II** (Figure 4). Weaker red spots correspond to bonds O—H···Cl in **I** and C—H···Cl for cations in both compounds. For the anion, red spots on the surface d_{norm} correspond to hydrogen bonds in **I** and **II**, and halogen bonds in **I**.

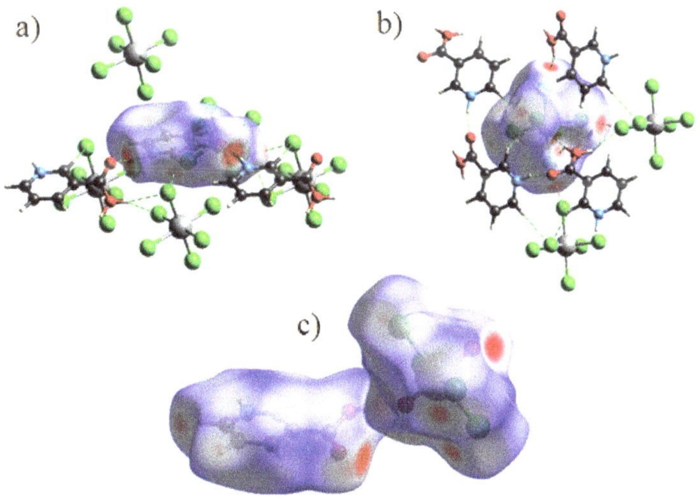

Figure 4. Hirshfeld surface mapper over d_{norm} for (**a**) cation and (**b**) for anion to visualize the interactions in compound **I** and in **II** (**c**).

π-stacking interactions are absent in **I** and **II**, which can be seen from the absence of characteristic red and blue triangles on the shape-index surface (Figure 5).

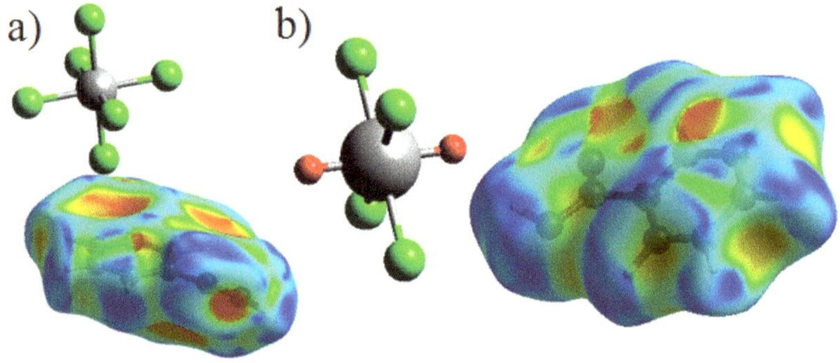

Figure 5. Hirshfeld surface mapper shape-index of cation in **I** (**a**) and **II** (**b**).

On the Hirschfeld surface d_{norm} above the six-membered ring, a weak red spot corresponds to the C···Cl interaction, which is also seen from the bright red spot on the shape index surface. The C6···Cl distance in **I** is 3.299 Å, which is shorter than the sum of the van der Waals radii of C and Cl. The distance between the center of the ring and the chlorine atom Cl1 in **I** at 100 K is 3.431 Å (3.489 Å at 296 K), and the angle α (Figure 6) is 70.97° (70.92° at 296 K). This contact can be called an anion-π interaction since the distance between the center of the ring and the anion is less than 5 Å, and the angle α is greater than 50 ° [3,50,51].

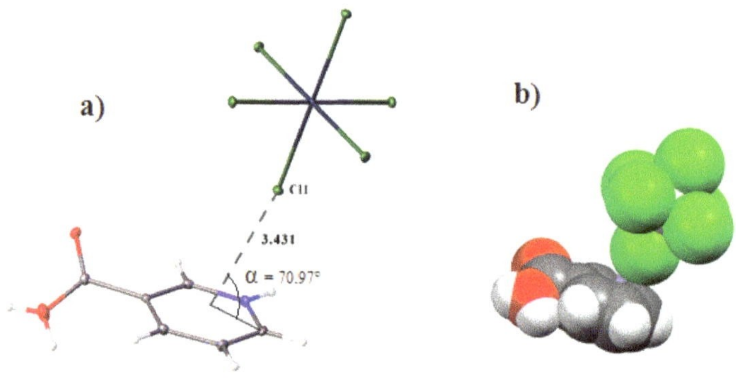

Figure 6. View showing halogen-π interactions in the structure **I** at 100 K (a) and "spacefill" image (b).

Since there are analogies with $(HNic)_2[SiF_6]$, the Hirschfeld surfaces were also constructed and compared for this derivative of the nicotinium cation. Fingerprint plots for compounds **I–III** are shown in Figures S1–S6 and can be used to highlight specific short interactions. Separately, 2D branchings for cations and anions were built. It should be noted that the shape of fingerprint graphs for different types of contacts in compounds **I–III** is different. For example, the O···H/H···O hydrogen bonds, although there are two elongated peaks in **I** and **II**, they are sharper in **I** than in **II**, and in **III** they look like something in between. In this case, all contacts occupy approximately the same position on the fingerprint plots. The different anionic geometry can explain the different shapes of the fingerprint graphs in the compounds and the substitution of chlorine atoms by other atoms.

The temperature changes do not introduce significant changes in the percentage contribution of each type of contacts in cation **I**, as shown in Figure 7. In all compounds in cations, the main contribution to intermolecular interactions is made by hydrogen bonds (contacts H···Hal/Hal···H and O···H/H···O). The van der Waals interactions H···H and C···H/H···C also play an essential role in the cations of all compounds. The total contribution of hydrogen bonds of the type H···Hal/Hal···H and O···H/H···O in cations in **I** and **II** is the same and is approximately 50%. However, the proportion of van der Waals interactions changes from **I** to **II**. The share of H···H decreases, while the share of C···H/H···C, on the contrary, increases. The proportion of contacts of the C···Cl/Cl···C type decreases, which in compound **I** is explained by the presence of the anion–π interaction. The replacement of the anion leads to an increase in the proportion of H···Hal/Hal···H and O···H/H···O contacts in the cation in compound **III** compared to **I** and **II**.

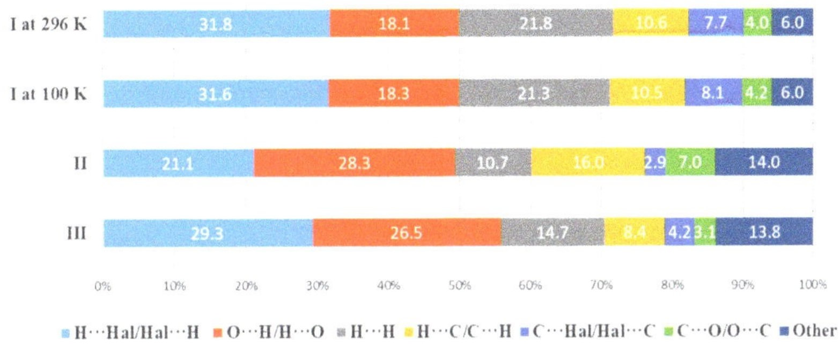

Figure 7. Percentage contributions to the Hirshfeld surface area for the various close intermolecular contacts for cations in **I–III**.

In the anions in compound **I**, a temperature change also does not affect the fraction of intermolecular interactions (Figure 8). In the transition from **I** to **II** in anions, the proportion of hydrogen bonds in the total increases. In **II**, the ratio of C···Cl/Cl···C and Cl···Cl interactions strongly decreases, which, as noted above, is due to the absence of anion–π interaction and halogen bonds in **II**. The replacement of a chlorine atom by fluorine leads to an increase in the proportion H···Hal/Hal···H, which may not all be called hydrogen bonds [52], but the virtual absence of other types of interactions. This may be because fluorine is not characterized by the formation of both hydrogen bonds and different types of non-valent interactions.

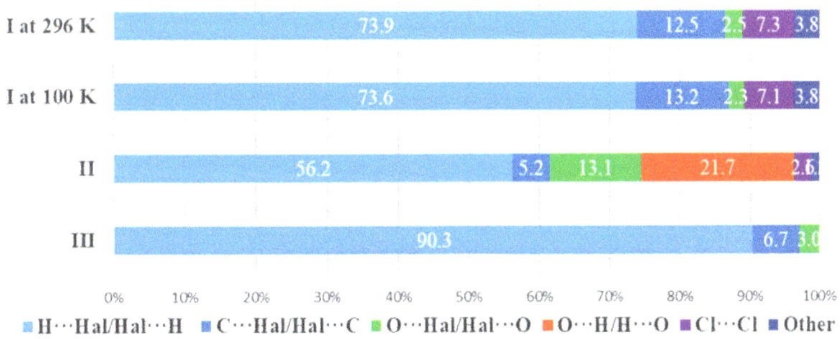

Figure 8. Percentage contributions to the Hirshfeld surface area for the various close intermolecular contacts for anions in **I–III**.

4. Conclusions

Although the compounds are similar (the same cation and octahedral or pseudooctahedral anion) and crystallize in the same space group, the compounds are not isostructural. They form a different system of hydrogen bonds and packing. For the first time, uranium and platinum compounds with protonated nicotinic acid were obtained, in which there are no M-N or M-O coordination bonds. These compounds are stabilized by classical hydrogen bonds of the N–H···Cl/O and O–H···Cl/O types, as well as by nonclassical C–H···Cl hydrogen bonds, halogen bonds, and π-anionic interactions in $(HNic)_2[PtCl_6]$. In $(HNic)_2[UO_2Cl_4]$, the structure is stabilized by N–H···Cl/O and O–H···Cl/O hydrogen bonds. Replacement of the anion in $(HNic)_2[SiF_6]$ leads to the formation of another type of packing.

Supplementary Materials: The following are available online at https://www.mdpi.com/article/10.3390/cryst12020271/s1, Figure S1. Experimental and theoretical powder XRD spectra of I at 296 K. Figure S2. The 2D fingerprint plots of interatomic interactions for cation in I at 100 K show the percentages of contacts contributed to the total Hirshfeld surface area of the molecules. Figure S3. The 2D fingerprint plots of interatomic interactions for cation in II at 100 K show the percentages of contacts contributed to the total Hirshfeld surface area of the molecules. Figure S4. The 2D fingerprint plots of interatomic interactions for cation in III at 100 K show the percentages of contacts contributed to the total Hirshfeld surface area of the molecules. Figure S5. The 2D fingerprint plots of interatomic interactions for anion in I at 100 K show the percentages of contacts contributed to the total Hirshfeld surface area of the molecules. Figure S6. The 2D fingerprint plots of interatomic interactions for anion in II at 100 K show the percentages of contacts contributed to the total Hirshfeld surface area of the molecules. Figure S7. The 2D fingerprint plots of interatomic interactions for anion in III at 100 K show the percentages of contacts contributed to the total Hirshfeld surface area of the molecules. Table S1. Bond lengths for I at 100 K. Table S2. Bond angles for I at 100 K. Table S3. Torsion angles for I at 100 K. Table S4. Bond lengths for I at 296 K. Table S5. Bond angles for I at 296 K. Table S6. Hydrogen bonds for I at 296 K. Table S7. Torsion angles for I at 296 K. Table S8. Bond lengths for II at 100 K. Table S9. Bond angles for II at 100 K. Table S10. Torsion angles for II at 100 K.

Author Contributions: Contributions: conceptualization, M.S.G. and A.V.S.; methodology, M.A.V., A.P.N.; validation, M.S.G.; investigation, M.A.V., A.P.N., A.V.S., M.S.G.; writing—original draft preparation, A.P.N., A.V.S., M.S.G.; writing—review and editing, A.P.N., A.V.S. visualization, A.P.N.; supervision, M.S.G.; project administration, M.S.G. All authors have read and agreed to the published version of the manuscript.

Funding: The study was supported by the Ministry of Science and Higher Education of the Russian Federation (program no. 122011300061-3).

Acknowledgments: X-ray diffraction experiments were performed at the Center for Shared Use of Physical Methods of Investigation at the Frumkin Institute of Physical Chemistry and Electrochemistry, RAS.

Conflicts of Interest: The authors declare no conflict of interest.

References

1. Nicotinic Acid | C5H4NCOOH–PubChem. Available online: https://pubchem.ncbi.nlm.nih.gov/compound/Nicotinic-acid (accessed on 29 January 2022).
2. Mahmudov, K.T.; Kopylovich, M.N.; Guedes da Silva, M.F.C.; Pombeiro, A.J.L. Non-covalent interactions in the synthesis of coordination compounds: Recent advances. *Coord. Chem. Rev.* **2017**, *345*, 54–72. [CrossRef]
3. Lucas, X.; Bauzá, A.; Frontera, A.; Quiñonero, D. A thorough anion–π interaction study in biomolecules: On the importance of cooperativity effects. *Chem. Sci.* **2016**, *7*, 1038–1050. [CrossRef] [PubMed]
4. Lu, W.; Chan, M.C.W.; Zhu, N.; Che, C.M.; He, Z.; Wong, K.Y. Structural Basis for Vapoluminescent Organoplatinum Materials Derived from Noncovalent Interactions as Recognition Components. *Chem. A Eur. J.* **2003**, *9*, 6155–6166. [CrossRef] [PubMed]
5. Zhuang, W.R.; Wang, Y.; Cui, P.F.; Xing, L.; Lee, J.; Kim, D.; Jiang, H.L.; Oh, Y.K. Applications of π-π stacking interactions in the design of drug-delivery systems. *J. Control. Release* **2019**, *294*, 311–326. [CrossRef]
6. Dudenko, D.V.; Yates, J.R.; Harris, K.D.M.; Brown, S.P. An NMR crystallography DFT-D approach to analyse the role of intermolecular hydrogen bonding and π–π interactions in driving cocrystallisation of indomethacin and nicotinamide. *CrystEngComm* **2013**, *15*, 8797–8807. [CrossRef]
7. Svenson, J.; Karlsson, J.G.; Nicholls, I.A. 1H Nuclear magnetic resonance study of the molecular imprinting of (−)-nicotine: Template self-association, a molecular basis for cooperative ligand binding. *J. Chromatogr. A* **2004**, *1024*, 39–44. [CrossRef]
8. Gokula, R.P.; Mahato, J.; Tripathi, A.; Singh, H.B.; Chowdhury, A. Self-Assembly of Nicotinic Acid-Conjugated Selenopeptides into Mesotubes. *ACS Appl. Bio. Mater.* **2021**, *4*, 1912–1919. [CrossRef]

9. Trzesowska-Kruszynska, A. On construction of lead coordination polymers derived from N′-(2-hydroxybenzylidene)nicotinohyd razide via covalent and non-covalent interactions. *J. Coord. Chem.* **2014**, *67*, 120–135. [CrossRef]
10. Lu, J.; Zhao, K.; Fang, Q.R.; Xu, J.Q.; Yu, J.H.; Zhang, X.; Bie, H.Y.; Wang, T.G. Synthesis and characterization of four novel supramolecular compounds based on metal zinc and cadmium. *Cryst. Growth Des.* **2005**, *5*, 1091–1098. [CrossRef]
11. Gianneschi, N.C.; Tiekink, E.R.T.; Rendina, L.M. Dinuclear platinum complexes with hydrogen-bonding functionality: Noncovalent assembly of nanoscale cyclic arrays. *J. Am. Chem. Soc.* **2000**, *122*, 8474–8479. [CrossRef]
12. Pahade, P.; Katolkar, P. A systematic review on metal complexes of some medicinal compounds. *Int. J. Appl. Adv. Sci. Res.* **2020**, *5*, 9–21. [CrossRef]
13. Fong, C.W. Platinum anti-cancer drugs: Free radical mechanism of Pt-DNA adduct formation and anti-neoplastic effect. *Free Radic. Biol. Med.* **2016**, *95*, 216–229. [CrossRef]
14. Cohen, G.L.; Ledner, J.A.; Bauer, W.R.; Ushay, H.M.; Caravana, C.; Lippard, S.J. Sequence dependent binding of cis-dichlorodiammineplatinum(II) to DNA. *J. Am. Chem. Soc.* **1980**, *102*, 2487–2488. [CrossRef]
15. Crisp, M.G.; Pyke, S.M.; Rendina, L.M. Dinuclear organoplatinum(II)-methyldiphenylphosphine complexes of nicotinic acid. *J. Organomet. Chem.* **2000**, *607*, 222–226. [CrossRef]
16. Fedorov, B.S.; Fadeev, M.A.; Kozub, G.I.; Aldoshin, S.M.; Aliev, Z.G.; Atovmyan, L.O.; Konovalova, N.P.; Sashenkova, T.E.; Kondrat'Eva, T.A.; Blokhina, S.V. Synthesis and antimetastatic activity of metal complexes based on substituted pyridinecarboxylic acid amides and platinum tetrachloride. *Pharm. Chem. J.* **2009**, *43*, 134–138. [CrossRef]
17. Durbin, P.W.; Kullgren, B.; Ebbe, S.N.; Xu, J.; Raymond, K.N. Chelating agents for uranium(vi): 2. Efficacy and toxicity of tetradentate catecholate and hydroxypyridinonate ligands in mice. *Health Phys.* **2000**, *78*, 511–521. [CrossRef] [PubMed]
18. Safonov, A.V.; Perepelov, A.V.; Babich, T.L.; Popova, N.M.; Grouzdev, D.S.; Filatov, A.V.; Shashkov, A.S.; Demina, L.I.; Nazina, T.N. Structure and gene cluster of the O-polysaccharide from Pseudomonas veronii A-6-5 and its uranium bonding. *Int. J. Biol. Macromol.* **2020**, *165*, 2197–2204. [CrossRef] [PubMed]
19. Raditzky, B.; Schmeide, K.; Sachs, S.; Geipel, G.; Bernhard, G. Interaction of uranium(VI) with nitrogen containing model ligands studied by laser-induced fluorescence spectroscopy. *Polyhedron* **2010**, *29*, 620–626. [CrossRef]
20. Joseph, C.; Raditzky, B.; Schmeide, K.; Geipel, G.; Bernhard, G. Complexation of uranium by sulfur and nitrogen containing model ligands in aqueous solution. *Uranium Min. Hydrogeol.* **2008**, *2008*, 539–548. [CrossRef]
21. Thuéry, P.; Harrowfield, J. Recent advances in structural studies of heterometallic uranyl-containing coordination polymers and polynuclear closed species. *Dalt. Trans.* **2017**, *46*, 13660–13667. [CrossRef]
22. Fortier, S.; Hayton, T.W. Oxo ligand functionalization in the uranyl ion (UO22+). *Coord. Chem. Rev.* **2010**, *254*, 197–214. [CrossRef]
23. Kannan, S.; Kumar, M.; Sadhu, B.; Jaccob, M.; Sundararajan, M. Unusual intramolecular CH···O hydrogen bonding interaction between a sterically bulky amide and uranyl oxygen. *Dalt. Trans.* **2017**, *46*, 16939–16946. [CrossRef] [PubMed]
24. Carter, K.P.; Kalaj, M.; Cahill, C.L. Harnessing uranyl oxo atoms via halogen bonding interactions in molecular uranyl materials featuring 2,5-diiodobenzoic acid and N-donor capping ligands. *Inorg. Chem. Front.* **2017**, *4*, 65–78. [CrossRef]
25. Andrews, M.B.; Cahill, C.L. Utilizing hydrogen bonds and halogen–halogen interactions in the design of uranyl hybrid materials. *Dalt. Trans.* **2012**, *41*, 3911–3914. [CrossRef]
26. Natrajan, L.S. Developments in the photophysics and photochemistry of actinide ions and their coordination compounds. *Coord. Chem. Rev.* **2012**, *256*, 1583–1603. [CrossRef]
27. Baldoví, J.J.; Cardona-Serra, S.; Clemente-Juan, J.M.; Coronado, E.; Gaita-Ariño, A. Modeling the properties of uranium-based single ion magnets. *Chem. Sci.* **2013**, *4*, 938–946. [CrossRef]
28. Fox, A.R.; Bart, S.C.; Meyer, K.; Cummins, C.C. Towards uranium catalysts. *Nature* **2008**, *455*, 341–349. [CrossRef]
29. Arnold, P.L. Uranium-mediated activation of small molecules. *Chem. Commun.* **2011**, *47*, 9005–9010. [CrossRef]
30. *SAINT-Plus*; Version 7.68; Shuker AXS Inc.: Madison, WI, USA, 2012.
31. Sheldrick, G.M. *SADABS*; Bruker AXS: Madison, WI, USA, 2008.
32. Sheldrick, G.M. A short history of SHELX. *Acta Crystallogr. Sect. A Found. Crystallogr.* **2008**, *64*, 112–122. [CrossRef]
33. Sheldrick, G.M. Crystal structure refinement with SHELXL. *Acta Crystallogr. Sect. C Struct. Chem.* **2015**, *71*, 3–8. [CrossRef]
34. Dolomanov, O.V.; Bourhis, L.J.; Gildea, R.J.; Howard, J.A.K.; Puschmann, H. OLEX2: A complete structure solution, refinement and analysis program. *J. Appl. Crystallogr.* **2009**, *42*, 339–341. [CrossRef]
35. Groom, C.R.; Bruno, I.J.; Lightfoot, M.P.; Ward, S.C. The Cambridge Structural Database. *Acta Crystallogr. Sect. B Struct. Sci. Cryst. Eng. Mater.* **2016**, *72*, 171–179. [CrossRef] [PubMed]
36. Liu, M.; Yin, C.; Chen, P.; Zhang, M.; Parkin, S.; Zhou, P.; Li, T.; Yu, F.; Long, S. sp2CH···Cl hydrogen bond in the conformational polymorphism of 4-chloro-phenylanthranilic acid. *CrystEngComm* **2017**, *19*, 4345–4354. [CrossRef]
37. Aakeröy, C.B.; Evans, T.A.; Seddon, K.R.; Pálinkó, I. The C–H···Cl hydrogen bond: Does it exist? *New J. Chem.* **1999**, *23*, 145–152. [CrossRef]
38. Novikov, A.P.; Ryagin, S.N.; Grigoriev, M.S.; Safonov, A.V.; German, K.E. 5,5-Dichloro-6-hydroxydihydropyrimidine-2,4(1H,3H)-dione: Molecular and crystal structure, Hirshfeld surface analysis and the new route for synthesis with Mg(ReO4)2 as a new catalyst. *Acta Crystallogr. Sect. E Crystallogr. Commun.* **2020**, *76*, 1557–1561. [CrossRef] [PubMed]
39. Gel'mbol'dt, V.O.; Minacheva, L.K.; Ganin, E.V.; Sergienko, V.S. Crystal structure of bis(3-carboxypyridinium) hexafluorosilicate. *Russ. J. Inorg. Chem.* **2008**, *53*, 875–878. [CrossRef]

40. Macazaga, M.J.; Rodríguez, J.; Quiroga, A.G.; Peregina, S.; Carnero, A.; Navarro-Ranninger, C.; Medina, R.M. Platinum(IV) complexes of 3- and 4-picolinic acids containing ammine or isopropylamine ligands—Synthesis, characterization, x-ray structures, and evaluation of their cytotoxic activity against cancer cell lines. *Eur. J. Inorg. Chem.* **2008**, *2008*, 4762–4769. [CrossRef]
41. Medina, R.M.; Rodríguez, J.; Quiroga, A.G.; Ramos-Lima, F.J.; Moneo, V.; Carnero, A.; Navarro-Ranninger, C.; Macazaga, M.J. Influence of (Hydroxymethyl)pyridine and Pyridine-carboxylic Acids, in trans-Position to the Isopropylamine and Ammine Ligands, on the Cytotoxicity of Platinum Complexes. *Chem. Biodivers.* **2008**, *5*, 2090–2100. [CrossRef]
42. Mit'kovskaya, E.V.; Mikhailov, Y.N.; Gorbunova, Y.; Serezhkina, L.; Serezhkin, V. X-ray diffraction study of [UO2SO4{NH 2CON(CH3)2}2]. *Russ. J. Inorg. Chem.* **2004**, *49*, 1923–1927.
43. Spackman, M.A.; Jayatilaka, D. Hirshfeld surface analysis. *CrystEngComm* **2009**, *11*, 19–32. [CrossRef]
44. Piña, J.J.; Gil, D.M.; Pérez, H. Revealing new non-covalent interactions in polymorphs and hydrates of Acyclovir: Hirshfeld surface analysis, NCI plots and energetic calculations. *Comput. Theor. Chem.* **2021**, *1197*, 113133. [CrossRef]
45. Marek, P.H.; Urban, M.; Madura, I.D. The study of interactions with a halogen atom: Influence of NH2 group insertion on the crystal structures of meta-bromonitrobenzene derivatives. *Acta Crystallogr. Sect. C Struct. Chem.* **2018**, *74*, 1509–1517. [CrossRef] [PubMed]
46. Novikov, A.P.; Bezdomnikov, A.A.; Grigoriev, M.S.; German, K.E. Synthesis, crystal structure and Hirshfeld surface analysis of 2-(perfluorophenyl)acetamide in comparison with some related compounds. *Acta Crystallogr. Sect. E Crystallogr. Commun.* **2022**, *78*, 80–83. [CrossRef]
47. Psycharis, V.; Dermitzaki, D.; Raptopoulou, C.P. The Use of Hirshfeld Surface Analysis Tools to Study the Intermolecular Interactions in Single Molecule Magnets. *Crystals* **2021**, *11*, 1246. [CrossRef]
48. Tan, S.L.; Jotani, M.M.; Tiekink, E.R.T. Utilizing Hirshfeld surface calculations, non-covalent inter-action (NCI) plots and the calculation of inter-action energies in the analysis of mol-ecular packing. *Acta Crystallogr. Sect. E Crystallogr. Commun.* **2019**, *75*, 308–318. [CrossRef]
49. Spackman, P.R.; Turner, M.J.; McKinnon, J.J.; Wolff, S.K.; Grimwood, D.J.; Jayatilaka, D.; Spackman, M.A. CrystalExplorer: A program for Hirshfeld surface analysis, visualization and quantitative analysis of molecular crystals. *J. Appl. Crystallogr.* **2021**, *54*, 1006–1011. [CrossRef]
50. Novikov, A.P.; Volkov, M.A.; Safonov, A.V.; Grigoriev, M.S.; Abkhalimov, E.V. Synthesis and Characterization of New Guanine Complexes of Pt(IV) and Pd(II) by X-Ray Diffraction and Hirshfeld Surface Analysis. *Crystals* **2021**, *11*, 1417. [CrossRef]
51. Savastano, M.; García, C.; López de la Torre, M.D.; Pichierri, F.; Bazzicalupi, C.; Bianchi, A.; Melguizo, M. Interplay between salt bridge, hydrogen bond and anion-π interactions in thiocyanate binding. *Inorganica Chim. Acta* **2018**, *470*, 133–138. [CrossRef]
52. Howard, J.A.K.; Hoy, V.J.; O'Hagan, D.; Smith, G.T. How good is fluorine as a hydrogen bond acceptor? *Tetrahedron* **1996**, *52*, 12613–12622. [CrossRef]

crystals

MDPI

Article

An Enhanced Oxidation of Formate on PtNi/Ni Foam Catalyst in an Alkaline Medium

Antanas Nacys *, Dijana Šimkūnaitė, Aldona Balčiūnaitė, Aušrinė Zabielaitė, Daina Upskuvienė, Benjaminas Šebeka, Vitalija Jasulaitienė, Vitalij Kovalevskij, Eugenijus Norkus and Loreta Tamašauskaitė-Tamašiūnaitė *

Center for Physical Sciences and Technology, Sauletekio Avenue 3, LT-10257 Vilnius, Lithuania; dijana.simkunaite@ftmc.lt (D.Š.); aldona.balciunaite@ftmc.lt (A.B.); ausrine.zabielaite@ftmc.lt (A.Z.); daina.upskuviene@ftmc.lt (D.U.); benjaminas.sebeka@ftmc.lt (B.Š.); vitalija.jasulaitiene@ftmc.lt (V.J.); vitalij.kovalevskij@ftmc.lt (V.K.); eugenijus.norkus@ftmc.lt (E.N.)

* Correspondence: antanas.nacys@ftmc.lt (A.N.); loreta.tamasauskaite@ftmc.lt (L.T.-T.)

Abstract: In this study, a platinum-coated Ni foam catalyst (denoted PtNi/Ni foam) was investigated for the oxidation of the formate reaction (FOR) in an alkaline medium. The catalyst was fabricated via a two-step procedure, which involved an electroless deposition of the Ni layer using sodium hypophosphite as a reducing agent and the subsequent electrodeposition of the platinum layer. The PtNi/Ni foam catalyst demonstrated enhanced electrocatalytic activity for the FOR in an alkaline medium compared to the Ni/Ni foam catalyst and pure Pt electrode. Moreover, the PtNi/Ni foam catalyst promoted the FOR at more negative potentials than the Pt electrode. This contributed to a significant negative shift in the onset potential, indicating the high activity of the catalyst. Notably, in alkaline media with the PtNi/Ni foam catalyst, the FOR proceeds via a direct pathway mechanism without significant accumulation of poisonous carbonaceous species on the PtNi/Ni foam catalyst.

Keywords: platinum; nickel foam; electroless deposition; formate; oxidation

Citation: Nacys, A.; Šimkūnaitė, D.; Balčiūnaitė, A.; Zabielaitė, A.; Upskuvienė, D.; Šebeka, B.; Jasulaitienė, V.; Kovalevskij, V.; Norkus, E.; Tamašauskaitė-Tamašiūnaitė, L. An Enhanced Oxidation of Formate on PtNi/Ni Foam Catalyst in an Alkaline Medium. *Crystals* **2022**, *12*, 362. https://doi.org/10.3390/cryst12030362

Academic Editor: Stefano Carli

Received: 2 February 2022
Accepted: 7 March 2022
Published: 8 March 2022

1. Introduction

The large-scale commercialization of formic acid (FA)/formate-based fuel cells is challenging because of the unavailability of highly active, durable, and low-cost catalysts for anodic oxidation reactions. The most promising are based on Pt or Pt-group materials (PGMs) [1–5]. Pd and Pd-based catalysts are the most widely studied and are currently regarded as state-of-the-art electrocatalysts for FA/formate oxidation because of their high catalytic activity and good anti-poisoning properties in acidic and alkaline media [6–8]. Combining Pd with other metals or metal nanoparticles, such as Ag [8–10], Cu [10,11], Ni [9], CeO_2 [12], In [13], and B [14], improves the characteristics of the catalysts while simultaneously reducing the cost. In some cases, especially under alkaline conditions, Pd-based catalysts are even more active and less prone to deactivation by carbon monoxide (CO) than Pt-based catalysts [15]; however, their insufficient stability remains a major concern. Meanwhile, significantly less attention has been paid to Pt-based electrodes for the FOR. However, further development is required to mitigate the high cost and low performance of Pt-based catalysts owing to the fast deactivation of the in situ-generated poisonous carbonaceous intermediates, such as CO, and to develop CO-tolerant catalysts. Some efficient ways to overcome the present drawbacks are to decrease the Pt loading and fabricate nanostructured materials with Pt to exploit its mass activity and stability, or decorate and modify catalysts with Pt heteroatoms for better poison resistance [8,16,17]. Another way to enhance Pt electrocatalysis is to control the acidity of fuel solutions by maintaining a low concentration of FA and a high concentration of formate, which results in a smaller volume of FA being adsorbed. This contributes to a lower level of severe Pt poisoning by the adsorbed CO [18]. However, decreasing FA concentration is closely related to decreases in the current density of FA oxidation. Recently, free formate was demonstrated to promote anode oxidation in acidic solutions and suppress the poisoning pathway [18]. The FOR demonstrates faster reaction kinetics, a lower overpotential, and less poisoning of the anode catalyst in an alkaline media than in an acidic media [19–22].

Several conventional design principles for catalysts have been applied to improve their poisoning resistance. The following two strategies have been used: (1) Pt is coupled with other metals such as Ni [23–26], Au [8,27], Bi [28,29], Sb [30], and Rh [31], to prevent CO adsorption on the Pt surface through so-called ensemble and electronic effects, and (2) oxidative removal of adsorbed CO from the Pt surface is facilitated at low potentials by enriching the surface with oxygen-containing species via a bifunctional mechanism. This may be achieved by alloying Pt with oxophilic transition metals or metal oxides, such as NiO_x [32–35], CoO_x [32], Cu_2O [36], FeO_x [37], MnO_x [38,39], and SnO_2 [40]. The presence of oxophilic materials, including Ni-oxy species, in the catalyst composition can assist with the electrochemical dissociation of water to form OH^- ions at a more negative potential than that required for OH^- to be formed in the presence of bare Pt [32,41–43]. This phenomenon aids in the removal of adsorbed surface CO (CO_{ads}) by oxidation, allowing for the development of CO-tolerant catalysts. In addition, the β-NiOOH phase, which originates from the oxidation of the lower oxidation state Ni or NiO through $Ni(OH)_2$, is considered to be the active anti-poison site, and can act as a catalytic mediator through a reversible Ni (II)/Ni (III) redox system to facilitate a charge transfer for the FA oxidation reaction [32]. Moreover, Ni-based oxides/hydroxides are of significant interest as electrocatalysts because of their low price, nontoxicity, and earth-abundant resources. Recently, the spontaneous deposition of Pt on commercial Ni foams (Ni_{foam}) was used to produce Pt-modified 3D Ni electrodes, which are perfect candidates for efficient catalysis owing to the high surface area and synergistic effects between the Pt nanoparticles and the supporting material [44]. This unique three-dimensional structure benefits significantly from the use of noble metals. As a result, Ni foams with a particular architecture and excellent conductivity have been successively used as supports for the preparation of catalysts for various electrochemical reactions, such as hydrogen evolution [45], water splitting in alkaline media [46], direct ammonia oxidation [47], methanol oxidation [48], ethanol oxidation [49], sodium borohydride hydrolysis [50], borohydride oxidation [51], hydrazine oxidation [52], and urea oxidation [53]. However, there are no reports of Ni foam as a support for FA oxidation.

Large surfaces are important for the good spacious dispersion of heterogeneous catalysts, as they provide an increased contact area for reactants to facilitate catalysis. Therefore, we deposited a Ni layer on the Ni foam surface (Ni/Ni_{foam}) using the electroless deposition technique. Subsequently, a thin Pt layer was electrochemically deposited on the surface. The electrocatalytic activity of the $PtNi/Ni_{foam}$ catalysts for FOR was evaluated in alkaline media. Owing to the synergy between the active sites of Pt, the porous structure of the Ni-oxy species layer on the Ni foam, and the extremely high specific surface area of the catalyst, the prepared catalyst demonstrated advanced activity in alkaline media and pronounced resistance to poisoning by carbonaceous species.

2. Materials and Methods

The Ni foam with 20 pores/cm, a bulk density of 0.45 g cm^{-3}, and a thickness of 1.6 mm was purchased from GoodFellow GmbH Supplier (Hamburg, Germany). The thin Ni layer was deposited on Ni foam, using sodium hypophosphite as a reducing agent described in detail in Ref. [50].

The Pt thin layer was electroplated on Ni/Ni_{foam} using the electrolyte containing $PtCl_2(NH_3)_2$, NH_4NO_3, NH_4OH, and $NaNO_2$ (pH 8) at the current density of 1 A dm^{-2} for 40 min. The temperature of the electrolyte was kept at a temperature of 95 °C.

The chemical composition of the samples was analyzed using the X-ray photoelectron spectroscopy method employing Kratos AXIS Supra+ spectrometer (Kratos Analytical, Manchester, UK, 2019) with monochromatic Al Kα (1486.6 eV) X-ray radiation powered at 225 W. The base pressure in the analysis chamber was less than 1×10^{-8} mbar and a low electron flood gun was used as a charge neutralizer. The survey spectra for each sample were recorded at a pass energy of 80 eV with a 1 eV energy step and high-resolution spectra (pass energy—10 eV, in 0.1 eV steps) over individual element peaks. The binding energy scale was calibrated by setting the adventitious carbon peak at 284.8 eV. XPS data were converted to VAMAS format and processed using the Avantage Software (Thermo Scientific, East Grinstead, UK).

The morphology and composition of the prepared coatings were investigated by scanning electron microscopy (SEM) using a SEM/FIB workstation Helios NanoLab 650 with an energy dispersive X-ray (EDX) spectrometer INCA Energy 350 X-Max 20. Pt loading was determined by inductively coupled plasma optical emission spectrometry (ICP-OES). It was determined that the Pt loading in the PtNi/Ni_{foam} catalyst was ca. 71 μg cm^{-2}.

The oxidation of formate (FOR) was investigated using a Zennium electrochemical workstation (ZAHNER-Elektrik GmbH & Co.KG, Kronach, Germany). A conventional three-electrode cell was used for electrochemical measurements. The Ni/Ni_{foam} and PtNi/Ni_{foam} catalysts with a geometric area of 2.45 cm^2 were employed as working electrodes. An Ag/AgCl/KCl (3 M KCl) electrode was used as a reference, Pt sheet with a geometric area of 4 cm^2 was used as a counter electrode. For comparison, the Pt bulk electrode with a geometric area of 1 cm^2. Cyclic voltammograms (CVs) were recorded at a potential scan rate of 50 mV s^{-1} from the open-circuit potential value in the anodic voltammetric scan up to +0.6 V unless otherwise stated in a 1 M NaOH solution containing FA concentration in the range of 0.1–0.7 M at a temperature of 25 °C. All reported potential values are referred to as "Ag/AgCl".

Chronoamperometric measurements were carried out at a constant electrode potential of −0.6 V for 2600 s.

3. Results

3.1. Characterization of the Electrodes

X-ray photoelectron spectroscopy (XPS) was performed to analyze the electronic state of the surface composition of the as-prepared PtNi/Ni_{foam}. The Pt 4f spectra reveal a doublet consisting of high-energy (Pt 4f5/2) and low-energy (Pt 4f7/2) bands (Figure 1a). Deconvolution of the latter reveals two peaks centered at 70.9 and 72.4 eV, indicating that Pt is in two different oxidation states, Pt (0) and Pt (II), respectively, and the lower binding energy (BE) peak at 70.9 eV is clearly dominant. This indicates that almost all Pt species grown on Ni/Ni_{foam} exist in a metallic state. The higher BE peak (72.4 eV) represents PtO or $Pt(OH)_2$ [54].

The Ni $2p_{3/2}$ XPS spectrum splits into three resolved peaks centered at 852.3 eV, 853.9, and 855.8 eV, indicating the presence of Ni, NiO, and $Ni(OH)_2$ species, respectively, on the Ni_{foam} surface (Figure 1b). The dominant BE peak at 855.8 eV indicates that the $Ni(OH)_2$ species predominates in the catalyst composition. Moreover, the BE peak of Pt $4f_{7/2}$ for the PtNi/Ni_{foam} catalyst shifted from 71.20 to 70.91 eV when compared to that for metallic Pt, suggesting that Pt undergoes electronic structural changes when it is coupled with $Ni(OH)_2$ [55]. This shift implies that the Ni atoms occupy the Pt lattice and that the metallic grains are intermixed with $Ni(OH)_2$ [56,57]. Furthermore, the O 1s XPS spectrum reveals three resolved peaks centered at 529.8, 531.3, and 532.8 eV (Figure 1c). The lowest energy contributions at 529.8 and 531.3 eV were assigned to oxide/hydroxide species such as NiO and $Ni(OH)_2$, respectively [45]. The highest BE value (532.8 eV) generally corresponds to physically adsorbed water molecules [58,59]. The coexistence of Pt and $Ni(OH)_2$ species was determined by XPS analysis and is supported by cyclic voltammetry measurements.

The cyclic voltammograms (CVs) of the bare Pt and PtNi/Ni_{foam} electrodes in 1 M NaOH solution measured at a potential scan rate of 50 mV s^{-1} are displayed in Figure 2. The existence of a pair of coupled peaks at ca. +0.46 V is generally attributed to the reversible transformation of Ni (II)/Ni (III) species via the redox reaction between $Ni(OH)_2$ and NiOOH, according to the following Equation (1) [60]:

$$Ni(OH)_2 + OH^- \leftrightarrow NiOOH + H_2O + e^- \quad (1)$$

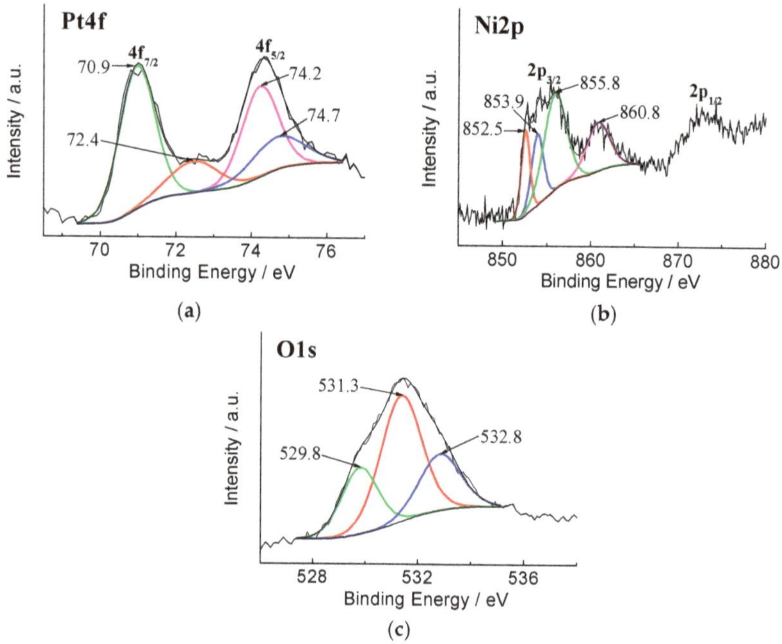

Figure 1. High-resolution XPS spectra of PtNi/Ni_{foam}: (**a**) Pt 4f; (**b**) Ni 2p; (**c**) O1s.

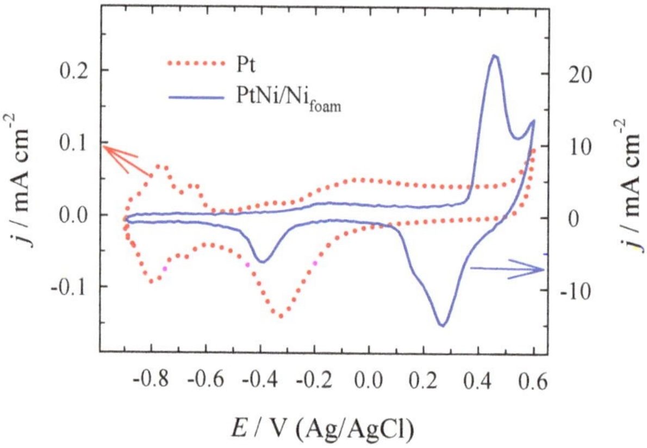

Figure 2. CVs of Pt (red dotted line) and PtNi/Ni_{foam} (solid blue line) were recorded in 1 M NaOH solution at a scan rate of 50 mV s^{-1}.

The presence of Pt on the surface of the Ni/Ni_{foam} electrode was confirmed by the cathodic peak located at ca. −0.4 V in the alkaline 1 M NaOH solution. The electrochemically active surface areas (ECSAs) of the prepared Pt and PtNi/Ni_{foam} catalysts were determined from the CVs recorded in a deaerated 0.5 M H_2SO_4 solution at a scan rate of 50 mV s^{-1} (Figure S1, Supplementary Material) by calculating the charge associated with hydrogen adsorption at 210 μC cm^{-2} [61]. The ECSA for the PtNi/Ni_{foam} electrode is 71 cm^2, and that for the bare Pt substrate is 1.5 cm^2. The above results reveal that the PtNi/Ni_{foam} electrode has significantly higher current values and a larger surface area than the bare Pt substrate, and it outperforms the Pt electrode ca. 47-fold.

The electrochemical performance of the Pt and PtNi/Ni_{foam} catalysts was evaluated by progressively increasing the anodic reverse potential limit (E_{ra}). A series of CVs for Pt and PtNi/Ni_{foam} catalysts were recorded in a 1 M NaOH solution at different E_{ra} values at a scan rate of 50 mV s^{-1} (Figure 3). The CV curves of the bare Pt electrodes (Figure 3a) are consistent with those typically found for bare Pt electrodes in an alkaline solution. They are characterized by the following three potential regions: (a) a hydrogen adsorption/desorption region ($-0.9 < E < -0.6$ V), (b) a double layer region ($-0.6 < E < -0.45$ V), and (c) a hydroxide–oxide formation region ($-0.45 < E < +0.6$ V) [62–64]. The applied potential region can generate two types of oxygen-containing species, namely, PtOH and PtO [65]. This common potential domain, up to ca. −0.10 V, is attributed to OH^- ion electroabsorption with charge transfer and the formation of surface hydroxides ($OH^- \rightarrow OH_{ad} + e^-$ in alkaline solutions) [63,66,67]. This potential domain is associated with reversible and weak OH_{ad} adsorption before the oxide layer formation. A more positive E domain represents the only somewhat reversible formation of strongly bound OH_{ad} species and high-valent oxides. The most positive potential region up to +0.6 V is attributed to stoichiometrically different Pt-OH species that transform into Pt-O^-, Pt=O, or Pt_2=O, depending on the surface coverage of strongly adsorbed OH_{ad}.

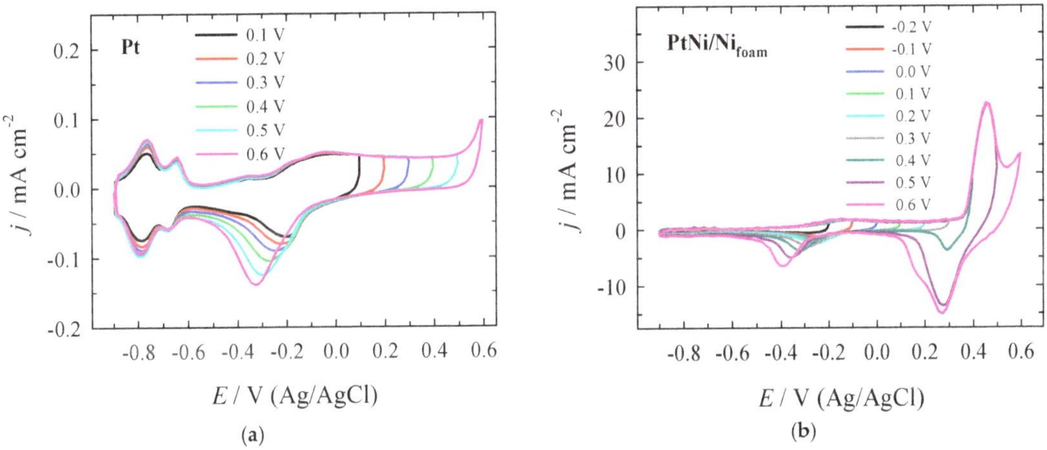

Figure 3. The CVs of Pt (**a**) and PtNi/Ni_{foam} (**b**) catalysts at the different anodic potential limit (E_{ra}) were recorded in 1 M NaOH solution at a scan rate of 50 mV s^{-1}.

The cathodic peak generated in the backward (cathodic voltammetric) scan when the electrode potential was scanned at progressively more positive values was attributed to the reduction of Pt oxides, which increased and shifted to more negative potential values, indicating the formation of irreversible oxide species.

Meanwhile, the hydrogen adsorption/desorption peaks at the bare Pt electrode changed significantly less in response to the applied E_{ra} than those of Pt oxides. However, the processes mentioned above in an alkaline medium are not well separated in potential ranges, as in the case of an acidic medium [68,69]. The coupled H and OH adsorption/desorption processes in alkaline media are thought to occur reversibly up to ca. +0.65 V vs. RHE for Pt (100) and Pt (110), respectively. The OH^- adsorption on Pt in an alkaline medium begins quite low at (ca. +0.35 V vs. RHE) [68,69], as soon as the desorption of hydrogen ceases. Notably, the presence of small quantities of adsorbed OH_{ad} species on the Pt surface was confirmed, even in the H_{upd} region in alkaline solutions [64,66]. Thus, the coupled competing adsorption/desorption processes of H_{upd} and OH_{ad} proceeded in the H_{upd} region.

Similarly, CVs were recorded for the PtNi/Ni_{foam} electrode with a gradual increase in the upper E_{ra} (up to +0.6 V, Figure 3b). In contrast to the CVs determined for the Pt electrode, those for PtNi/Ni_{foam} reveal a pronounced redox current peak pair at ca. +0.46 V, which was attributed to the reversible transformation of Ni (II)/Ni (III) species during the anodic and cathodic scans. With an increase in the upper E_{ra}, a slight negative shift of the Pt oxide formation/reduction potentials was observed, together with an increase in the corresponding current peak values during both anodic and cathodic scans. Meanwhile, almost complete suppression of the peak current was observed in the hydrogen adsorption/desorption domain, unlike in bulk Pt. The corresponding Pt oxide reduction current density peak values (j_{pc}) for the Pt and PtNi/Ni_{foam} catalysts determined in a 1 M NaOH solution with increasing E_{ra} are presented in Table 1. The results clearly reveal that j_{pc} shifted to more negative potential regions. The reduction current density peak values for the PtNi/Ni_{foam} catalysts increased by ca. 39.78 and 46.49 times compared to those for the bare Pt electrode, with increases in E_{ra} of +0.5 or +0.6 V, respectively. These data indicate that when the oxidation of Ni^{2+} to Ni^{3+} is achieved, the reduction of more Pt oxides to metallic Pt are revealed on the backward scan, demonstrating the synergy between the Pt and Ni species.

Table 1. Comparison of electrochemical activities for the bare Pt and PtNi/Ni_{foam} catalysts in a 1 M NaOH solution with respect to different anodic potential limits (E_{ra}).

E_{ra}, V	E_{pc}, V	j_{pc}, mA cm^{-2}	E_{pc}, V	j_{pc}, mA cm^{-2}	j_{pc} PtNi/Ni_{foam}/ j_{pc} Pt
	Pt		PtNi/Ni_{foam}		
+0.1	−0.205	−0.069	−0.235	−2.151	30.77
+0.2	−0.221	−0.079	−0.256	−2.508	31.39
+0.3	−0.252	−0.090	−0.288	−2.988	33.13
+0.4	−0.281	−0.105	−0.321	−3.711	35.32
+0.5	−0.303	−0.125	−0.356	−5.004	39.78
+0.6	−0.332	−0.139	−0.389	−6.489	46.49

3.2. *Electrocatalytic Evaluation for Formate Oxidation*

The electrocatalytic activity of the Pt and PtNi/Ni_{foam} catalysts in a 1 M NaOH solution containing 0.3 M or 0.5 M FA with different E_{ra} was evaluated by the CV measurements at a potential scan rate of 50 mV s^{-1} (Figure 4). Stabilized CVs of FOR for the Pt catalyst plotted in Figure 4a,b reveal one broad peak on the anodic scan, followed by another on the backward scan, regardless of the E_{ra} applied. Meanwhile, the first scans presented in Figure S2 (Supplementary Material) for different FA concentrations are significantly different from the stabilized ones. Three current peaks (labeled Peak I, Peak II, and Peak III) can be observed in the positive-going potential scan (solid lines), and one more peak (labeled Peak IV, dashed lines) appears in the negative-going potential scan. The nature of those peaks can be interpreted regarding the fact that all FA solutions with different concentrations studied in our work are strongly alkaline, and that the pKa of HCOOH is well above ~3.8; therefore, it is inevitable that a formate ($HCOO^-$) oxidation reaction will occur [19]. A dual-pathway mechanism, analogous to FAO in acidic media, takes place for formate oxidation in alkaline media on polycrystalline Pt at pH $\approx$ 14 [19]. Formate typically undergoes oxidation through a reactive intermediate either directly to CO_2, or indirectly through a strongly chemisorbed intermediate (CO_{ad}) that is further oxidized to CO_2 at higher potentials. Additionally, a third independent pathway, involving formate oxidation via the formation of less-active surface formate ($HCOO_{ads}$), and its oxidation to CO_2 on the Pt surface at significantly higher potentials, should be considered [70].

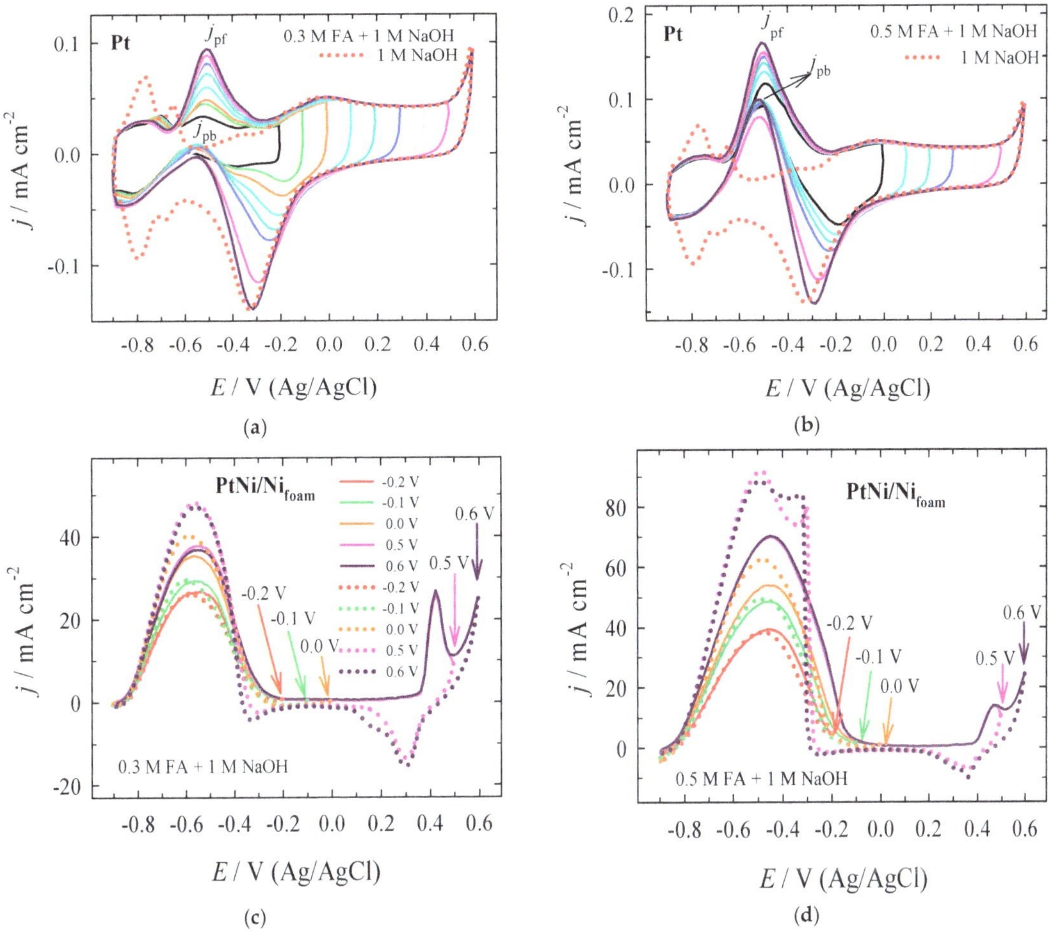

Figure 4. Stabilized CVs of the Pt (**a**,**b**) and PtNi/Ni$_{foam}$ (**c**,**d**) catalysts at different E_{ra} recorded in a 1 M NaOH solution containing different FA concentrations of 0.3 M (**a**,**c**) or 0.5 M (**b**,**d**) at a scan rate of 50 mV s^{-1}. (**a**,**b**) The dotted lines indicate the stabilized CV curve of Pt recorded in a 1 M NaOH solution at a scan rate of 50 mV s^{-1}. (**c**,**d**) Solid lines and dotted lines represent the anodic and cathodic potential scans, respectively, in the CVs.

The first peak (I) at ca. −0.5 V in Figure S2 was assigned to the direct oxidation of formate, and a second hardly discernible peak (II) at ca. −0.35 V was attributed to the indirect oxidation pathway via the strongly chemisorbed intermediate CO_{ads} [19]. The third peak (III) at ca. +0.00 V indicates an independent pathway of formate oxidation via the formation of the less active surface formate ($HCOO_{ads}$) and its oxidation to CO_2 [70]. At ca. −0.5 V, the last peak (IV) on the reverse potential scan developed in the electrode potential region where no surface hydrogen adsorption occurs. The oxide was reduced almost to completion. The properties of peak IV depend on several factors, such as the reduction of surface oxides, CO adsorption on the electrode surface, and mass transfer limitation [70]. As the catalyst active sites were cleaned of adsorbed intermediates at high potentials and the Pt oxides/hydroxides were reduced, this peak is most likely related to the direct oxidation pathway (dehydrogenation). Moreover, the presence of this peak and the potential region of peak I both suggest a similar reaction pathway. Peak IV might have been influenced by the oxidation of adsorbed residual species in this potential region, but this does not seem likely under the conditions of our experiment.

An increase in the number of applied scan cycles leads to a change in the slight shape transformation of the stabilized CVs, resulting in a broader formate oxidation current density peak on the bare Pt electrode (Figure 4a,b). Current density peaks for FO on forward (anodic voltammetric) and backward scans are denoted (j_{pf}) and (j_{pb}), respectively. It is impossible to propose an unambiguous treatment of the origin of the oxidation peak obtained in the case of stabilized CVs on the bare Pt electrode, as the overlapping of the peaks is not excluded, and the merging of direct and indirect FO pathways is presumable.

The stabilized CVs (Figure 4a,b) clearly reveal that a progressive increase in the upper anodic potential limit accelerates the formation of Pt oxides and results in higher Pt oxide reduction peak values on the negative-going potential scans and higher FO current density peak values on the forward-going potential scans of the Pt electrode. These results are similar to those obtained when the FA concentration is increased from 0.3 to 0.5 M.

The increase in the current density peak value for the Pt oxide reduction to metallic Pt occurs with a reduction potential shift to a more negative value. This indicates a more difficult reduction of Pt oxides. Meanwhile, the FO potential for both j_{pf} and j_{pb} remains almost unchanged regardless of the applied E_{ra} value. The j_{pf} value triples if E_{ra} is changed from −0.2 to +0.6 V for FO at the Pt electrode in 0.3 M FA + 1 M NaOH solution (Figure 4a) and is ca. 1.40 times higher in 0.5 M FA + 1 M NaOH solution (Table 2).

Table 2. Summary of electrochemical measurements at Pt catalyst in 1 M NaOH + 0.5 M FA from the data in Figure 4b.

E_{ra}, V	E_{pf}, V	j_{pf}, mA cm^{-2}	E_{pb}, V	j_{pb}, mA cm^{-2}	j_{pf}/j_{pb}	j_{pf} at X V/j_{pf} at +0.0 V
+0.0	−0.493	0.1185	−0.497	0.0913	1.30	1.00
+0.1	−0.490	0.1322	−0.500	0.0976	1.35	1.12
+0.2	−0.500	0.1423	−0.507	0.0978	1.46	1.20
+0.3	−0.495	0.1501	−0.501	0.0943	1.59	1.27
+0.4	−0.499	0.1537	−0.507	0.0875	1.76	1.30
+0.5	−0.493	0.1549	−0.509	0.0785	1.97	1.31
+0.6	−0.507	0.1661	−0.512	0.0989	1.68	1.40

The presence of an extremely clean metallic Pt surface thermodynamically improves the oxidation reaction of CO_{ads}. However, the decrease in j_{pb} with increasing E_{ra} in the forward scan in 0.3 M FA + 1 M NaOH solution is associated with the oxidation of residual incompletely oxidized carbon species on the catalyst surface rather than the oxidation of freshly chemisorbed species. Meanwhile, the absence of a clear correlation between the change in j_{pb} value and the E_{ra} in 0.5 M FA + 1 M NaOH solution could be explained by the competitive oxidation between residual incompletely oxidized carbon species and freshly chemisorbed species due to the activation of the surface by the reduction of Pt oxides and/or hydroxides. The presence of Pt oxides and/or hydroxide oxides (such as PtO_2) on the catalyst surface is thought to help oxidize the CO_{ad}, and thus, reduce catalyst poisoning, similar to the case of PdO_2 and Au_2O_3 in an alkaline media [6,70–72]. Nevertheless, the scope of the results indicates that the increase in the upper E_{ra} appears to favor the direct oxidation pathway in all cases for peaks I and IV.

Similarly, stabilized CVs were recorded for the PtNi/Ni_{foam} catalyst with increasing E_{ra} (Figure 4c,d). One broad current density peak was generated in the negative potential region of the PtNi/Ni_{foam} catalyst, characterized by FO via a direct pathway that was followed by a clear current density peak at positive potentials corresponding to the Ni(OH)/NiOOH surface transformation. The formate oxidation current density peak values increased tremendously with increases in the positively going potential limit. Current density peak values ca. 400 times higher than those obtained with the bare Pt electrode were observed, confirming that the ECSAs of $PtNi/Ni_{foam}$ are much larger and that there are many more active sites that facilitate significantly improved FO catalytic reactions owing to the synergy between the Pt and the porous Ni-coated foam substrate. The increase in FA concentration contributes similarly to current growth. A summary of the electrochemical measurements of the $PtNi/Ni_{foam}$ catalyst in 1 M NaOH with 0.3 or 0.5 M FA (Figure 4c,d) is presented in Tables 3 and 4. The j_{pf} value increased ca. 1.4 or 1.8 times for the $PtNi/Ni_{foam}$ catalyst in 1 M NaOH containing 0.3 M or 0.5 M FA, respectively, when the E_{ra} was changed from −0.2 to +0.6 V. The j_{pf}/j_{pb} ratio varied from ca. 1.0–0.8, confirming that the incorporation of Ni oxy-species effectively improves the electrocatalytic kinetics of the Pt-based catalysts and can prevent the accumulation of incompletely oxidized carbonaceous species (CO_{ads}), directing the FOR toward the dehydrogenation pathway.

Table 3. Summary of electrochemical measurements at $PtNi/Ni_{foam}$ catalyst in 1 M NaOH + 0.3 M FA from the data in Figure 4c.

E_{ra}, V	E_{pf}, V	j_{pf}, mA cm^{-2}	E_{pb}, V	j_{pb}, mA cm^{-2}	j_{pf}/j_{pb}	j_{pf} at X V/j_{pf} at −0.2 V
−0.2	−0.530	26.75	−0.591	25.99	1.03	1.00
−0.1	−0.563	29.30	−0.597	29.57	0.99	1.10
+0.0	−0.570	35.38	−0.582	40.10	0.88	1.32
+0.5	−0.549	37.77	−0.565	48.04	0.79	1.41
+0.6	−0.552	36.80	−0.557	47.22	0.78	1.38

Table 4. Summary of electrochemical measurements at $PtNi/Ni_{foam}$ catalyst in 1 M NaOH + 0.5 M FA for the data in Figure 4d.

E_{ra}, V	E_{pf}, V	j_{pf}, mA cm^{-2}	E_{pb}, V	j_{pb}, mA cm^{-2}	j_{pf}/j_{pb}	j_{pf} at X V/j_{pf} at −0.2 V
−0.2	−0.461	39.48	−0.478	38.40	1.03	1.00
−0.1	−0.446	48.73	−0.490	49.55	0.98	1.23
+0.0	−0.454	54.00	−0.483	62.53	0.86	1.37
+0.5	−0.451	69.91	−0.481	91.95	0.76	1.77
+0.6	−0.449	70.32	−0.496	88.61	0.79	1.78

The influence of the first applied potential scan cycles recorded for the $PtNi/Ni_{foam}$ catalyst for different E_{ra} values in alkaline FA was investigated. For the sake of simplicity, only the positive-going potential CVs with two different E_{ras} (−0.2 and +0.6 V) for the $PtNi/Ni_{foam}$ catalyst in a 1 M NaOH solution, containing different FA concentrations of 0.1, 0.3, and 0.5 M, are presented in Figure 5. Successive cycling in the potential region with an E_{ra} of −0.2 V resulted in an apparent j_{pf} decrease at all FA concentrations applied to the $PtNi/Ni_{foam}$ catalyst. Meanwhile, when the E_{ra} was decreased to −0.6 V to include the potential window for the oxidation of Ni^{2+} to Ni^{3+}, the j_{pf} increased (unlike when E_{ra} = −0.2 V) and increased further with increasing FA concentration, indicating an improvement in FO. Changing the E_{ra} from −0.2 to +0.6 V resulted in increases of ca. 1.3, 1.4, and even 1.8 times for the I_{pf} values, for 0.1, 0.3, and 0.5 M FA containing alkaline solutions, respectively, on the $PtNi/Ni_{foam}$ catalyst. An increase in FA concentration from 0.1 to 0.5 M led to ca. 4.4–6.1-fold increased j_{pf} values when the upper potential limit was −0.2 V or +0.6 V. Furthermore, a decrease in j_{pb} values can be observed on the negative-going potential scans presented in Figure S3 (Supplementary Material), regardless of the E_{ra} value applied. Notably, when E_{ra} = −0.6 V, the forward and backward-going current density peak values become closer to each other, indicating better catalyst tolerance to the accumulation of carbonaceous species.

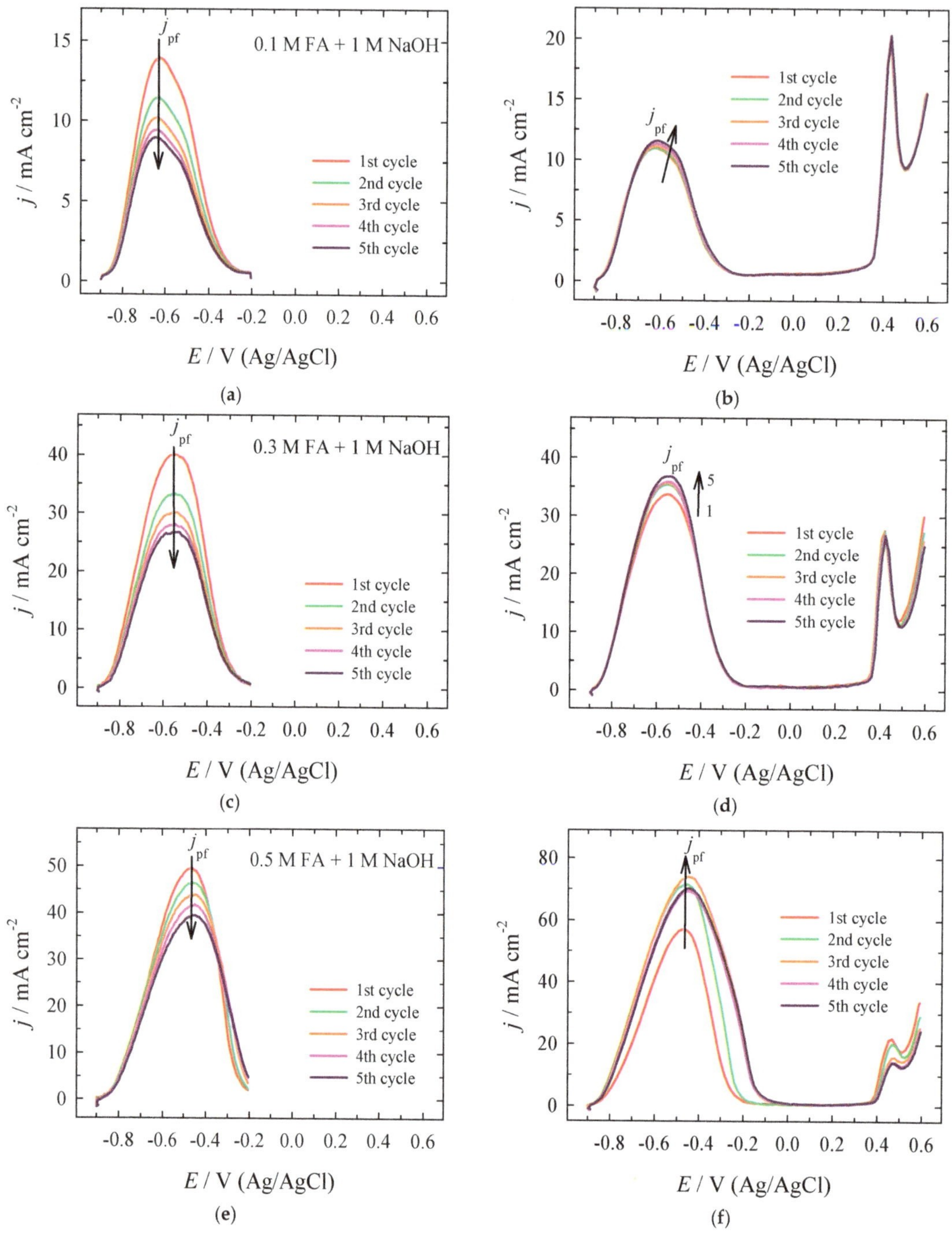

Figure 5. Positive-going potential scans of PtNi/Ni_{foam} recorded at anodic potential limit of +0.2 (**a**,**c**,**e**) and +0.6 V (**b**,**d**,**f**) in a 1 M NaOH solution, containing 0.1 (**a**,**b**), 0.3 (**c**,**d**) and 0.5 (**e**,**f**) M FA, at a scan rate of 50 mV s^{-1}.

FO improves with increasing (more positive) E_{ra} because of the impact of the oxophilic transition metal component in the catalyst composition, which involves the potential region in which the Ni(OH)/NiOOH surface transformation takes place and provides an additional quantity of oxygen-containing species at low potentials, which contributes to the oxidation of adsorbed CO, thus decreasing the catalytic poisoning of the catalyst. In general, nickel hydroxide nanostructures play a crucial role in facilitating the dissociative adsorption of water molecules in alkaline media and the subsequent oxidative removal of carbonaceous poisons from neighboring platinum sites [33–35].

The electrocatalytic activity of the investigated catalysts toward FO on FA concentrations is shown in Figure 6. The insets in Figure 6a,b represent the linear relationship between j_{peak} and C_{FA}, indicating the irreversible electrochemical process kinetics on both Pt and PtNi/Ni_{foam} catalysts [73]. Moreover, the peak potential slightly shifted to the positive potential values with increasing FA concentration (Figure 6a,b). The reaction order of FO oxidation was determined by the slopes of lnj_{peak} vs. lnC_{FA} (Figure 6c) using Equation (2) [73].

$$j_{peak} = zC^{\beta}{}_{FA} \quad (2)$$

where j_{peak} is the peak current density (mA cm^{-2}), z is a constant, C refers to the FA concentration (mol L^{-1}), and β the reaction order concerning C_{FA}. The value of β is obtained as 1.11 on the PtNi/Ni_{foam} catalyst and 1.19 on the bare Pt electrode (Figure 6c). The determined values corresponded to a first-order reaction.

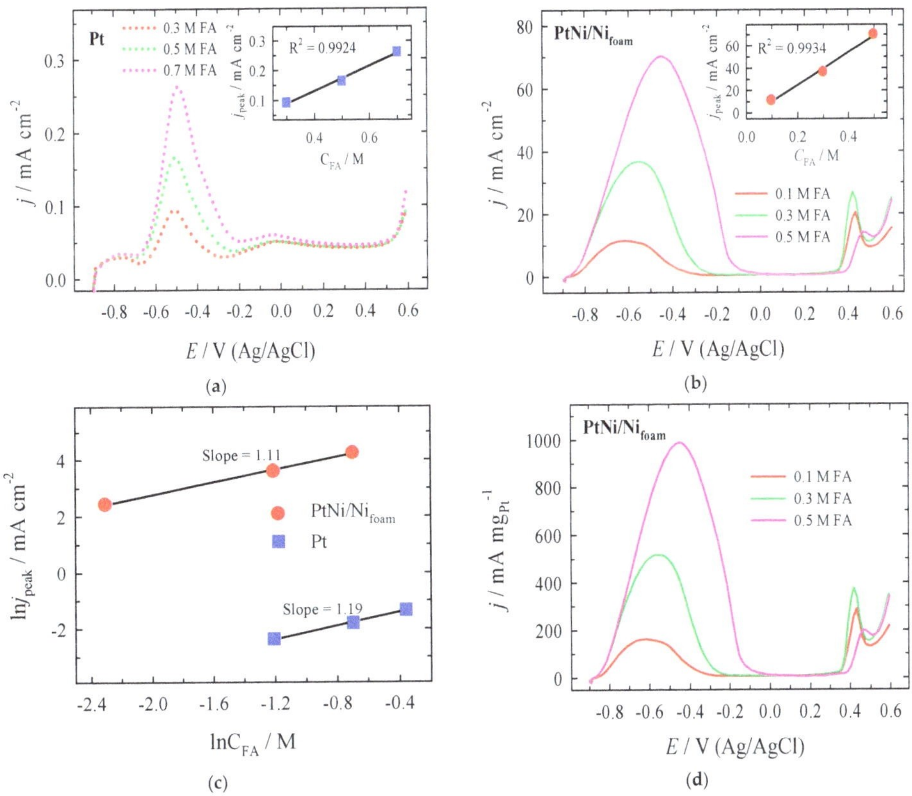

Figure 6. Positive-going potential scans of Pt (**a**) and PtNi/Ni_{foam} (**b**) for different FA concentrations with the corresponding j_{peak} vs. E (the insets) and lnj_{peak} vs. lnC_{FA} (**c**). (**d**) The current density peak values normalized with reference to the Pt loading for PtNi/Ni_{foam}.

Nevertheless, increasing the FA concentration from 0.3 to 0.5 M led to a ca. 1.8-times increase in j_{pf} on both catalysts (Figure 6a,b). However, at all FA concentrations, the j_{pf} on PtNi/Ni_{foam} notably outperforms those on the bare Pt electrode. This result was more than ca. 390–420 times higher in alkaline media on PtNi/Ni_{foam} than on bare Pt. The onset potential of the PtNi/Ni_{foam} (ca. −0.9 V) electrode shifted significantly to a more negative potential region than that of the bare Pt electrode (ca. −0.7 V). The large difference in E_{onset} indicates higher PtNi/Ni_{foam} electrocatalytic activity than that of Pt. Moreover, the obtained characteristics of the PtNi/Ni_{foam} catalyst are consistent with those of the state-of-the-art PGM catalysts for formate oxidation. A comparison of their properties with those of Pt-and Pt-based electrocatalysts used for formate oxidation in alkaline media is presented in Table 5. The results reveal that the developed PtNi/Ni_{foam} catalyst is competitive with state-of-the-art PGM catalysts and outperforms them in many cases. Such favorable results are attributed to the synergy between the Pt and the Ni-coated porous structure of the Ni_{foam} substrate, which prevented the accumulation of incompletely oxidized carbonaceous species (CO_{ads}), directing the FO reaction toward the dehydrogenation pathway.

Table 5. Comparison of Pt and the Pt-based electrocatalysts for formic acid/formate oxidation at ambient conditions in an alkaline media with respect to the onset potential (E_{onset}) and maximum current density (I_{max}, normalized by electrode geometric area) obtained on anodic peaks (at a peak potential (E_{peak})). The data are adopted from [4,8,9,11,12,14,22,70,74–77].

Catalyst	Conditions of Experiment	E_{onset} (V vs. RHE)	I_{max} (mA cm^{-2}) at E_{peak} (V vs. RHE)	Ref.
Pt disk	0.2 M HCOONa + 1 M NaOH (pH ≈ 14.0) 20 mV s^{-1}	+0.3	0.1 mA cm^{-2} (+0.5 V vs. RHE)	[19]
Pt disk	0.4 M $HCOO^-$ + 1 M KNO_3 (pH 13) 5 mV s^{-1}	+0.5	3.1 mA cm^{-2} (+1.1 V vs. RHE)	[74]
Pt nanoparticles (Pt black)	0.5 M HCOOK + 0.5 M KOH (pH ≈ 13.7) 50 mV s^{-1}	+0.2	0.2 mA cm^{-2} (+0.5 V vs. RHE)	[70]
Pt (40%)/ Vulcan carbon	1 M HCOOK + 1 M KOH (pH ≈ 14.0) 20 mV s^{-1}	+0.4	14.6 mA cm^{-2} (+0.6 V vs. RHE)	[22]
Pt (50%)/C	0.5 M HCOOH + 0.5 M KOH (pH ≈ 13.7) 50 mV s^{-1}	+0.8	33.9 mA cm^{-2} (+1.2 V vs. RHE)	[11]
PtAg alloy nanoballoon nanoassembly	1 M HCOOK + 1 M KOH (pH ≈ 14.0) 50 mV s^{-1}	+0.2	32.6 mA cm^{-2} (+0.7 V vs. RHE)	[17]
Pd black	0.5 M HCOOH + 0.5 M KOH (pH ≈ 13.7) 50 mV s^{-1}	+0.8	27.3 mA cm^{-2} (+1.2 V vs. RHE)	[11]
Pd/C	1 M HCOOK + 1 M KOH (pH ≈ 14.0) 50 mV s^{-1}	+0.1	108.8 mA cm^{-2} (+0.7 V vs. RHE)	[4]
Pd (20%)-H/Vulcan carbon	0.5 M HCOOK + 1 M KOH (pH ≈ 14.0) 20 mV s^{-1}	+0.2	71.0 mA c cm^{-2} (+0.8 V vs. RHE)	[76]
Pd (20%)/C	1 M HCOONa + 1 M NaOH (pH ≈ 14.0) 20 mV s^{-1}	+0.2	40.0 mA cm^{-2} (+0.8 V vs. RHE)	[75]
Pd (25%)/Vulcan carbon	1 M HCOOK + 1 M KOH (pH ≈ 14.0) 50 mV s^{-1}	+0.2	4.6 mA cm^{-2} (+0.7 V vs. RHE)	[8]
Pd (40%)/Vulcan carbon	1 M HCOOK + 1 M KOH (pH ≈ 14.0) 20 mV s^{-1}	+0.2	102.0 mA cm^{-2} (+1.0 V vs. RHE)	[22]
Pd/Vulcan carbon	1 M HCOOK + 1 M KOH (pH ≈ 14.0) 50 mV s^{-1}	+0.4	23.0 mA cm^{-2} (+0.8 V vs. RHE)	[77]
Pd (20%)/Reduced graphene oxide	1 M HCOONa + 1 M NaOH (pH ≈ 14.0) 20 mV s^{-1}	+0.2	57.0 mA cm^{-2} (+0.8 V vs. RHE)	[75]

Table 5. *Cont.*

Catalyst	Conditions of Experiment	E_{onset} (V vs. RHE)	I_{max} (mA cm^{-2}) at E_{peak} (V vs. RHE)	Ref.
$Pd_{54}Ag_{46}$ (core-shell)	1 M HCOOK + 1 M KOH (pH ≈ 14.0) 50 mV s^{-1}	+0.2	31.0 mA cm^{-2} (+0.7 V vs. RHE)	[8]
$Pd_{70}Cu_{30}$/Vulcan carbon	1 M HCOOK + 1 M KOH (pH ≈ 14.0) 50 mV s^{-1}	+0.2	4.3 mA cm^{-2} (+0.6 V vs. RHE)	[12]
$Pd_{72}Ce_{28}$/Vulcan carbon	1 M HCOOK + 1 M KOH (pH ≈ 14.0) 50 mV s^{-1}	+0.1	19.4 mA cm^{-2} (+0.6 V vs. RHE)	
$Pd_{2.3}Co$/Vulcan carbon	1 M HCOOK + 1 M KOH (pH ≈ 14.0) 50 mV s^{-1}	+0.3	38.0 mA cm^{-2} (+0.8 V vs. RHE)	
PdNi/Vulcan carbon	1 M HCOOK + 1 M KOH (pH ≈ 14.0) 50 mV s^{-1}	0.2	74.0 mA cm^{-2} (+0.8 V vs. RHE)	
PdNi/Ketjen carbon	1 M HCOOK + 1 M KOH (pH ≈ 14.0) 50 mV s^{-1}	0.2	117.0 mA cm^{-2} (+0.8 V vs. RHE)	[77]
$Pd_{60}Ag_{20}Ni_{20}$ (alloyed)	1 M HCOOK + 1 M KOH (pH ≈ 14.0) 50 mV s^{-1}	0.2	99.6 mA cm^{-2} (+0.8 V vs. RHE	[9]
Pd_2Ag_1 aerogel	0.5 M HCOOK + 0.5 M KOH (pH ≈ 13.7) 50 mV s^{-1}	0.2	27.5 mA cm^{-2} (+0.8 V vs. RHE)	[8]
$Pd_2Ag_1Pt_{0.25}$ aerogel	0.5 M HCOOK + 0.5 M KOH (pH ≈ 13.7) 50 mV s^{-1}	0.1	60.0 mA cm^{-2} (+0.7 V vs. RHE)	
Pd (interstitial B)	0.5 M HCOOK + 1 M KOH (pH ≈ 14.0) 100 mv s^{-1}	0.2	90 mA cm^{-2} (+0.8 V vs. RHE)	[14]
PdCuPt (hierarchical zigzag-branched urchin-like superstructure	0.5 M HCOOH + 0.5 M KOH (pH ≈ 13.7) 50 mV s^{-1}	0.5	102.4 mA cm^{-2} (+1.2 V vs. RHE)	[11]
Pt	0.5 M HCOOH + 1 M NaOH (pH ≈ 14.0) 50 mV s^{-1}	0.34	0.17 mA cm^{-2} (+0.53 V vs. RHE)	Our work
PtNi/Ni_{foam}	0.5 M HCOOH + 1 M NaOH (pH ≈ 14.0) 50 mV s^{-1}	0.14	70.32 mA cm^{-2} (+0.59 V vs. RHE	Our work

To evaluate the mass activity of the investigated PtNi/Ni_{foam} catalyst toward FO, the current density values for different FA concentrations in alkaline media were normalized with reference to the Pt loading for the PtNi/Ni_{foam} catalysts (Figure 6d). The highest FO mass activity of the PtNi/Ni_{foam} catalyst is ca. 990 mA mg_{Pt}^{-1}, larger than Pt/C (ca. 225 mA mg_{Pt}^{-1}) [40], Pt/C (391.2 mA mg_{Pt}^{-1}) [11], and $PtSnO_2$/C (ca. 600 mA mg_{Pt}^{-1}) [40].

The stability of the PtNi/Ni_{foam} catalyst was examined by chronoamperometry in 1 M NaOH solutions containing 0.1, 0.3, and 0.5 M FA (Figure 7). As illustrated, the FOR current density for the PtNi/Ni_{foam} catalyst was the highest in 0.5 M FA + 1 M NaOH media; meanwhile, the current density disappeared completely after only 2000 s in a solution containing a lower concentration of FA. This indicates that an increase in the FA concentration increases the stability of the PtNi/Ni_{foam} catalyst.

Additionally, the morphological measurements confirmed the stability of the PtNi/Ni_{foam} catalyst. Figure S4 (Supplementary Material) shows an SEM view of a thin Ni layer deposited on the Ni_{foam} substrate, using sodium hypophosphite as a reducing agent before the electrochemical deposition of Pt. The Ni layer accurately replicates the three-dimensional network structure of the Ni_{foam} backbone with micro-open cages and wide flow channels inside them (Figure S4a). The SEM image of the Pt layer on the Ni/Ni_{foam} surface after continuous electrode cycling in 0.5 M FA + 1 M NaOH in the potential region from –0.9 to +0.6 V vs. Ag/AgCl is displayed in Figure S4b,c (Supplementary Material). Bright spots of Pt the size of about 12–18 nm are evenly distributed over the entire Ni/Ni_{foam} surface, as depicted in Figure S4c (Supplementary Material). The presence of Pt is further supported by EDX data indicating that the amount of deposited Pt in the PtNi/Ni_{foam} catalyst reaches 39.63 at%. Meanwhile, the amount of Ni is 60.37 at%. Continuous cycling in

FA solution does not significantly change the surface morphology of the catalyst, indicating the high stability and activity of the PtNi/Ni_{foam} catalyst.

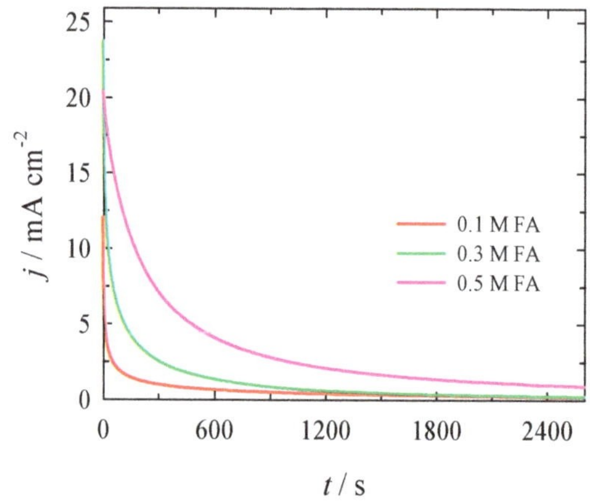

Figure 7. CA data for the PtNi/Ni_{foam} catalyst were recorded in a 1 M NaOH solution containing 0.1, 0.3 and 0.5 M FA, at a constant potential value $E = -0.6$ V for 2600 s.

To confirm the better tolerance toward catalyst poisoning by adsorbed carbonaceous species on the PtNi/Ni_{foam} catalyst compared to that on the bare Pt electrode, the CO stripping parameters were adjusted. The obtained data are shown in Figure 8.

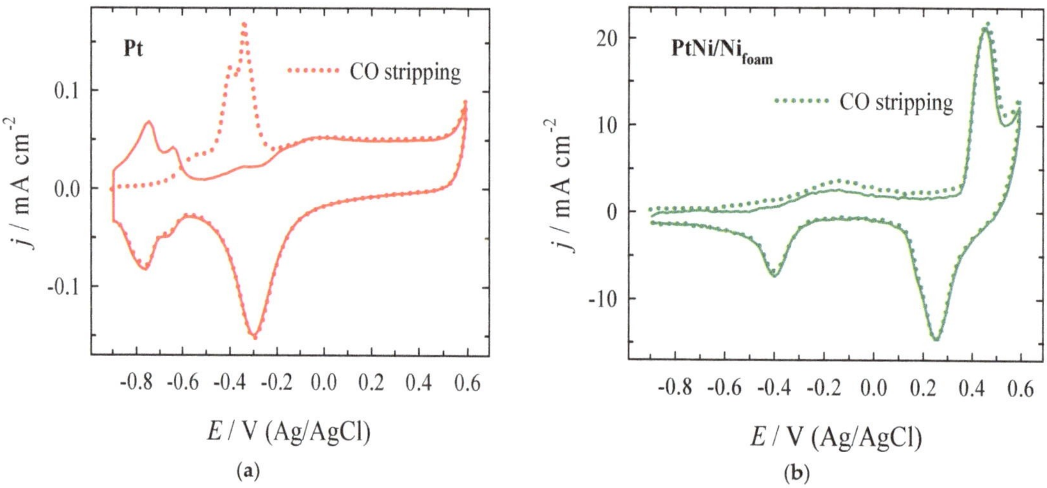

Figure 8. CVs for the oxidative CO stripping from the Pt (**a**) and PtNi/Ni_{foam} (**b**) catalysts surface in 1 M NaOH at a scan rate of 50 mV s^{-1}. CO was adsorbed at -0.9 V from 1 M NaOH for 15 min; a potential sweep was carried out in an N_2-saturated solution.

A prominent CO_{ads} oxidation current density peak was observed during the positive potential scan of the bare Pt electrode in 1 M NaOH. Concurrently, the same peak for the PtNi/Ni_{foam} catalyst completely disappeared, indicating that the PtNi/Ni_{foam} catalyst has almost total CO poisoning tolerance in an alkaline medium, unlike bare Pt. The results indicate that the oxidative removal of carbonaceous poisoning species from adjacent Pt sites by $Ni(OH)_2$ was facilitated by the presence of a stable and reversible Ni (II)/Ni (III) redox transformation in alkaline media, as shown in Equation (1). The improvement in the formate oxidation reaction could be explained by the presence of a Ni-oxy species that can act as catalytic mediators via the abovementioned reaction by facilitating a charge transfer during the direct oxidation of formate to CO_2, simultaneously oxidizing CO at a low potential throughout subsequent reactions [32,42,43].

4. Conclusions

A promising novel binary catalyst composed of a Pt-modified Ni layer coated on Ni_{foam} was proposed for efficient formate oxidation in an alkaline medium. The PtNi/Ni_{foam} catalyst was found to have a significantly high FO mass activity (990 mA mg_{Pt}^{-1}). Moreover, it exhibited enhanced electrocatalytic activity toward formate oxidation via a direct pathway in an alkaline medium, unlike the pure Pt electrode. The prepared PtNi/Ni_{foam} catalyst was tolerant to CO poisoning in an alkaline medium and demonstrated high poison removal ability. It shows rather negative onset potential on the bare Pt. The significant improvement in the E_{onset} to ca. −0.9 V for PtNi/Ni_{foam}, compared to ca. −0.7 V for bare Pt, indicates that the PtNi/Ni_{foam} electrode is competitive with the benchmark Pt and Pt-based electrocatalysts for formate oxidation in alkaline media. The improvement in electrocatalytic activity is a result of synergy between Pt and the porous structure of the Ni-oxy species layer on the Ni_{foam}, which is intended to assist in the oxidative removal of the accumulated carbonaceous species from the surface by acting as a catalytic mediator for charge transfer in the oxidation process.

The PtNi/Ni_{foam} catalyst appears to be a promising anode material for application in direct liquid FA fuel cells.

Supplementary Materials: The following supporting information can be downloaded at: https://www.mdpi.com/article/10.3390/cryst12030362/s1, Figure S1: CVs of Pt (dotted lines) and PtNi/Ni_{foam} (blue solid line) recorded in a 0.5 M H_2SO_4 solution at a scan rate of 50 mV s^{-1}; Figure S2: CVs at the Pt electrode recorded in a 1 M NaOH solution (red solid line) and that containing 0.3, 0.5, and 0.7 M FA (anodic potential scan—solid lines and cathodic potential scan—dotted lines) at a scan rate 50 mV s^{-1}; Figure S3: Negative-going potential CVs of PtNi/Ni_{foam} recorded at anodic potential limit of 0.2 (a,c,e) and 0.6 V (b,d,f) in a 1 M NaOH solution containing 0.1 (a,b), 0.3 (c,d), and 0.5 (e,f) M FA at a scan rate of 50 mV s^{-1}; Figure S4: SEM views of as-prepared Ni_{foam} substrate coated by a Ni layer (a) and PtNi/Nifoam electrode at different magnifications (b,c) after continuous cycling in 0.5 M FA + 1 M NaOH in a potential region from −0.9 to +0.6 V vs. Ag/AgCl at a scan rate of 50 mV s^{-1}. (d) The corresponding EDX spectra for PtNi/Ni_{foam} electrode.

Author Contributions: This study was conducted through the contributions of all authors. Conceptualization, L.T.-T., A.N. and E.N.; Methodology, D.U., V.K. and V.J.; Investigation, B.Š., A.B. and A.Z.; formal analysis, B.Š. and A.B.; Validation, A.Z., D.Š.; Writing—Original Draft Preparation, A.N. and D.Š.; Writing—Review and Editing, L.T.-T. and E.N.; Visualization, D.U., V.K. and V.J. All authors have read and agreed to the published version of the manuscript.

Funding: This research was funded by European Union Structural Funds project "Development of Doctoral Studies" under grant No. 09.3.3.-ESFA-V-711-01-0001.

Institutional Review Board Statement: Not applicable.

Informed Consent Statement: Not applicable.

Data Availability Statement: Not applicable.

Conflicts of Interest: The authors declare no conflict of interest.

References

1. Fang, Z.; Chen, W. Recent advances in formic acid electro-oxidation: From the fundamental mechanism to electrocatalysts. *Nanoscale Adv.* **2021**, *3*, 94–105. [CrossRef]
2. Folkman, S.J.; González-Cobos, J.; Giancola, S.; Sánchez-Molina, I.; Galán-Mascarós, J.R. Benchmarking Catalysts for Formic Acid/Formate Electrooxidation. *Molecules* **2021**, *26*, 4756. [CrossRef] [PubMed]

3. Shen, T.; Zhang, J.; Chen, K.; Deng, S.; Wang, D. Recent Progress of Palladium-Based Electrocatalysts for the Formic Acid Oxidation Reaction. *Energy Fuels* **2020**, *34*, 9137–9153. [CrossRef]
4. An, L.; Chen, R. Direct formate fuel cells: A review. *J. Power Sources* **2016**, *320*, 127–139. [CrossRef]
5. Uwitonze, N.; Chen, Y.X. The study of Pt and Pd based anode catalysis for formic acid fuel cell. *Chem. Sci. J.* **2017**, *8*, 167. [CrossRef]
6. Mao, H.; Huang, T.; Yu, A. Electrochemical surface modification on CuPdAu/C with extraordinary behavior toward formic acid/formate oxidation. *Int. J. Hydrogen Energy* **2016**, *41*, 13190–13196. [CrossRef]
7. Ding, J.; Liu, Z.; Liu, X.; Liu, J.; Deng, Y.; Han, X.; Zhong, C.; Hu, W. Mesoporous decoration of freestanding palladium nanotube arrays boosts the electrocatalysis capabilities toward formic acid and formate oxidation. *Adv. Energy Mater.* **2019**, *9*, 1900955. [CrossRef]
8. Wang, J.; Chen, F.; Jin, Y.; Guo, L.; Gong, X.; Wanga, X.; Johnston, R.L. In situ high-potential-driven surface restructuring of ternary AgPd-Pt dilute aerogels with record-high performance improvement for formate oxidation electrocatalysis. *Nanoscale* **2019**, *11*, 14174–14185. [CrossRef]
9. Wang, Q.; Chen, F.; Tang, Q.; Guo, L.; Gebremarian, T.; Jin, T.; Liu, H.; Kou, B.; Li, Z.; Bian, W. Transition from core-shell to janus segregation pattern in AgPd nanoalloy by Ni doping for the formate oxidation. *Appl. Catal. B Environ.* **2020**, *270*, 118861. [CrossRef]
10. Wang, Q.; Chen, F.; Guo, L.; Jin, T.; Liu, H.; Wang, X.; Gong, X.; Liu, Y. Nanoalloying effects on the catalytic activity of the formate oxidation reaction over AgPd and AgCuPd aerogels. *J. Mater. Chem. A* **2019**, *7*, 16122–16135. [CrossRef]
11. Chen, Y.; Niu, H.; Feng, Y.; Wu, J.; Wang, A.; Huang, H.; Feng, J. Three-dimensional hierarchical urchin-like PdCuPt nanoassembles with zigzag branches: A highly efficient and durable electrocatalyst for formic acid oxidation reaction. *Appl. Surf. Sci.* **2020**, *510*, 145480. [CrossRef]
12. Tang, Q.; Chen, F.; Jin, T.; Guo, L.; Wang, Q.; Liu, H. Alloying in inverse CeO_2/Pd nanoparticles to enhance the electrocatalytic activity for the formate oxidation reaction. *J. Mater. Chem. A* **2019**, *7*, 22996–23007. [CrossRef]
13. Hwang, E.; Park, H.; Kim, H.; Ahn, S.H.; Kim, S.K. Electrochemically fabricated Pd–In catalysts for carbon dioxide- formate/formic acid inter-conversion. *Bull. Korean Chem. Soc.* **2017**, *38*, 607–613. [CrossRef]
14. Hong, S.; Chung, S.; Park, J.; Hwang, J.; Lee, C.; Uhm, S.; Bong, S.; Lee, J. Contribution of interstitial boron in a boron-incorporated palladium catalyst toward formate oxidation in an alkaline direct formate fuel cell. *ACS Catal.* **2021**, *11*, 4722–4729. [CrossRef]
15. Yang, L.; Li, G.; Chang, J.; Ge, J.; Liu, C.; Vladimir, F.; Wang, G.; Jin, Z.; Xing, W. Modification of palladium nanocrystals with single atom platinum via an electrochemical self-catalysis strategy for efficient formic acid electrooxidation. *Appl. Catal. B* **2020**, *260*, 118200. [CrossRef]
16. da Silva, S.G.; Silva, J.C.M.; Buzzo, G.S.; Neto, A.O.; Assumpção, M.H.M.T. Use of PtAu/C electrocatalysts toward formate oxidation: Electrochemical and fuel cell considerations. *Mater. Renew. Sustain. Energy* **2016**, *5*, 15. [CrossRef]
17. Han, S.-H.; Liu, H.-M.; Bai, J.; Tian, X.L.; Xia, B.Y.; Zeng, J.-H.; Jiang, J.-X.; Chen, Y. Platinum-silver alloy nanoballoon nanoassemblies with super catalytic activity for the formate electrooxidation. *ACS Appl. Energy Mater.* **2018**, *1*, 1252–1258. [CrossRef]
18. Yang, Z.; Wang, Y.; Dong, T.; Yuan, X.; Lv, L.; Wei, X.; Wang, J. Formate: A possible replacement for formic acid in fuel cells. *Austral. J. Chem.* **2017**, *70*, 757–763. [CrossRef]
19. John, J.; Wang, H.; Rus, E.D.; Abruna, H.D. Mechanistic studies of formate oxidation on platinum in alkaline medium. *J. Phys. Chem. C* **2012**, *116*, 5810–5820. [CrossRef]
20. Joo, J.; Uchida, T.; Cuesta, A.; Koper, M.T.M.; Osawa, M. Importance of acid-base equilibrium in electrocatalytic oxidation of formic acid on platinum. *J. Am. Chem. Soc.* **2013**, *135*, 9991–9994. [CrossRef]
21. Li, Y.; Feng, Y.; Sun, X.; He, Y. A sodium-ion-conducting direct formate fuel cell: Generating electricity and producing base. *Angew. Chem.* **2017**, *129*, 5828–5831. [CrossRef]
22. Yu, X.; Manthiram, A. Catalyst-selective, scalable membraneless alkaline direct formate fuel cells. *Appl. Catal. B Environ.* **2015**, *165*, 63–67. [CrossRef]
23. Themsirimongkon, S.; Pongpichayakul, N.; Fang, L.; Jakmunee, J.; Saipanya, S. New catalytic designs of Pt on carbon nanotube-nickel-carbon black for enhancement of methanol and formic acid oxidation. *J. Electroanal. Chem.* **2020**, *876*, 114518. [CrossRef]
24. Kiani, M.; Zhang, J.; Luo, Y.; Chen, Y.; Chen, J.; Fan, J.; Wang, G.; Wang, R. Facile synthesis and enhanced catalytic activity of electrochemically dealloyed platinum-nickel nanoparticles towards formic acid electro-oxidation. *J. Energy Chem.* **2019**, *35*, 9–16. [CrossRef]
25. Zhang, B.-W.; Zhang, Z.-C.; Liao, H.-G.; Gong, Y.; Gu, L.; Qu, X.-M.; You, L.-X.; Liu, S.; Huang, L.; Tian, X.-C.; et al. Tuning Pt-skin to Ni-rich surface of Pt_3Ni catalysts supported on porous carbon for enhanced oxygen-reduction reaction and formic electro-oxidation. *Nano Energy* **2016**, *19*, 198–209. [CrossRef]
26. Pei, A.; Ruan, L.; Liu, B.; Chen, W.; Lin, S.; Chen, B.; Liu, Y.; Zhu, L.H.; Chen, B.H. Ultra-low Au decorated PtNi alloy nanoparticles on carbon for high-efficiency electro-oxidation of methanol and formic acid. *Int. J. Hydrogen Energy* **2020**, *45*, 22893–22905. [CrossRef]
27. Lee, J.; Yoo, J.K.; Lee, H.; Kim, S.H.; Sohn, Y.; Rhee, C.K. Formic acid oxidation on Pt deposit model catalysts on Au: Single-layered Pt deposits, plateau-type Pt deposits, and conical Pt deposits. *Electrochim. Acta* **2019**, *310*, 38–44. [CrossRef]
28. Perales-Rondón, J.V.; Ferre-Vilaplana, A.; Feliu, J.M.; Herrero, E. Oxidation mechanism of formic acid on the bismuth adatom-modified Pt(111) surface. *J. Amer. Chem. Soc.* **2014**, *136*, 13110–13113. [CrossRef]

29. Wang, C.Y.; Yu, Z.-Y.; Li, G.; Song, Q.-T.; Li, G.; Luo, C.-X.; Yin, S.-H.; Lu, B.-A.; Xiao, C.; Xu, B.-B.; et al. Intermetallic PtBi nanoplates with high catalytic activity towards electrooxidation of formic acid and glycerol. *ChemElectroChem* **2020**, *7*, 239–245. [CrossRef]
30. Zhang, Y.; Qiao, M.; Huang, Y.; Zou, Y.; Liu, Z.; Tao, L.; Li, Y.; Dong, C.L.; Wang, S. In situ exfoliation and Pt deposition of antimonene for formic acid oxidation via a predominant dehydrogenation pathway. *Research* **2020**, *2020*, 5487237. [CrossRef]
31. Sawy, E.N.E.; Pickup, P.G. Carbon monoxide and formic acid oxidation at Rh@Pt nanoparticles. *Electrochim. Acta* **2019**, *302*, 234–240. [CrossRef]
32. El-Nagar, G.A.; Mohammad, A.M.; El-Deab, M.S.; El-Anadouli, B.E. Electrocatalysis of formic acid electro-oxidation at platinum nanoparticles modified surfaces with nickel and cobalt oxides nanostructures. In *Progress in Clean Energy*; Dincer, I., Colpan, C., Kizilkan, O., Ezan, M., Eds.; Springer International Publishing: Cham, Switzerland, 2015; Volume 1. [CrossRef]
33. El-Nagar, G.A.; Mohammad, A.M. Enhanced electrocatalytic activity and stability of platinum, gold, and nickel oxide nanoparticles based ternary catalyst for formic acid electro-oxidation. *Int. J. Hydrogen Energy* **2014**, *39*, 11955–11962. [CrossRef]
34. Mohammad, A.M.; El-Nagar, G.A.; Al-Akraa, I.M.; El-Deab, M.S.; El-Anadouli, B.E. Towards improving the catalytic activity and stability of platinum-based anodes in direct formic acid fuel cells. *Int. J. Hydrogen Energy* **2015**, *40*, 7808–7816. [CrossRef]
35. El-Refaei, S.M.; El-Nagar, G.A.; Mohammad, A.M.; El-Deab, M.S.; El-Anadouli, B.E. Electrocatalytic activity of NiOx nanostructured modified electrodes towards oxidation of small organic molecules. In *Springer Proceedings in Energy, Proceedings of the 2nd International Congress on Energy Efficiency and Energy Related Materials (ENEFM2014), 16–19 October 2014*; Oral, A., Bahsi Oral, Z., Ozer, M., Eds.; Springer: Cham, Switzerland, 2015. [CrossRef]
36. El-Nagar, G.A.; Mohammad, A.M.; El-Deab, M.S.; El-Anadouli, B.E. Propitious dendritic Cu_2O-Pt nanostructured anodes for direct formic acid, Fuel cells. *ACS Appl. Mater. Interf.* **2017**, *9*, 19766–19772. [CrossRef]
37. Ali Al-Qodami, B.; Farrag, H.H.; Sayed, S.Y.; Allam, N.K.; El-Anadouli, B.E.; Mohammad, A.M. Bifunctional tailoring of platinum surfaces with earth abundant iron oxide nanowires for boosted formic acid electro-oxidation. *J. Nanotechnol.* **2018**, *2018*, 4657040. [CrossRef]
38. Mohammad, A.M.; Al-Akraa, I.M.; El-Deab, M.S. Superior electrocatalysis of formic acid electrooxidation on a platinum, gold and manganese oxide nanoparticle-based ternary catalyst. *Int. J. Hydrogen Energy* **2018**, *43*, 139–149. [CrossRef]
39. Asal, Y.M.; Al-Akraa, I.M.; Mohammad, A.M.; El-Deab, M.S. A competent simultaneously co-electrodeposited Pt-MnO_x nanocatalyst for enhanced formic acid electro-oxidation. *J. Taiwan Inst. Chem. Eng.* **2019**, *96*, 169–175. [CrossRef]
40. Santos, A.O.; Silva, J.C.M.; Antoniassi, E.R.M.; Ponzio, A.; Alves, O.C. The formate electrooxidation on Pt/C and $PtSnO_2$/C nanoparticles in alkaline media: The effect of morphology and SnO_2 on the platinum catalytic activity. *Int. J. Hydrogen Energy* **2020**, *45*, 33895–33905. [CrossRef]
41. Subbaraman, R.; Tripkovic, D.; Chang, K.-C.; Strmcnik, D.; Paulikas, A.P.; Hirunsit, P.; Chan, M.; Greeley, J.; Stamenkovic, V.; Markovic, N.M. Trends in activity for the water electrolyser reactions on 3d M(Ni Co, Fe, Mn) hydr(oxy)oxide catalysts. *Nat. Mater.* **2012**, *11*, 550–557. [CrossRef]
42. El-Nagar, G.A.; Mohammad, A.M.; El-Deab, M.S.; El-Anadouli, B.E. Facilitated electrooxidation of formic acid at nickel oxide nanoparticles modified electrodes. *J. Electrochem. Soc.* **2012**, *159*, F249–F254. [CrossRef]
43. El-Nagar, G.A.; Mohammad, A.M.; El-Deab, M.S.; El-Anadouli, B.E. Electrocatalysis by design: Enhanced electrooxidation of formic acid at platinum nanoparticles–nickel oxide nanoparticles binary catalysts. *Electrochim. Acta* **2013**, *94*, 62–71. [CrossRef]
44. Fiameni, S.; Herraiz-Cardona, I.; Musiani, M.; Pérez-Herranz, V.; Vázquez-Gómez, L.; Verlato, E. The HER in alkaline media on Pt-modified three-dimensional Ni cathodes. *Int. J. Hydrogen Energy* **2012**, *37*, 10507–10516. [CrossRef]
45. Gu, Y.; Wang, Y.; Shi, J.; Yang, M.; Rui, Y.; An, W.; Men, Y. Well-dispersed Pt nanodots interfaced with $Ni(OH)_2$ on anodized nickel foam for efficient hydrogel evolution reaction. *Int. J. Hydrogen Energy* **2020**, *45*, 27067–27077. [CrossRef]
46. Yuan, S.; Cui, L.; He, X.; Zhang, W.; Asefa, T. Nickel foam-supported Fe, Ni-Polyporphyrin microparticles: Efficient bifunctional catalysts for overall water splitting in alkaline media. *Int. J. Hydrogen Energy* **2020**, *45*, 28860–28869. [CrossRef]
47. Wang, R.; Liu, H.; Zhang, K.; Zhang, G.; Lan, H.; Qu, J. Ni(II)/Ni(III) redox couple endows Ni foam-supported Ni_2P with excellent capability for direct ammonia oxidation. *Chem. Eng. J.* **2021**, *404*, 126795. [CrossRef]
48. Yuan, G.; Wang, L.; Zhang, X.; Wang, Q. Self-supported Pt nanoflakes-doped amorphous $Ni(OH)_2$ on Ni foam composite electrode for efficient and stable methanol oxidation. *J. Colloid Interf. Sci.* **2019**, *536*, 189–195. [CrossRef]
49. Sheng, S.; Song, Y.; Sha, L.; Ye, K.; Zhu, K.; Gao, Y.; Yan, J.; Wang, G.; Cao, D. Simultaneous hydrogen evolution and ethanol oxidation in alkaline medium via a self-supported bifunctional electrocatalyst of Ni-Fe phosphide/Ni foam. *Appl. Surf. Sci.* **2021**, *561*, 150080. [CrossRef]
50. Tamašauskaitė-Tamašiūnaitė, L.; Zabielaitė, A.; Balčiūnaitė, A.; Šebeka, B.; Stanionienė, I.; Buzas, V.; Mačiulis, L.; Tumonis, L.; Norkus, E. Deposition of Pt nanoparticles on Ni foam via galvanic displacement. *J. Electrochem. Soc.* **2017**, *164*, D53–D56. [CrossRef]
51. Song, C.; Wang, G.; Li, B.; Miao, C.; Ma, K.; Zhu, K.; Cheng, K.; Ye, K.; Yan, J.; Cao, D.; et al. A novel electrode of ternary CuNiPd nanoneedles decorated Ni foam and its catalytic activity toward $NaBH_4$ electrooxidation. *Electrochim. Acta* **2019**, *299*, 395–404. [CrossRef]
52. Lei, Y.; Liu, Y.; Fan, B.; Mao, L.; Yu, D.; Huang, Y.; Guo, F. Facile fabrication of hierarchically porous Ni foam@Ag-Ni catalyst for efficient hydrazine oxidation in alkaline medium. *J. Taiwan Inst. Chem. Eng.* **2019**, *105*, 75–84. [CrossRef]

53. Chen, N.; Wang, Y.; Du, X.; Zhang, X. Facile synthesis of Ni doped $CoWO_4$ nanoarrays grown on nickel foam substrates for efficient urea oxidation. *Int. J. Hydrogen Energy* **2021**, *46*, 25114–25120. [CrossRef]
54. NIST X-ray Photoelectron Spectroscopy Database. *NIST Standard Reference Database Number 20;* National Institute of Standards and Technology: Gaithersburg, MD, USA, 2000; p. 20899. [CrossRef]
55. Igarashi, H.; Fujino, T.; Zhu, Y.; Uchida, H.; Watanabe, M. CO tolerance of Pt alloy electrocatalysts for polymer electrolyte fuel cells and the detoxification mechanism. *Phys. Chem. Chem. Phys.* **2001**, *3*, 306–314. [CrossRef]
56. Zhao, Y.; Yifeng, E.; Fan, L.; Qiu, Y.; Yang, S. A new route for the electrodeposition of platinum-nickel alloy nanoparticles on multi-walled carbon nanotubes. *Electrochim. Acta* **2007**, *52*, 5873–5878. [CrossRef]
57. Kim, K.S.; Winograd, N. X-ray photoelectron spectroscopic studies of nickel oxygen surfaces using oxygen and argon ion-bombardment. *Surf. Sci.* **1974**, *43*, 625–643. [CrossRef]
58. Wittstok, G.; Strübing, A.; Szargan, R.; Werner, G. Glucose oxidation at bismuth-modified platinum electrodes. *J. Electroanal. Chem.* **1998**, *444*, 61–73. [CrossRef]
59. Blasini, D.R.; Rochefort, D.; Fachini, E.; Aldena, L.R.; DiSalvo, J.F.; Cabrera, C.R.; Abruna, H.D. Surface composition of ordered intermetallic compounds PtBi and PtPb. *Surf. Sci.* **2006**, *600*, 2670–2680. [CrossRef]
60. Eiler, K.; Krawiec, H.; Kozina, I.; Sort, J.; Pellicer, E. Electrochemical characterisation of multifunctional electrocatalytic mesoporous Ni-Pt thin films in alkaline and acidic media. *Electrochim. Acta* **2020**, *359*, 136952. [CrossRef]
61. Trasatti, S.; Petrii, O.A. Real surface area measurements in electrochemistry. *Pure Appl. Chem.* **1991**, *63*, 711–734. [CrossRef]
62. Daubinger, P.; Kieninger, J.; Unmussig, T.; Urban, G.A. Electrochemical characteristics of nanostructured platinum electrodes A cyclic voltammetry study. *Phys. Chem. Chem. Phys.* **2014**, *16*, 8392–8399. [CrossRef]
63. Dražić, D.M.; Tripković, A.V.; Popović, K.D.; Lović, D.J. Kinetic and mechanistic study of hydroxyl ion electrosorption at the Pt(111) surface in alkaline media. *J. Electroanal. Chem.* **1999**, *466*, 155–164. [CrossRef]
64. Schmidt, T.J.; Ross, P.N.; Marković, N.M. Temperature-dependent surface electrochemistry on Pt single crystals in alkaline electrolyte: Part 1: CO oxidation. *J. Phys. Chem. B* **2001**, *105*, 12082–12086. [CrossRef]
65. Tilak, B.V.; Conway, B.E.; Angerstain-Kozlowska, H. The real condition of oxidized Pt electrodes: Part III. Kinetic theory of formation and reduction of surface oxides. *J. Electroanal. Chem.* **1973**, *48*, 1–23. [CrossRef]
66. Marković, N.M.; Schmidt, T.J.; Grgur, B.N.; Gasteiger, H.A.; Ross, P.N.; Behm, R.J., Jr. Effect of temperature on surface processes at the Pt(111)-liquid interface: Hydrogen adsorption, oxide formation, and CO oxidation. *J. Phys. Chem. B.* **1999**, *103*, 8568–8577. [CrossRef]
67. Tripković, A.V.; Popović, J.K.D.; Lović, D.J. The influence of the oxygen-containing species on the electrooxidation of the C_1–C_4 alcohols at some platinum single crystal surfaces in alkaline solution. *Electrochim. Acta* **2001**, *46*, 3163–3173. [CrossRef]
68. Marković, N.M.; Gasteiger, H.A.; Ross, P.N. Oxygen reduction on platinum low-index single-crystal surfaces in alkaline solution: Rotating ring disk$_{Pt(hkl)}$ studies. *J. Phys. Chem. B* **1996**, *100*, 6715–6721. [CrossRef]
69. Marković, N.M.; Sarraf, S.T.; Gasteiger, H.A.; Ross, P.N. Hydrogen electrochemistry on platinum low-index single-crystal surfaces in alkaline solution. *J. Chem. Soc., Faraday Trans.* **1996**, *92*, 3719–3725. [CrossRef]
70. Jiang, J.; Scott, J.; Wieckowski, A. Direct evidence of triple-path mechanism of formate electrooxidation on Pt black in alkaline media at varying temperature. Part I: The electrochemical studies. *Electrochim. Acta* **2013**, *104*, 124–133. [CrossRef]
71. Yan, L.; Yao, S.; Chang, J.; Liu, C.; Xing, W. Pd oxides/hydrous oxides as highly efficient catalyst for formic acid electrooxidation. *J. Power Sources* **2014**, *250*, 128–133. [CrossRef]
72. Okamoto, H.; Kon, W.; Mukouyama, Y. Five current peaks in voltammograms for oxidations of formic acid, formaldehyde, and methanol on platinum. *J. Phys. Chem. B* **2005**, *109*, 15659–15666. [CrossRef]
73. Šljukić, B.; Milikić, J.; Santos, D.M.F.; Sequeira, C.A.C.; Macciò, D.; Saccone, A. Electrocatalytic performance of Pt–Dy alloys for direct borohydride fuel cells. *J. Power Sources* **2014**, *272*, 335–343. [CrossRef]
74. Han, L.; Gonzįlez-Cobos, J.; Sįnchez-Molina, I.; Giancola, S.; Folkman, S.; Tang, P.; Heggen, M.; Dunin-Borkowski, R.; Arbiol, J.; Giménez, S.; et al. Cobalt hexacyanoferrate as a selective and high current density formate oxidation electrocatalyst. *ACS Appl. Energy Mater.* **2020**, *3*, 9198–9207. [CrossRef]
75. Galvan, V.; Glass, D.E.; Baxter, A.F.; Surya Prakash, G.K. Reduced graphene oxide supported palladium nanoparticles for enhanced electrocatalytic activity toward formate electrooxidation in an alkaline medium. *ACS Appl. Energy Mater.* **2019**, *2*, 7104–7111. [CrossRef]
76. Song, S.; Hwang, H.; Kim, J.W.; Lee, J. The effect of synthesis temperature on Pd-H catalyst structure for alkaline direct formate fuel cells. *ECS Trans.* **2018**, *85*, 149–158. [CrossRef]
77. Sankar, S.; Anilkumar, G.M.; Tamaki, T.; Yamaguchi, T. Binary Pd–Ni nanoalloy particles over carbon support with superior alkaline formate fuel electrooxidation performance. *ChemCatChem* **2019**, *11*, 4731–4737. [CrossRef]

crystals

MDPI

Article

Synthesis and Characterization of Graphite Intercalation Compounds with Sulfuric Acid

Gintare Rimkute [1,*], Mantvydas Gudaitis [2], Jurgis Barkauskas [1], Aleksej Zarkov [1,*], Gediminas Niaura [2] and Justina Gaidukevic [1]

1 Institute of Chemistry, Faculty of Chemistry and Geosciences, Vilnius University, Naugarduko Str. 24, LT-03225 Vilnius, Lithuania; jurgis.barkauskas@chf.vu.lt (J.B.); justina.gaidukevic@chf.vu.lt (J.G.)

2 Department of Organic Chemistry, Center for Physical Sciences and Technology (FTMC), Saulėtekio Av. 3, LT-10257 Vilnius, Lithuania; mantvydas.gudaitis@chgf.stud.vu.lt (M.G.); gediminas.niaura@ftmc.lt (G.N.)

* Correspondence: gintare.rimkute@chgf.vu.lt (G.R.); aleksej.zarkov@chf.vu.lt (A.Z.)

Abstract: In this work, graphite intercalation compounds (GICs) were synthesized using three different oxidizers: $(NH_4)_2S_2O_8$, $K_2S_2O_8$, and CrO_3 with and without P_2O_5 as a water-binding agent. Furthermore, the samples obtained were heat-treated at 800 °C. Specimens were characterized by optical microscopy, Raman spectroscopy, Fourier transform infrared spectroscopy (FTIR), X-ray powder diffraction (XRD), and scanning electron microscopy (SEM). The correlation between different characteristic parameters of the Raman analysis has shown that the use of CrO_3 results in a much higher structural disorder compared to the products obtained using persulfate oxidizers. Narrowing the correlation set revealed that minimal defect concentration can be reached by using $K_2S_2O_8$, while the use of $(NH_4)_2S_2O_8$ causes a slightly higher concentration of defects. It was also established that the additional use of P_2O_5 can help to achieve more effective intercalation and has a positive effect on the formation of the stage I GIC phase. After heat treatment, the intercalated products mostly return to a graphite-like structure; however, the samples obtained with CrO_3 stand out with the most significant changes in their surface morphology. Therefore, analysis suggests that GICs obtained using persulfate oxidizers and P_2O_5 could be a candidate to produce high-quality graphene or graphene oxide.

Keywords: graphite intercalation compounds; graphite bisulfate; staging in graphite intercalation compounds; structural characterization; thermal treatment

Citation: Rimkute, G.; Gudaitis, M.; Barkauskas, J.; Zarkov, A.; Niaura, G.; Gaidukevic, J. Synthesis and Characterization of Graphite Intercalation Compounds with Sulfuric Acid. *Crystals* **2022**, *12*, 421. https://doi.org/10.3390/cryst12030421

Academic Editor: Sawanta S. Mali

Received: 3 March 2022
Accepted: 16 March 2022
Published: 18 March 2022

1. Introduction

A special group of compounds known as graphite intercalation compounds (GICs) was first mentioned in 1840 by German chemist C. Schafhaeutl [1]. In 1855, B. C. Brodie described the synthesis of GIC using a mixture of sulfuric and nitric acids [2]. This work is considered to be the beginning of GIC research. Nowadays, research in the area of GICs is the focus, especially due to their use in Li-ion batteries [3]. Furthermore, GICs can form numerous groups of compounds. A variety of GICs can be generally divided into two groups: covalent and ionic. The number of compounds in the latter group far exceeds that in the former [4]. Ionic GICs, such as graphite salts, graphite-halogen compounds, graphite-metal chloride compounds, and graphite alkali-metal compounds, have received more recognition than covalent GICs due to the change in the electronic properties of graphite, which is attributed to the π-bonds that can donate or accept electrons [5]. Therefore, ionic GICs can be further divided into donor-type (Lewis base guests, e.g., alkali metals) and acceptor-type (Lewis acid guests, e.g., strong mineral acids), while unique redox amphoteric characteristics enable graphite to host both cations and anions.

The significance of GICs is not limited to their use in energy storage systems. Some of them reveal the superconductivity phenomenon at low temperatures, while others are used as reagents and catalysts in organic synthesis [6,7]. Another relevant area where GICs play a key role is the production of graphene and graphene oxide. High-quality, low-defective graphene can be obtained using the GIC as a direct precursor [8]. Likewise, GIC, known as graphite bisulfate (GBS), forms as an intermediate in the process of the preparation of graphene oxide, which is widely used for the further production and processing of graphene [9]. GBS is the GIC, where graphite layers are intercalated by HSO_4^- ions and H_2SO_4 molecules. The reaction of GBS formation implies oxidation of the graphite matrix conjugated with insertion of the molecules and ions in the presence of an oxidant (Ox):

$$\begin{cases} nC_{24} \overset{[Ox]}{\rightarrow} C_{24n}{}^{+} \\ C_{24n}{}^{+} + HSO_4{}^{-} + mH_2SO_4 \rightarrow C_{24n}{}^{+}{\cdot}HSO_4{}^{-}{\cdot}mH_2SO_4, \end{cases} \quad (1)$$

where m/n ratio stands for the staging index, i.e., the number of graphite layers between two intercalant layers [10]. Formula $C_{24}{}^{+}{\cdot}HSO_4{}^{-}{\cdot}2H_2SO_4$ represents the stage I compound, where a single layer of graphene is alternated regularly with intercalated species. In subsequent stages (stage II, stage III, etc.), the corresponding number of graphene layers (2, 3, etc.) are separated by layers of HSO_4^- ions and H_2SO_4 molecules [11].

GBS is also the most widely used precursor for manufacturing exfoliated graphite, which is valued for compactness, flexibility, high electrical and thermal conductivity, and corrosion resistance. For this purpose, GBS is treated with microwaves or thermal shock conditions. Under these conditions, graphene layers are separated and form a characteristic cellular structure. When pressed without binder, exfoliated graphite can be formed into disks, coatings, fabric, etc., and used as a fire retardant, absorbent, electrode material, photo-catalyst, and many similar cases [12].

Although GICs have been known for a long time and are extensively used nowadays, the knowledge in this field so far is not sufficient. Thermodynamics of GIC formation do not always correspond to the theoretical model, and it is still difficult to predict the potential of the system that includes GIC. Simultaneously, very limited structural information is given in the literature concerning these systems, at most due to their instability. Since the properties of GIC with the same chemical composition may vary depending on graphite precursor, preparation conditions, and other factors, it is important to investigate as many different synthesis variations as possible [13].

In this research, GBS compounds were chemically synthesized using three different oxidizing agents as well as P_2O_5 as a water binding agent. All obtained samples were heat-treated at 800 °C in order to get exfoliated graphite. The obtained samples were investigated using optical microscopy, Raman spectroscopy, Fourier transform infrared spectroscopy (FTIR), X-ray powder diffraction (XRD), and scanning electron microscopy (SEM).

2. Materials and Methods

2.1. Materials

In the synthesis, all reagents were used without further purification. Extra pure graphite powder (<50 μm (≥99.5%)) was purchased from Merck KGaA (Darmstadt, Germany), H_2SO_4 (98%), $(NH_4)_2S_2O_8$ (98%), $K_2S_2O_8$ (99.99%), CrO_3 (99.99%), and P_2O_5 (99%)—from Sigma-Aldrich (Darmstadt, Germany).

2.2. Synthesis of GBS Products

The synthesis of GBS was performed according to the protocol described by Dimiev et al. [14]. Three different oxidizing agents (ammonium persulfate—$(NH_4)_2S_2O_8$, potassium persulfate—$K_2S_2O_8$, and chromium trioxide—CrO_3) were used. The oxidizing mixture was prepared in a 50 mL Erlenmeyer flask with a ground glass joint by adding 1.6 g of an oxidant to 10 mL of H_2SO_4 (≥98%) at constant swirling. After 5 min, 0.25 g of graphite powder was slowly added to the solution of the oxidizing mixture. The Erlenmeyer flask was covered with a ground glass stopper, and the swirling procedure was continued for 1 week. The procedure of GBS synthesis was repeated with all three oxidizing agents, adding 2 g of P_2O_5 to the oxidizing mixture. Abbreviations of obtained 6 GBS products are summarized in Table 1. Additional synthesis in the mixture of CrO_3 and H_2SO_4 was repeated following the protocol, but at the end stage, 40 mL of deionized water was added, ensuring that the temperature of the mixture would not rise above 60 °C. The GBS product of this synthesis is also introduced in Table 1. After synthesis, the GBS products were filtered through a sintered glass filter and stored in a desiccator.

Table 1. Synthesized and heat-treated GBS products.

Oxidizer	Addition of P_2O_5	Addition of H_2O	Thermal Treatment	Abbreviation in Text
$(NH_4)_2S_2O_8$	–	–	–	A000
$K_2S_2O_8$	–	–	–	P000
CrO_3	–	–	–	C000
$(NH_4)_2S_2O_8$	+	–	–	A100
$K_2S_2O_8$	+	–	–	P100
CrO_3	+	–	–	C100
CrO_3	–	+	–	C010
$(NH_4)_2S_2O_8$	–	–	+	A001
$K_2S_2O_8$	–	–	+	P001
CrO_3	–	–	+	C001
$(NH_4)_2S_2O_8$	+	–	+	A101
$K_2S_2O_8$	+	–	+	P101
CrO_3	+	–	+	C101
CrO_3	–	+	+	C011

All GBS samples obtained by the synthesis described above were heat-treated under thermal shock conditions. For this purpose, samples were placed into ceramic combustion boats, which were inserted into a quartz glass tube. The tube was sealed and placed in a furnace preheated to 800 °C (heating rate was 5 $°C\cdot min^{-1}$) and annealed in argon ambient for 15 min. The flow rate of argon was maintained at 15 $mL\cdot min^{-1}$. Samples reached 800 °C approximately in a minute. After a set time, the tube was taken out of the furnace and cooled to room temperature in 15 min (cooling rate was about 50 $°C\cdot min^{-1}$). The heat-treated samples were stored in a desiccator. The thermally treated products are summarized in Table 1. In the table, "+" means that the P_2O_5 or H_2O was added to the reaction mixture or that the obtained products were thermally treated.

2.3. Characterization

Optical images were obtained using a BX51, Olympus optical microscope (Tokyo, Japan) at a magnification of 50×. Raman spectra were recorded using an inVia, Renishaw spectrometer (Wotton-under Edge, UK) equipped with an optical microscope at a magnification of 20× and an objective numerical aperture of 0.40, a CCD camera, and 1800 $grooves\cdot mm^{-1}$ grating. The laser excitation wavelength was 532 nm, beam concentration area—2 μm^2, and integration time—100 s. To analyze the band changes, Raman peaks were fitted with a pseudo-Voigt function—a linear combination of Lorentzian and Gaussian functions [15]. FTIR measurements were carried out using a Frontier PerkinElmer FTIR spectrometer (Seattle, WA, USA) in the spectral range of 4000–800 cm^{-1} with 4 cm^{-1} resolution and 25 scans. The samples were prepared using the KBr pellet technique. For this purpose, sample powder (0.5% by wt.) was mixed with pure KBr powder. Then, for 5 min, the mixtures were pressed into transparent pellets using a CrushIR PIKE hydraulic press (Madison, WI, USA) with 8 $ton\cdot cm^{-2}$ pressure. Background correction was made using a reference blank KBr pellet. The wet samples were analyzed by applying a thin layer of the substance on the surface of a pure KBr pellet. XRD measurements were performed in the 2θ range between 5–60° with a Miniflex II, Rigaku diffractometer (Neu-Isenburg,

Germany) (Cu K_α radiation with a graphite monochromator). The interlayer distance d was calculated according to Bragg's equation [16]:

$$n\lambda = 2d \cdot \sin\theta, \quad (2)$$

where n is a positive integer, λ is the wavelength of the X-ray, d is the interlayer distance and θ is the scattering angle. Crystallite size L was calculated using the Scherrer equation [17]:

$$L = \frac{0.89 \cdot \lambda}{\beta \cdot \cos\theta}, \quad (3)$$

where λ is the wavelength of the X-ray, β is a full width at half maximum (FWHM) and θ is the scattering angle. The morphology of the samples was examined using a SU-70, Hitachi microscope (Tokyo, Japan) at an accelerating voltage of 10 kV and magnification of 5000×.

3. Results and Discussion

Synthesized and heat-treated GBS products were characterized using optical and SEM microscopy, FTIR and Raman spectroscopy, and XRD measurements. GBS products are unstable (sensitive to moisture and higher temperatures) and have a significant corrosive effect. For this reason, SEM and XRD analyses were applied only to those products (treated at higher temperatures or washed with H_2O), from which the H_2SO_4 intercalant was removed.

The effect of the synthesis duration of GBS products was examined using optical microscopy. The specimens for analysis were collected after 2 h and 24 h from the beginning of the synthesis, respectively. The optical micrographs of the GBS samples are presented in Figures 1 and 2.

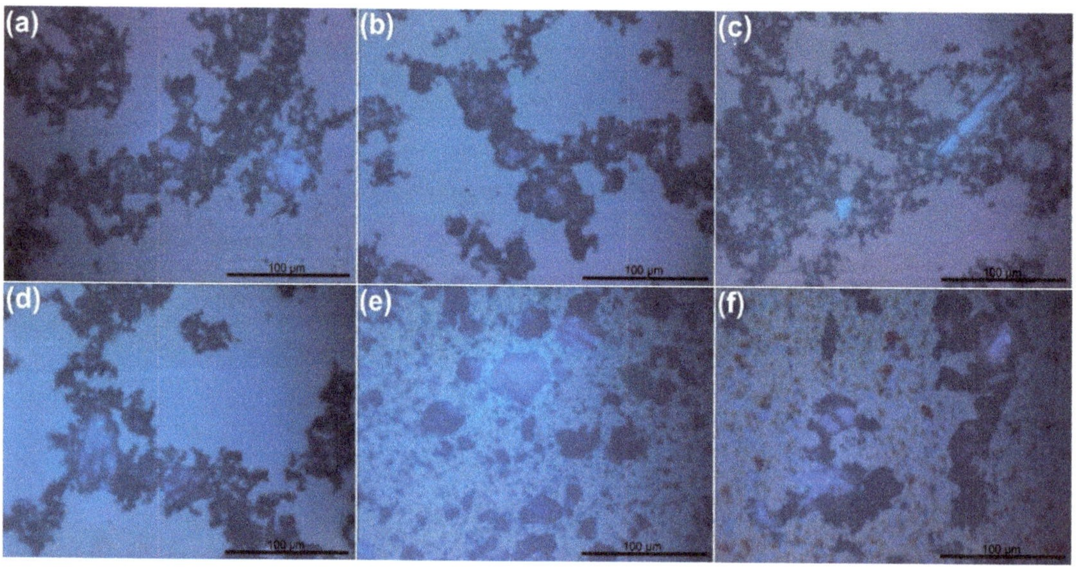

Figure 1. Optical micrographs of GBS obtained after 2 h from the start of the synthesis—A000 (**a**), A100 (**b**), P000 (**c**), P100 (**d**), C000 (**e**), and C100 (**f**).

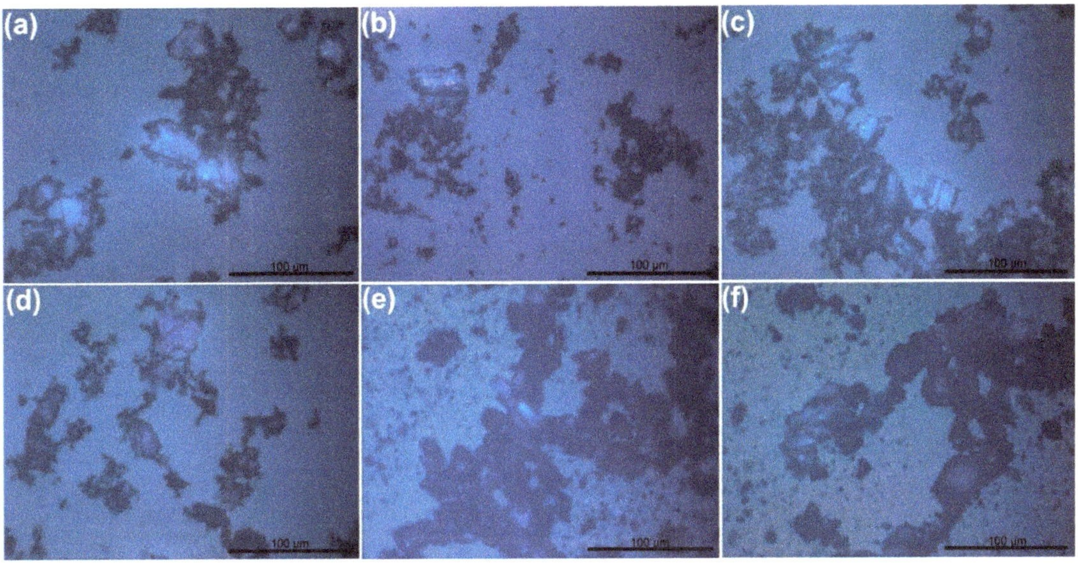

Figure 2. Optical micrographs of GBS obtained after 24 h from the start of the synthesis—A000 (**a**), A100 (**b**), P000 (**c**), P100 (**d**), C000 (**e**), and C100 (**f**).

Many stage I and stage II GICs are colored. A rich blue color is observed in optical microscopy pictures both 2 h and 24 h from the start of the synthesis, indicating the formation of stage I GBS [18,19]. The abundance of blue zones is approximately the same and does not depend on the duration of the synthesis. Therefore, we can conclude that the GBS of stage I represented by the formula $C_{24}^{+}\cdot HSO_4^{-}\cdot 2H_2SO_4$ is stable under the synthesis conditions in a medium of conc. H_2SO_4. We can also note that the GBS of stage I is more likely to form in the central part of larger crystallites, irrespective of the oxidizing agent. Blue areas are commonly surrounded by crystallites of darker color, which vary from black to brown-yellowish tint. These crystallites might be attributed both to non-intercalated graphite and GBS with a lower staging index [19,20]. Single inserts of this color can be detected in the micrographs of all GBS samples. However, the shape and tint of the brown-yellowish inserts of the GBS samples obtained using CrO_3 (Figure 1e,f and Figure 2f) are slightly different: some of them are lighter in color with more expressed yellow tint and more regular shape. Such a difference may occur due to the presence of either orange-yellowish CrO_3 crystals or miscellaneous Cr(VI) compounds [21].

Raman spectroscopy may help reveal the defectiveness of the carbon materials, staging indices, and the amount of stacked graphene sheets in a layered graphene-like material [19,22–24]. Raman spectra of all samples are provided below (Figures 3–5), and data obtained from the analysis are presented in Table 2 and Figure 6.

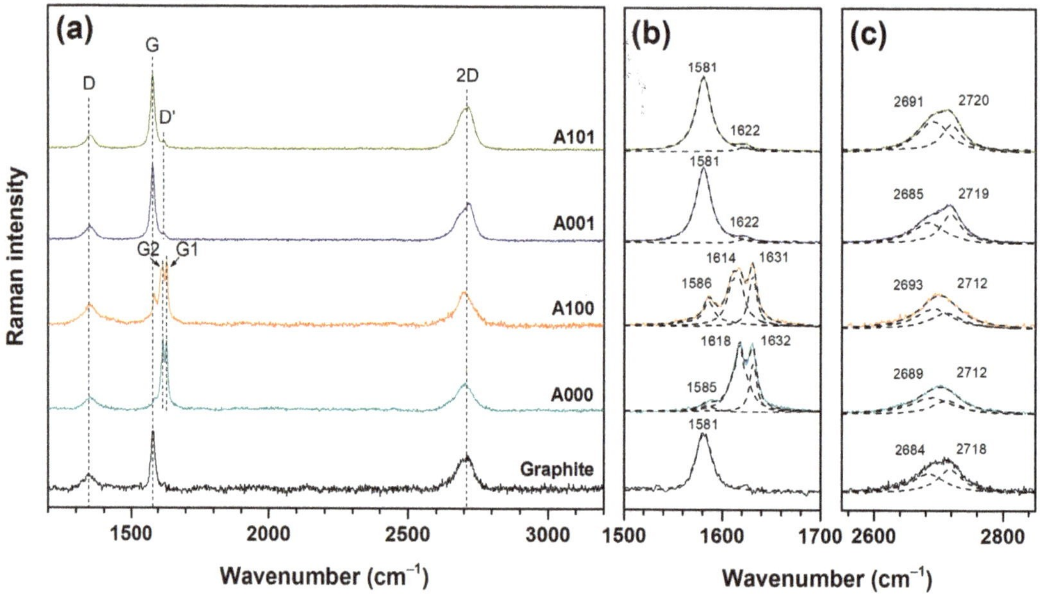

Figure 3. Raman spectra of GBS products obtained using $(NH_4)_2S_2O_8$ oxidizing agent (**a**), G band frequencies (**b**), and 2D band frequencies (**c**).

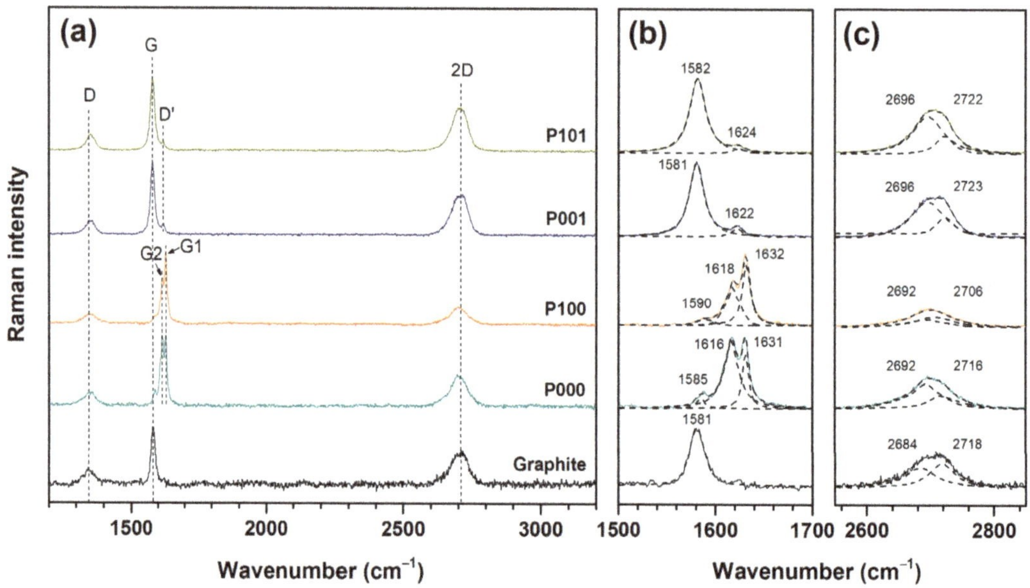

Figure 4. Raman spectra of GBS products obtained using $K_2S_2O_8$ oxidizing agent (**a**), G band frequencies (**b**), and 2D band frequencies (**c**).

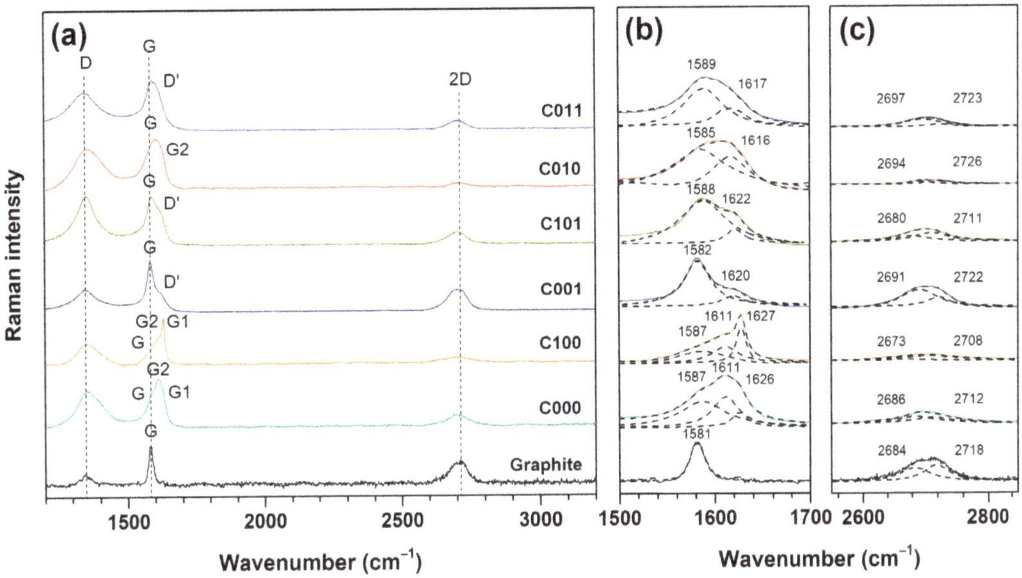

Figure 5. Raman spectra of GBS products obtained using CrO_3 oxidizing agent (**a**), G band frequencies (**b**), and 2D band frequencies (**c**).

Table 2. Amount (%) of stage I, stage II, and non-intercalated graphite phases in GBS samples determined according to the intensities of the G1, G2, and G peaks, respectively.

Sample	Stage I (%)	Stage II (%)	Non-Intercalated Graphite (%)
A000	45.62	46.78	7.60
A100	40.96	38.79	20.25
P000	44.28	45.47	10.25
P100	56.60	37.14	6.26
C000	30.15	31.81	30.05
C100	50.85	29.34	19.81
C010	–	49.63	50.37

Three prominent bands were observed in the resonance Raman spectra of graphite. The D band at 1345 cm^{-1} is caused by defects of the graphene layer (A_{1g} symmetry mode), the G band at 1581 cm^{-1} is assigned to the first-order scattering of E_{2g} mode arising from the sp^2 bonded carbon, and the 2D band at 2709 cm^{-1} is an overtone of the D mode; however, contrary to the D band, this mode is always Raman-allowed and the presence of defects or disorders is not required for activation of this band in the resonance Raman spectrum [22,23,25–29]. In addition to these peaks, at about 1622 cm^{-1} a low intensity band named D′ can be seen in Raman spectra of annealed samples. The D′ band is the second Raman-forbidden band, which appears as a shoulder at the high-frequency side of the G peak due to the presence of disorder. The D′ band appears when randomly distributed impurities interact with extended graphene phonon modes [30]. However, the main attribute observed in the Raman spectra of GBS products is splitting of the G band (see Figures 3–5). Splitting can occur when charged graphite layers adjacent to the intercalated layers are differentiated from those uncharged next to the empty galleries [31]. The effect of splitting is clearly noticeable using the persulfate oxidizing agents ($(NH_4)_2S_2O_8$ and $K_2S_2O_8$; Figures 3 and 4; samples A000, A100, P000, and P100). Simultaneously, the blue shift occurs when the graphene layers are charged with intercalant molecules [18]. This shift, also accompanied by the splitting of the G band into three peaks, evidences the process of graphite intercalation [32]. The smaller G bands at approximately 1586 cm^{-1}, which remain in the samples A000, A100, P000, and P100, represent the part of graphite that must have remained non-intercalated [13]. The other two peaks, G1 at about 1632 cm^{-1} and G2 at about 1618 cm^{-1} represent stage I and stage II GBS phases, respectively, present in the samples A000, A100, P000, and P100 [33]. The Raman spectra of G bands of GBS phases obtained using both persulfate oxidizing agents are quite similar (Figures 3b and 4b); more differences can be found between the samples with and without using P_2O_5 additive. Meantime, the Raman spectra of the GBS phases obtained using CrO_3 oxidizing agent (Figure 5) show a significant difference from those obtained using the persulfate oxidizing agents. Splitting of the G band for C000 and C100 is implicit, also different from that in the case of the A000, A100, P000, and P100 samples. The G2 peak at 1611 cm^{-1} can be assigned to the GBS stage II phase, and the peak G1 at about 1626 cm^{-1} indicates the stage I GBS compound. In the spectrum of C010, two peaks may be hidden in the amid G band region. The peak in the lower wavenumbers corresponds to graphite, and the G2 peak at about 1616 cm^{-1} is characteristic of stage II GIC. Moreover, it can be seen that the G peaks of the samples synthesized with CrO_3 are clearly broadened. FWHM(G) (full width at half maximum of the G band) represents the degree of the graphitization of carbon materials, which is the grade to which the carbon atoms form a hexagonal close-packed graphite crystalline structure [28]. Therefore, high FWHM(G) values reveal that the use of CrO_3 oxidizer could result in an almost amorphous structure, the formation of vacancies, sp^3 defects as well as reduction in particle size [34].

A comparison of the intensities of peaks G, G1, and G2, evaluated by the fit procedure, was used as a relative indicator of the quantity of different intercalation phases in the samples [13]. The amount of stage I, stage II phases, and non-intercalated graphite according to the intensities of the G1, G2, and G peaks, respectively, are presented in Table 2. Samples A000, P000, and C000 have the majority of stage II GBS compounds, while the stage I GBS is formed mostly when P_2O_5 is added. The GBS sample obtained using CrO_3 oxidizing agent and water at the end of the reaction (sample C010) shows a different case. In this spectrum, the G1 peak cannot be found, which suggests that the stage I GBS phase was not formed in this compound. For this reason, the value of stage I for C010 is blank in Table 2. However, Raman analysis reveals that stage II phase remains in the sample even after washing it with water. Nevertheless, non-intercalated graphite makes up the majority of this compound. Apparently, the addition of P_2O_5 can help to get a higher proportion of intercalated compound, while the addition of H_2O gives the opposite results. In this work, P_2O_5 was used as a dehydration agent. When dissociation of concentrated sulfuric acid occurs, H_3O^+ ion, which is responsible for the acidity, is obtained. When there is no water present, the dissociation of sulfuric acid proceeds in a different way. In that case, $H_3SO_4{}^+$ dominates among ions and it is a much stronger acid than H_3O^+. This results in a higher oxidation state of graphite. For this reason, the addition of P_2O_5 and dehydration can help to reach a higher oxidation state of graphite and, therefore, a higher degree of intercalation [35].

The D band is present in Raman spectra of graphite, all GBS samples, and their annealed products (Figures 3–5). Although the I(D)/I(G) ratio is the most commonly used Raman marker to evaluate the concentration of defects in graphite and graphene-like materials, the splitting of the G band due to the presence of intercalated phases raises the question of whether this ratio can be directly applied as a defect marker. Furthermore, a parameter named FWHM(D) (full width at half maximum of the band D) shows high sensitivity to all types of defects [36]. We used both parameters to evaluate the concentration of defects in the GBS products and to assess the correlation between them (Figure 6a). For I(D)/I(G) calculation, the maximum peak at the G band was normalized to be 1 in an arbitrary unit scale. As one can see in Figure 6a, a positive correlation exists between these two parameters: the higher I(D)/I(G) ratio, as well as the wider FWHM(D), results in a higher concentration of defects. In the correlation chart, a sample C101 stands out; the R^2 value including C101 reaches 0.6256. After excluding C101 from the correlation data set, R^2

rises to 0.8598. The Raman spectrum of C101 is rather different from the other spectra of that series; the D peak is extremely high, and the G band includes a pronounced shoulder of the D′ peak. Since the D′ peak appears as a result of defects that are different from those caused by the broadening of the D band [36], the high concentration of specific defects may be the reason for non-fitting of the C101 sample to the correlation data set. Figure 6a also shows that the concentration of defects in GBS products obtained using ammonium and potassium persulfate oxidizing agents is much lower compared to those obtained using CrO_3. Meanwhile, no significant differences can be detected between these two persulfate oxidizing agents. Similar results were obtained by measuring the correlation between the FWHM of two main bands in the Raman spectra of GBS products: FWHM(D) and FWHM(G) (see Figure 6b). While the higher values of both FWHM(D) and FWHM(G) mean the higher defect concentration, two characteristic zones, one for the samples oxidized with CrO_3 and another for the samples oxidized with $((NH_4)_2S_2O_8$ and $K_2S_2O_8$, can be distinguished in the correlation chart. As in the previous case, no significant differences were observed between the action of $(NH_4)_2S_2O_8$ and $K_2S_2O_8$.

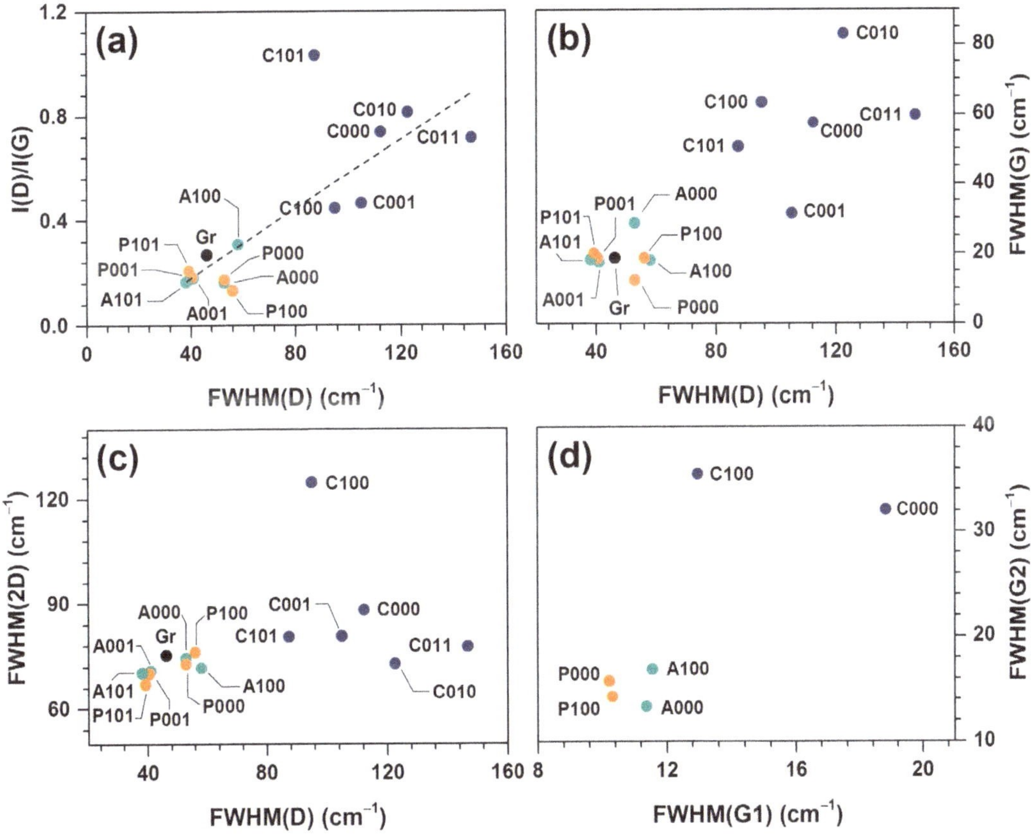

Figure 6. Correlation between FWHM data of Raman spectra of GBS products obtained using different oxidizing agents: FWHM(D) and I(D)/I(G) ratio representing defect concentration in GBS products (**a**), FWHM(D) and FWHM(G) representing defect and sp^2 bonded carbon vibrations (**b**), FWHM(D) and FWHM(2D) representing defect and second mode defect vibrations (**c**), FWHM(G1) and FWHM(G2) representing the staging index (stage I and stage II) (**d**).

Another important Raman marker is the FWHM(2D) (full width at half maximum of the 2D band), which is frequently used to identify the amount of graphene layers [27]. Single layer graphene has FWHM(2D) values around 30 cm^{-1}. When the number of layers increases, the FWHM values also show an increase [33]. In addition to this, the 2D Raman band shape also helps to distinguish single and bilayer graphene from the multilayer. Monolayer graphene has a single 2D component, bilayer graphene is fitted to four components, and the 2D mode in bulk graphite can be decomposed into two components. In this work, calculated FWHM(2D) values of all samples exceed 60 cm^{-1} and can be best fitted with two components (see Figures 3c, 4c and 5c). Thus, it can be concluded that the materials obtained feature a multilayer structure. The correlation between FWHM(D) and FWHM(2D) shows the presence of two zones, as in previous cases: one for the samples oxidized with CrO_3, another—for the samples oxidized with persulfates (Figure 6c).

Previously discussed correlations between the parameters in Raman spectra of GBS products do not allow us to distinguish any significant difference between the action of $(NH_4)_2S_2O_8$ and $K_2S_2O_8$. Such a result can be treated as an evident outcome, since the samples in the correlation set include completely different compounds (e.g., intercalates and annealed graphene-like phases). We tried to narrow the correlation set and compared the features of the Raman spectra of intercalated GBS phases (see Figure 6d). By comparing the FWHM(G1) and FWHM(G2), which are sensitive to the presence of defects in the corresponding GBS phase (stage I and stage II correspondingly), one can see that the quality of the phase is extremely sensitive to the use of CrO_3—it results in the evident increase of defect concentration [36]. Meanwhile, the minimal defect concentration in GBS phases can be reached by using $K_2S_2O_8$, while the use of $(NH_4)_2S_2O_8$ causes a slightly higher concentration of defects. Supposedly, the effect of NH_4^+ ion may be related to its higher redox activity in comparison with the K^+ ion, which is significant in the reaction of carbocation formation (see Equation (1)).

FTIR spectra were employed to determine the functional groups that emerged on the surface of the GBS products obtained using three oxidizing agents: $(NH_4)_2S_2O_8$ (Figure 7), $K_2S_2O_8$ (Figure 8), and CrO_3 (Figure 9). Additionally, the spectrum of the graphite precursor is presented (Figure 7), where the low intensity of the peaks represents a low concentration of surface functional groups in pristine graphite. The most characteristic features are the following: the absorption band at 3435 cm^{-1} (stretching vibrations of the O–H bond), band at 1637 cm^{-1} (stretching vibrations of the C=C bond), and peak at 1418 cm^{-1} (deformation vibrations of the C–OH bond) [28,37]. The addition of an oxidant increases both the intensity and variety of peaks in the FTIR spectra. Some newly appeared peaks may correspond to the SO_2 asymmetric stretching vibrations (1293 cm^{-1}) and S=O (1070 cm^{-1}) groups [13,38]. Peaks at 1231 cm^{-1} and 1167 cm^{-1} may be related to C–O vibrations [28,39,40]; the peak at 1004 cm^{-1} can be assigned to graphitic domains C–O vibrations [41,42], while 882 cm^{-1} and 850 cm^{-1} may correspond to out-of-plane vibrations of the C–H bond [42]. The broad feature in the vicinity of 2700−3100 cm^{-1} might be related to O−H stretching vibrations of strongly hydrogen-bonded carboxylate groups or adsorbed water molecules and C−H stretching motions [27]. In addition to these bands, the peak at 1117 cm^{-1} (asymmetric SO_4^{2-} vibrations) appears in the FTIR spectrum of the P001 sample [43]. In the spectra of GBS products obtained using CrO_3 (C010 and C011), minor bands that indicate the vibrations in the Cr=O group (950 cm^{-1} and 914 cm^{-1}), are also introduced [44]. After thermal treatment, absorption bands related to oxygen-containing groups in the fingerprint area mostly disappear, though a band in the vicinity of about 1070 cm^{-1}, which indicates the presence of S=O group, remains. As can be seen, during the intercalation of graphite, some of the carbon double bonds can be oxidized by strong acids, leading to the formation of oxygen-containing functional groups, and sulfur derivatives can be intercalated into the graphite structure [45]. In addition, analysis showed that the functional groups introduced during the intercalation process were not influenced using different persulfate oxidizing agents; however, using CrO_3 oxidizer determined that a small part of Cr=O groups was formed during the synthesis reaction.

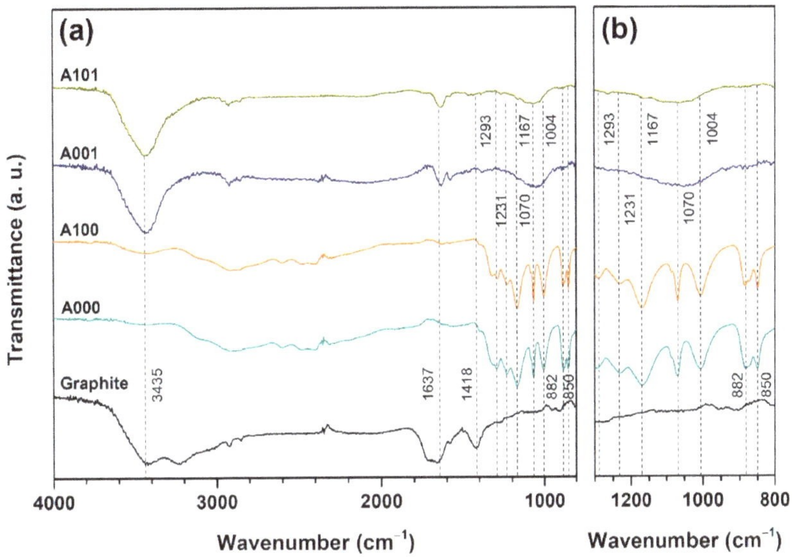

Figure 7. FTIR spectra of GBS products obtained using $(NH_4)_2S_2O_8$ oxidizing agent in the region of 4000–800 cm^{-1} (**a**) and 1300–800 cm^{-1} (**b**).

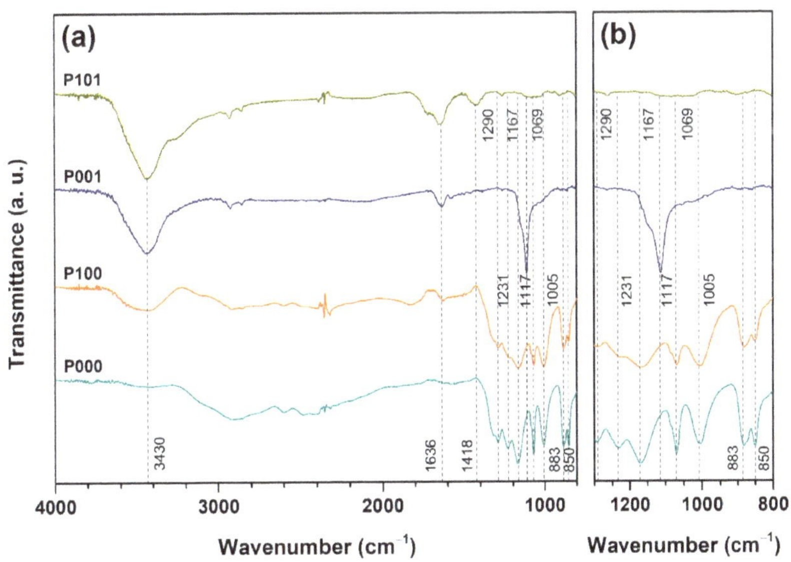

Figure 8. FTIR spectra of GBS products obtained using $K_2S_2O_8$ oxidizing agent in the region of 4000–800 cm^{-1} (**a**) and 1300–800 cm^{-1} (**b**).

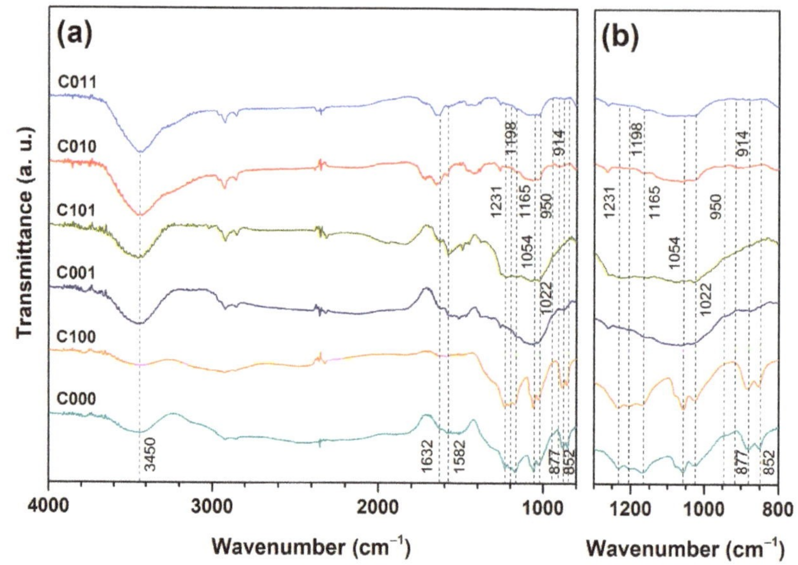

Figure 9. FTIR spectra of GBS products obtained using CrO_3 oxidizing agent in the region of 4000–800 cm^{-1} (**a**) and 1300–800 cm^{-1} (**b**).

The annealed GBS products, together with the GBS sample washed with H_2O (C010), were examined using XRD analysis. The XRD patterns are presented in Figure 10. The XRD pattern of pristine graphite is also presented for comparison. The determined values of interplanar spacing d_{002} and crystallite size L are presented in Table 3.

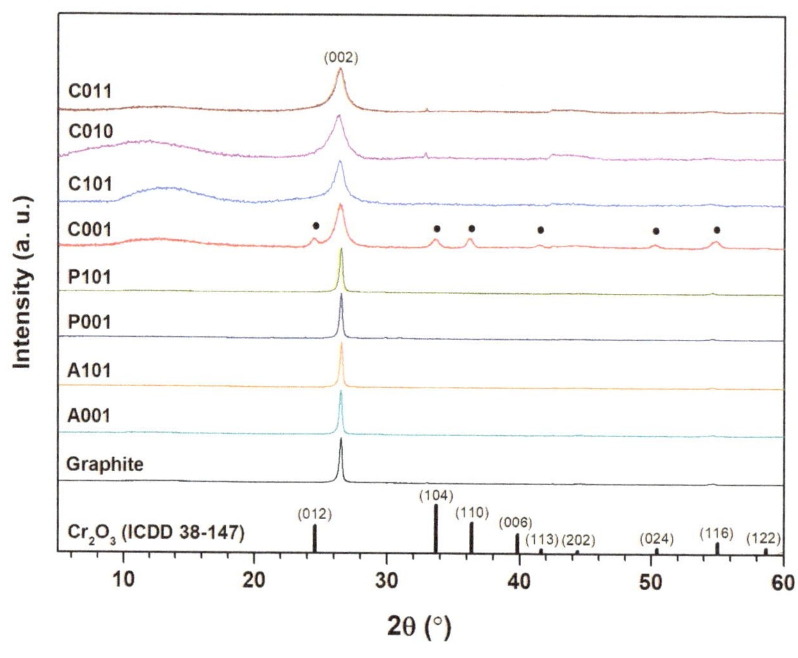

Figure 10. XRD patterns of GBS products, graphite, and Cr_2O_3 (ICDD 38-147; marked with •).

Table 3. XRD peak positions, interplanar spacing d_{002} and crystallite size of the annealed and water-treated GBS products.

Sample	2θ (deg)	d_{002} (nm)	*L* (nm)
Graphite	26.53	0.335	27.87
A001	26.55	0.335	27.40
P001	26.54	0.335	27.58
C001	26.48	0.336	7.34
A101	26.61	0.334	27.37
P101	26.52	0.335	23.71
C101	26.41	0.337	4.49
C010	26.41	0.337	4.36
C011	26.43	0.337	5.87

The use of $(NH_4)_2S_2O_8$ and $K_2S_2O_8$ oxidizing agents made it possible to obtain GBS products which, after annealing, return to the graphite-like structure: the peak positions, crystallite sizes and interplanar spacing of A001, A101, P001, and P101 closely coincide with that of pristine graphite (Table 3). Different results were obtained using the CrO_3 oxidizing agent. The peak (002) is clearly broadened compared to that of pure graphite. This is most likely caused by the drastic decrease in average particle size. As in most cases, smaller particles lead to broader reflections as estimated by the Scherrer equation. Such a result is observed not only for the annealed GBS products, but also for sample C010, which is obtained by adding water at the end of the intercalation reaction. A broadened peak with the maximum position at about 11.5° in the XRD pattern of the C010 sample supposedly indicates the presence of the structure, which is similar to that of graphite oxide. Furthermore, compared to the graphite, the (002) peak is shifted to lower 2θ angles and the interplanar spacing is increased from 0.335 nm up to 0.337 nm in GBS products obtained with CrO_3. A shift to lower angles can occur due to the presence of defects in the crystal lattice [13]. This observation, together with the Raman analysis, indicates that the GBS produced with CrO_3 can have a significant number of defects. Additional peaks (marked with dots) were also observed in the XRD patterns, where CrO_3 was used. These peaks may indicate the presence of Cr_2O_3. This compound can be formed when CrO_3 remains in the mixture and decomposes into Cr_2O_3 during thermal treatment [46]. The Cr_2O_3 reference pattern (ICDD 38-147) is shown in Figure 10 for comparison.

SEM micrographs were used to assess the surface morphology of the annealed GBS samples and compare it to that of pristine graphite (Figure 11).

The variable surface morphology of GBS products is evident. The SEM image of the pristine graphite sample (Figure 11a) revealed the presence of lamellar structure and flat surfaces with regular edges. The use of $(NH_4)_2S_2O_8$ and $K_2S_2O_8$ oxidizing agents produces particles with an expanded lamellar structure and less regular edges (Figure 11b–e). Significant differences in the morphology of GBS products appear after the use of CrO_3 oxidizing agent (Figure 11f–i); the lamellar structure becomes much less pronounced, while the edges of the particles turn highly jagged and irregular. This observation is in good agreement with the results discussed previously, as the use of CrO_3 in the GBS synthesis results in the most affected graphite structure.

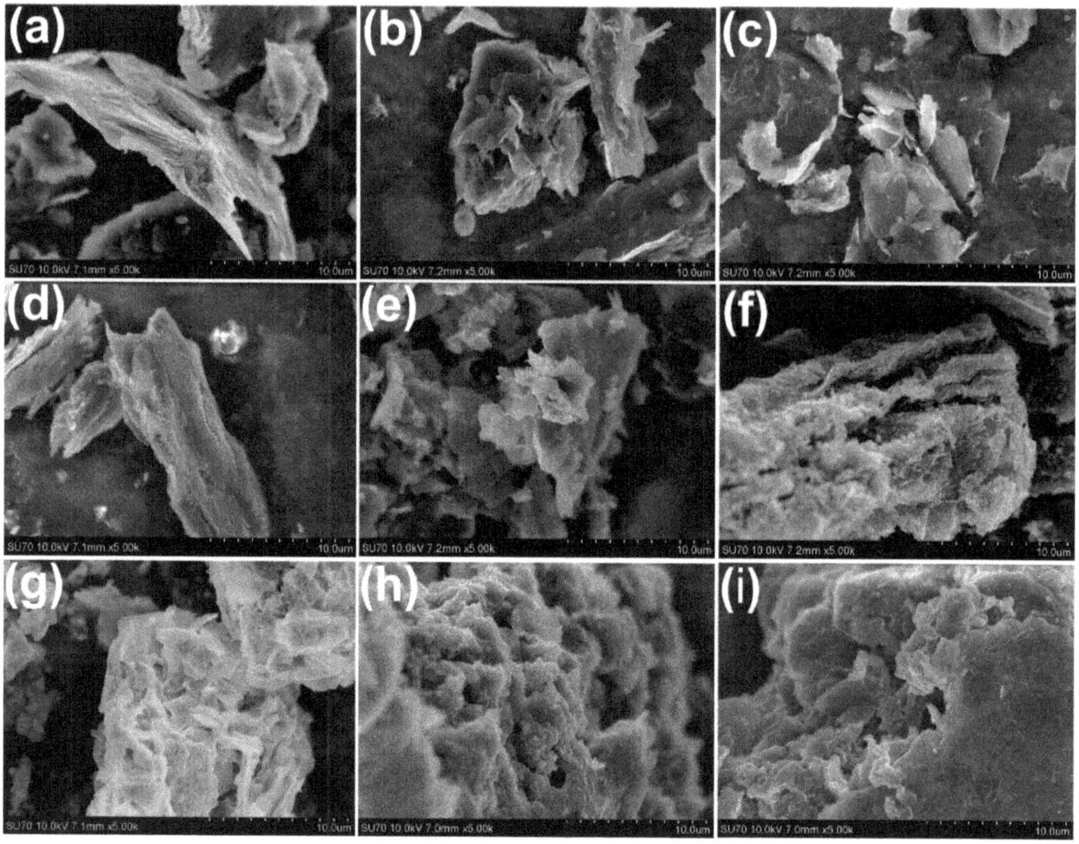

Figure 11. SEM micrographs of pristine graphite and GBS products: graphite (**a**), A001 (**b**), A101 (**c**), P001 (**d**), P101 (**e**), C001 (**f**), C101 (**g**), C010 (**h**), and C011 (**i**).

4. Conclusions

In this work, samples of GBS were synthesized using three different oxidizing agents: $(NH_4)_2S_2O_8$, $K_2S_2O_8$, and CrO_3 with and without using P_2O_5 as the water-binding agent. In addition, all the specimens obtained were heat-treated at 800 °C. Optical microscopy pictures taken after 2 and 24 h from the start of the synthesis show blue GBS particles, which correspond to stage I of the GBS compound, confirming that graphite has been successfully intercalated and is stable under used synthesis conditions in a medium of conc. H_2SO_4. Raman spectroscopy analysis revealed differences between the GBS phases and annealed GBS products when using all three oxidizing agents and P_2O_5. It was established that when P_2O_5 acts as a dehydration agent, the intercalation system can reach a higher acidity, resulting in a higher oxidation state of graphite and a more effective intercalation. Correlation between different characteristic parameters of the Raman analysis has shown that the use of CrO_3 results in a much higher defect concentration compared to the GBS products obtained using persulfate oxidizers. The correlation among all GBS products did not allow us to distinguish any significant difference between the action of $(NH_4)_2S_2O_8$ and $K_2S_2O_8$. However, narrowing the correlation set only to the characteristic parameters of GBS phases (stage I and stage II) revealed that the difference between the action of $K_2S_2O_8$ and $(NH_4)_2S_2O_8$ is still noticeable. The minimal defect concentration in the GBS phases can be reached by using $K_2S_2O_8$, while the use of $(NH_4)_2S_2O_8$ causes a slightly higher concentration of defects. Supposedly, the effect of the NH_4^+ ion may be related to its higher redox activity in comparison with the K^+ ion, which is significant in the reaction of carbocation formation. FTIR analysis has shown that during graphite intercalation, part of the carbon double bonds is oxidized, and sulfur derivatives are intercalated into the graphite structure. XRD analysis revealed that after heat treatment, GBS products mostly returned to a graphite-like structure; however, several differences were observed in samples with CrO_3. These compounds are characterized by the largest interplanar distances and the smallest crystallite sizes. SEM pictures confirm that the most significant changes in surface morphology were observed in the GBS products obtained using the CrO_3 oxidizing agent.

Author Contributions: Conceptualization, G.R., M.G., J.B., A.Z., J.G. and G.N.; methodology, J.B., J.G. and G.N.; formal analysis, J.B. and G.R.; investigation, M.G. and G.R.; resources, J.B. and J.G.; data curation, G.R., M.G., J.B., J.G. and G.N.; writing—original draft preparation, G.R.; writing—review and editing, G.R., M.G., J.B., A.Z., J.G. and G.N.; supervision, J.B. and J.G. All authors have read and agreed to the published version of the manuscript.

Funding: This research received no external funding.

Institutional Review Board Statement: Not applicable.

Informed Consent Statement: Not applicable.

Data Availability Statement: Data are contained within the article.

Conflicts of Interest: The authors declare no conflict of interest.

References

1. Schafhaeutl, C. Ueber Die Verbindungen Des Kohlenstoffes Mit Silicium, Eisen Und Anderen Metallen, Welche Die Verschiedenen Gallungen von Roheisen, Stahl Und Schmiedeeisen Bilden. *J. Prakt. Chem.* **1840**, *21*, 129–157. [CrossRef]
2. Brodie, B. Note Sur Un Nouveau Procédé Pour La Purification et La Désagrégation Du Graphite. *Ann. Chim. Phys.* **1855**, *45*, 351–353.
3. Brédas, J.-L.; Buriak, J.M.; Caruso, F.; Choi, K.-S.; Korgel, B.A.; Palacín, M.R.; Persson, K.; Reichmanis, E.; Schüth, F.; Seshadri, R.; et al. An Electrifying Choice for the 2019 Chemistry Nobel Prize: Goodenough, Whittingham, and Yoshino. *Chem. Mater.* **2019**, *31*, 8577–8581. [CrossRef]
4. Li, Y.; Lu, Y.; Adelhelm, P.; Titirici, M.-M.; Hu, Y.-S. Intercalation Chemistry of Graphite: Alkali Metal Ions and Beyond. *Chem. Soc. Rev.* **2019**, *48*, 4655–4687. [CrossRef]
5. Xu, J.; Dou, Y.; Wei, Z.; Ma, J.; Deng, Y.; Li, Y.; Liu, H.; Dou, S. Recent Progress in Graphite Intercalation Compounds for Rechargeable Metal (Li, Na, K, Al)-Ion Batteries. *Adv. Sci.* **2017**, *4*, 1700146. [CrossRef] [PubMed]
6. Setton, R.; Beguin, F.; Piroelle, S. Graphite Intercalation Compounds as Reagents in Organic Synthesis. An Overview and Some Recent Applications. *Synth. Met.* **1982**, *4*, 299–318. [CrossRef]
7. Smith, R.P.; Weller, T.E.; Howard, C.A.; Dean, M.P.M.; Rahnejat, K.C.; Saxena, S.S.; Ellerby, M. Superconductivity in Graphite Intercalation Compounds. *Phys. C Supercond. Its Appl.* **2015**, *514*, 50–58. [CrossRef]
8. Abdelkader, A.M.; Cooper, A.J.; Dryfe, R.A.W.; Kinloch, I.A. How to Get between the Sheets: A Review of Recent Works on the Electrochemical Exfoliation of Graphene Materials from Bulk Graphite. *Nanoscale* **2015**, *7*, 6944–6956. [CrossRef]
9. Li, C.; Shi, Y.; Chen, X.; He, D.; Shen, L.; Bao, N. Controlled Synthesis of Graphite Oxide: Formation Process, Oxidation Kinetics, and Optimized Conditions. *Chem. Eng. Sci.* **2018**, *176*, 319–328. [CrossRef]
10. Sorokina, N.E.; Nikol'skaya, I.V.; Ionov, S.G.; Avdeev, V.V. Acceptor-Type Graphite Intercalation Compounds and New Carbon Materials Based on Them. *Russ. Chem. Bull.* **2005**, *54*, 1749–1767. [CrossRef]
11. Bonaccorso, F.; Lombardo, A.; Hasan, T.; Sun, Z.; Colombo, L.; Ferrari, A.C. Production and Processing of Graphene and 2d Crystals. *Mater. Today* **2012**, *15*, 564–589. [CrossRef]
12. Chung, D.D.L. A Review of Exfoliated Graphite. *J. Mater. Sci.* **2015**, *51*, 554–568. [CrossRef]
13. Salvatore, M.; Carotenuto, G.; de Nicola, S.; Camerlingo, C.; Ambrogi, V.; Carfagna, C. Synthesis and Characterization of Highly Intercalated Graphite Bisulfate. *Nanoscale Res. Lett.* **2017**, *12*, 167. [CrossRef]
14. Dimiev, A.M.; Bachilo, S.M.; Saito, R.; Tour, J.M. Reversible Formation of Ammonium Persulfate/Sulfuric Acid Graphite Intercalation Compounds and Their Peculiar Raman Spectra. *ACS Nano* **2012**, *6*, 7842–7849. [CrossRef] [PubMed]
15. Lünsdorf, N.K. Raman Spectroscopy of Dispersed Vitrinite—Methodical Aspects and Correlation with Reflectance. *Int. J. Coal Geol.* **2016**, *153*, 75–86. [CrossRef]
16. Zaaba, N.I.; Foo, K.L.; Hashim, U.; Tan, S.J.; Liu, W.W.; Voon, C.H. Synthesis of Graphene Oxide Using Modified Hummers Method: Solvent Influence. *Procedia Eng.* **2017**, *184*, 469–477. [CrossRef]
17. Ivanov, A.V.; Maksimova, N.V.; Kamaev, A.O.; Malakho, A.P.; Avdeev, V.V. Influence of Intercalation and Exfoliation Conditions on Macrostructure and Microstructure of Exfoliated Graphite. *Mater. Lett.* **2018**, *228*, 403–406. [CrossRef]

18. Dimiev, A.M.; Shukhina, K.; Behabtu, N.; Pasquali, M.; Tour, J.M. Stage Transitions in Graphite Intercalation Compounds: Role of the Graphite Structure. *J. Phys. Chem. C* **2019**, *123*, 19246–19253. [CrossRef]
19. Seiler, S.; Halbig, C.E.; Grote, F.; Rietsch, P.; Börrnert, F.; Kaiser, U.; Meyer, B.; Eigler, S. Effect of Friction on Oxidative Graphite Intercalation and High-Quality Graphene Formation. *Nat. Commun.* **2018**, *9*, 836. [CrossRef]
20. Dimiev, A.M.; Shukhina, K.; Khannanov, A. Mechanism of the Graphene Oxide Formation. The Role of Water, "Reversibility" of the Oxidation, and Mobility of the C–O Bonds. *Carbon* **2020**, *166*, 1–14. [CrossRef]
21. 3—Fixation. In *Chromosome Techniques*, 3rd ed.; Sharma, A.K.; Sharma, A. (Eds.) Butterworth-Heinemann: Boston, MA, USA, 1980; pp. 30–70, ISBN 978-0-408-70942-2.
22. Ferrari, A.C.; Robertson, J. Raman Spectroscopy of Amorphous, Nanostructured, Diamond-like Carbon, and Nanodiamond. *Philos. Trans. R. Soc. A Math. Phys. Eng. Sci.* **2004**, *362*, 2477–2512. [CrossRef] [PubMed]
23. Trusovas, R.; Ratautas, K.; Račiukaitis, G.; Niaura, G. Graphene Layer Formation in Pinewood by Nanosecond and Picosecond Laser Irradiation. *Appl. Surf. Sci.* **2019**, *471*, 154–161. [CrossRef]
24. ChacónChac, J.C.; Wirtz, L.; Pichler, T. Raman Spectroscopy of Graphite Intercalation Compounds: Charge Transfer, Strain, and Electron-Phonon Coupling in Graphene Layers. *Phys. Status Solidi B* **2014**, *251*, 2337–2355. [CrossRef]
25. Gurzęda, B.; Buchwald, T.; Krawczyk, P. Thermal Exfoliation of Electrochemically Synthesized Graphite Intercalation Compound with Perrhenic Acid. *J. Solid State Electrochem.* **2020**, *24*, 1363–1370. [CrossRef]
26. Eigler, S. Graphite Sulphate—A Precursor to Graphene. *Chem. Commun.* **2015**, *51*, 3162–3165. [CrossRef] [PubMed]
27. Pei, S.; Wei, Q.; Huang, K.; Cheng, H.; Ren, W. Green synthesis of graphene oxide by seconds timescale water electrolytic oxidation. *Nat. Commun.* **2018**, *9*, 145. [CrossRef] [PubMed]
28. Trusovas, R.; Račiukaitis, G.; Niaura, G.; Barkauskas, J.; Valušis, G.; Pauliukaite, R. Recent Advances in Laser Utilization in the Chemical Modification of Graphene Oxide and Its Applications. *Adv. Opt. Mater.* **2016**, *4*, 37–65. [CrossRef]
29. Ferrari, A.C. Raman Spectroscopy of Graphene and Graphite: Disorder, Electron-Phonon Coupling, Doping and Nonadiabatic Effects. *Solid State Commun.* **2007**, *143*, 47–57. [CrossRef]
30. Douda, J.; González Vargas, C.R.; Basiuk, E.V.; Díaz Cano, A.I.; Fuentes García, J.A.; Hernández Contreras, X.A. Optical Properties of Amine-Functionalized Graphene Oxide. *Appl. Nanosci.* **2019**, *9*, 567–578. [CrossRef]
31. Zou, J.; Sole, C.; Drewett, N.E.; Velický, M.; Hardwick, L.J. In Situ Study of Li Intercalation into Highly Crystalline Graphitic Flakes of Varying Thicknesses. *J. Phys. Chem. Lett.* **2016**, *7*, 4291–4296. [CrossRef]
32. Zhou, X.; Liu, Q.; Jiang, C.; Ji, B.; Ji, X.; Tang, Y.; Cheng, H.-M. Strategies towards Low-Cost Dual-Ion Batteries with High Performance. *Angew. Chem. Int. Ed.* **2020**, *59*, 3802–3832. [CrossRef]
33. Dimiev, A.M.; Ceriotti, G.; Behabtu, N.; Zakhidov, D.; Pasquali, M.; Saito, R.; Tour, J.M. Direct Real-Time Monitoring of Stage Transitions in Graphite Intercalation Compounds. *ACS Nano* **2013**, *7*, 2773–2780. [CrossRef] [PubMed]
34. Gaidukevic, J.; Aukstakojyte, R.; Navickas, T.; Pauliukaite, R.; Barkauskas, J. A Novel Approach to Prepare Highly Oxidized Graphene Oxide: Structural and Electrochemical Investigations. *Appl. Surf. Sci.* **2021**, *567*, 150883. [CrossRef]
35. Melezhyk, A.V.; Tkachev, A.G. Synthesis of graphene nanoplatelets from peroxosulfate graphite intercalation compounds. Наносистемы *физика* Химия Математика **2014**, *5*, 294–306.
36. Nagyte, V.; Kelly, D.J.; Felten, A.; Picardi, G.; Shin, Y.Y.; Alieva, A.; Worsley, R.E.; Parvez, K.; Dehm, S.; Krupke, R.; et al. Raman Fingerprints of Graphene Produced by Anodic Electrochemical Exfoliation. *Nano Lett.* **2020**, *20*, 3411–3419. [CrossRef]
37. Lavin-Lopez, M.P.; Patón-Carrero, A.; Muñoz-Garcia, N.; Enguilo, V.; Valverde, J.L.; Romero, A. The Influence of Graphite Particle Size on the Synthesis of Graphene-Based Materials and Their Adsorption Capacity. *Colloids Surf. A Physicochem. Eng. Asp.* **2019**, *582*, 123935. [CrossRef]
38. Martinez-Rubi, Y.; Gonzalez-Dominguez, J.M.; Ansón-Casaos, A.; Kingston, C.T.; Daroszewska, M.; Barnes, M.; Hubert, P.; Cattin, C.; Martinez, M.T.; Simard, B. Tailored SWCNT Functionalization Optimized for Compatibility with Epoxy Matrices. *Nanotechnology* **2012**, *23*, 285701. [CrossRef]
39. Bera, M.; Gupta, P.; Maji, P.K. Facile One-Pot Synthesis of Graphene Oxide by Sonication Assisted Mechanochemical Approach and Its Surface Chemistry. *J. Nanosci. Nanotechnol.* **2018**, *18*, 902–912. [CrossRef]
40. Ţucureanu, V.; Matei, A.; Avram, A.M. FTIR Spectroscopy for Carbon Family Study. *Crit. Rev. Anal. Chem.* **2016**, *46*, 502–520. [CrossRef]
41. Zhong, L.; Yun, K. Graphene Oxide—Modified ZnO Particles: Synthesis, Characterization and Application. *Int. J. Nanomed.* **2018**, *10*, 79–92.
42. He, X.; Liu, X.; Nie, B.; Song, D. FTIR and Raman Spectroscopy Characterization of Functional Groups in Various Rank Coals. *Fuel* **2017**, *206*, 555–563. [CrossRef]
43. Kusrini, E.; Ng, S.S.; Yulizar, Y.; Ismail, D. Effects of Monocarboxylic Acids and Potassium Persulfate on Preparation of Chitosan Nanoparticles. *Int. J. Technol.* **2015**, *6*, 10–21. [CrossRef]
44. Trivedi, M.K. Characterization of Physical, Thermal and Structural Properties of Chromium (VI) Oxide Powder: Impact of Biofield Treatment. *J. Powder Metall. Min.* **2015**, *4*, 128. [CrossRef]
45. Konwer, S.; Gogoi, J.P.; Kalita, A.; Dolui, S.K. Synthesis of Expanded Graphite Filled Polyaniline Composites and Evaluation of Their Electrical and Electrochemical Properties. *J. Mater. Sci. Mater. Electron.* **2011**, *22*, 1154–1161. [CrossRef]
46. Sajadi, S.A.A.; Khaleghian, M. Study of Thermal Behavior of CrO_3 Using TG and DSC. *J. Therm. Anal. Calorim.* **2014**, *116*, 915–921. [CrossRef]

crystals

MDPI

Article

The Impact of Full-Scale Substitution of Ca^{2+} with Ni^{2+} Ions on Brushite's Crystal Structure and Phase Composition

Mazen Alshaaer [1,2,*], Khalil Issa [3,*], Ahmed S. Afify [4], Moustapha E. Moustapha [5], Abdulaziz A. Alanazi [5], Ammar Elsanousi [1] and Talal F. Qahtan [1]

1 Department of Physics, College of Science and Humanities in Al-Kharj, Prince Sattam Bin Abdulaziz University, Al-Kharj 11942, Saudi Arabia; a.elsanousi@psau.edu.sa (A.E.); tfqahtan@gmail.com (T.F.Q.)
2 GeoBioTec Research Centre, Campus de Santiago, University of Aveiro, 3810-193 Aveiro, Portugal
3 Orthopedics Unit, Faculty of Medicine and Health Sciences, An-Najah National University, Nablus 00972, Palestine
4 Department of Basic Sciences, Higher Institute of Engineering and Automotive Technology and Energy, New Heliopolis 11829, Egypt; ahmed.afify@polito.it
5 Department of Chemistry, College of Science and Humanities in Al-Kharj, Prince Sattam Bin Abdulaziz University, Al-Kharj 11942, Saudi Arabia; m.moustapha@psau.edu.sa (M.E.M.); abdulaziz.alanazi@psau.edu.sa (A.A.A.)
* Correspondence: mazen.alshaaer@yahoo.com or m.alshaaer@psau.edu.sa (M.A.); k.issa@najah.edu (K.I.)

Abstract: Because the impact of the full-scale substitution of Ca^{2+} in brushite ($CaHPO_4 \cdot 2H_2O$) with Ni^{2+} ions has never been systematically explored, it is the focus of this investigation, as it holds potential for use in $Ca_xNi_{1-x}HPO_4 \cdot nH_2O$ production. These biomaterials have many beneficial characteristics that can be modified to suit diverse applications, including bone tissue regeneration and pharmaceutics. For the present study, $NaH_2PO_4 \cdot 2H_2O$, $Ca(NO_3)_2 \cdot 4H_2O$, and $Ni(NO_3)_2 \cdot 6H_2O$ were used in various molar concentrations to obtain the required starting solutions. Previous studies have shown that adding Ni ions in the initial solution below 20% results in the precipitation of monophasic brushite with slight changes in the crystal structure. However, this study confirms that when the Ni ions substitution increases to 20%, a mixture of phases from both brushite and hexaaquanickel(II) hydrogenphosphate monohydrate HNiP ($Ni(H_2O)_6 \cdot HPO_4 \cdot H_2O$) is formed. The results confirm that the full replacement (100%) of Ca ions by Ni ions results in a monophasic compound solely comprising orthorhombic HNiP nanocrystals. Therefore, a novel technique of HNiP synthesis using the precipitation method is introduced in this research work. These materials are subsequently analyzed utilizing powder X-ray diffraction (XRD), X-ray photoelectron spectroscopy (XPS), thermogravimetric analysis (TGA), and scanning electron microscopy (SEM). The obtained results confirm that the material microstructure is controlled by the Ni/Ca ratio in the starting solution and can be modified to obtain the desired characteristics of phases and crystals.

Keywords: brushite; hexaaquanickel(II) hydrogen phosphate monohydrate; biomaterials; XPS; kidney stone

Citation: Alshaaer, M.; Issa, K.; Afify, A.S.; Moustapha, M.E.; Alanazi, A.A.; Elsanousi, A.; Qahtan, T.F. The Impact of Full-Scale Substitution of Ca^{2+} with Ni^{2+} Ions on Brushite's Crystal Structure and Phase Composition. *Crystals* **2022**, *12*, 940. https://doi.org/10.3390/cryst12070940

Academic Editors: Aleksej Zarkov, Aivaras Kareiva and Loreta Tamasauskaite-Tamasiunaite

Received: 21 May 2022
Accepted: 30 June 2022
Published: 4 July 2022

1. Introduction

Owing to their beneficial mineralogical and biochemical characteristics, calcium phosphates (CaPs) already have diverse applications, including those in the environmental sciences and the field of engineering. However, their similarity with the mineral phases present in osseous tissue—as well as a low toxicity, high bioactivity, and excellent biocompatibility—makes them particularly useful candidates for medical applications [1–4]. Presently, CaPs serve as precursors for biocements and bioceramics, to be used in medicine and dentistry, and to obtain other types of advanced materials [4,5]. Available evidence further indicates that CaPs can be adopted for drug delivery or bone tissue engineering [5,6], in fertilizers [7], and in the construction industry [8–10].

Nickel is one of the trace elements in the human body, where the average amount in a healthy individual is about 10 mg [11,12]. The concentration of nickel per liter of serum is about 0.2 μg/L in a normal population [11]. Nickel is known to be harmful to human health, causing cancer, inflammation, skin allergies, lung fibrosis, asthma, and kidney diseases [13]. Yet later, it was found that nickel toxicity depends on the amount of the dose so that it can be used in vivo under certain conditions and modifications. In view of these facts, nickel has been adopted for biomedical applications involving stainless steel (15% Ni) or shape-memory alloys such as nitinol (50% Ni) [14,15]. Such Nickel-based alloys are now used in biomedical implants for clinical application [12].

Nickel ions and other d-block elements (Co, Cu, V, Ti, Cr) exhibit one very important property, which is the stabilization of HIF-1α; therefore, they can induce cellular VEGF secretion by creating intracellular hypoxia mimicking conditions. Nickel-doped nHAp having different percentages of Ni (1%, 5%, and 10% w/w) [12] is a proangiogenic–osteogenic material that could be used in bone tissue engineering, as doped samples proved bone-cell compatible and osteoconductive. Analysis showed that Ni^{2+} doped nHAp samples were potent inducers of the cellular Vascular Endothelial Growth Factor (VEGF), the expression of VEGF being directly proportional to the doping concentration of Ni^{2+} [12].

Several studies indicate that dicalcium phosphate dehydrate (DCPD)—commonly known as brushite, with the chemical formula $CaHPO_4{\cdot}2H_2O$—is one of the most widely used CaPs [8,11]. This mineral acts as a precursor of hydroxyapatite (HA), bone cement, and bioceramics. Hence, the brushite encountered in mineralized tissues can be useful in medicine, especially in bone cement formulations [16]. It shows high stability in environments characterized by pH 4.0–6.5 and temperatures below 60 °C [17–20]. Moreover, as it is usually metastable at pH ~7.4, it can be quickly resorbed in the human body to aid in bone remodeling [21–23].

The influence of ionic substitution on the morphology, structure, synthesis and hydrolysis process of brushite has been investigated [1]. The results of structural refinements indicate that the range of possible substitutions can reach different values based on the doped ions. This study showed that the Sr substitution can reach up to about 38%, whereas it is quite limited for Co, Mn, and Zn. A previous study [24] has shown the most complete incorporation of Ni^{2+} to the solid phase of brushite up to 15%. However, for larger Ni^{2+} concentrations in the initial solution, a mixture of phases has been detected [24,25].

Such findings have prompted us to study the crystal morphology, thermal properties, chemical composition, phases, and mineralogy of the produced compounds when Ca^{2+} in brushite is gradually substituted or replaced (from 20% to 100%) by Ni^{2+}. The findings fill an important research gap and may prove valuable for mineral synthesis to derive biomaterials for a variety of applications in the bone tissue engineering and the pharmaceutical industry.

2. Experimental Methodology

2.1. Materials

Sodium dihydrogen orthophosphate dihydrate ($NaH_2PO_4{\cdot}2H_2O$) was purchased from Techno Pharmchem, India; while calcium nitrate tetrahydrate ($Ca(NO_3)_2{\cdot}4H_2O$) and nickel (II) nitrate hexahydrate ($Ni(NO_3)_2{\cdot}6H_2O$) were procured from LOBA Chemie, India. Distilled water (0.055 μS/cm) was prepared using a water purification system (PURELAB option-Q, ELGA, Oxford, UK). A digital analytical balance (EX324N, OHAUS, Parsippany, NJ, USA) and a magnetic stirrer (ISOTEMP, Fisher Scientific, Shanghai, China) were used when required.

2.2. Synthesis of $Ca_xNi_{1-x}HPO_4{\cdot}nH_2O$ Compounds

A series of six $Ca_xNi_{1-x}HPO_4{\cdot}nH_2O$ compounds (with x = 0, 2, 4, 5, 6, and 10) was synthesized at room temperature (RT) based on the reaction given in Equation (1), which called for the use of $Na_2HPO_4{\cdot}2H_2O$, $Ca(NO_3)_2{\cdot}4H_2O$ and $Ni(NO_3)_2{\cdot}6H_2O$ 0.5 moL/L solutions in the molar proportions provided in Table 1.

$$\alpha Ni(NO_3)_2{\cdot}6H_2O + (1-\alpha)Ca(NO_3)_2 + Na_2HPO_4 + NH_3 + H_2O \rightarrow (Ca_{1-\alpha} + Ni_\alpha)HPO_4{\cdot}2H_2O + 2NaNO_3 \quad (1)$$

Table 1. Molar proportions of $NaH_2PO_4{\cdot}2H_2O$, $Ca(NO_3)_2{\cdot}4H_2O$, and $Ni(NO_3)_2{\cdot}6H_2O$, as well as Ni/Ca molar ratios used for the synthesis of $Ca_xNi_{1-x}HPO_4{\cdot}nH_2O$ compounds.

Product ID	$NaH_2PO_4{\cdot}2H_2O$	$Ca(NO_3)_2{\cdot}4H_2O$	$Ni(NO_3)_2{\cdot}6H_2O$	Ni/Ca Molar Ratio
BNi0	1	1	0	0
BNi2	1	0.8	0.2	0.25
BNi4	1	0.6	0.4	0.67
BNi5	1	0.5	0.5	1.0
BNi6	1	0.4	0.6	1.5
BNi10	1	0	1	-

To obtain pure brushite (denoted in Table 1 as BNi0), 100 mL of $Ca(NO_3)_2{\cdot}4H_2O$ solution was added at the ≈2 mL/min flow rate to the $Na_2HPO_4{\cdot}2H_2O$ solution using a glass funnel with a glass stopcock while stirring at 450 rpm until a 1.0 Ca/P molar ratio was achieved (which typically took about 60 min). The resulting solution was stirred at RT for 60 min to ensure that it was fully homogenized, after which the pH was adjusted to 6–6.5 by adding ammonia solution (≈15 moL/L, Labochemie, Mumbai, India) as required. The aim was to obtain a white precipitate that was then vacuum filtered using a Buchner funnel and a qualitative filter paper (45 μm, ∅12 cm, Double Rings, Shanghai, China). The filter cake was first washed three times with deionized water, which was followed by three washes in ethanol to prevent agglomeration [26–28]. Finally, the sample was placed on a watch glass and was left to dry in an oven set at 40 °C for one week (ED53/E2, Binder, Tuttlingen, Germany) [29].

To obtain BNi2, BNi4, BNi5, and BNi6 compounds, the $Ca(NO_3)_2{\cdot}4H_2O$ and $Ni(NO_3)_2{\cdot}6H_2O$ solutions were mixed in the molar ratios outlined in Table 1, after which 100 mL of the resulting solution was added at the ≈2 mL/min flow rate to 100 mL of $Na_2HPO_4{\cdot}2H_2O$ solution as described above. For BNi10, the same process was performed after mixing $NaH_2PO_4{\cdot}2H_2O$ and $Ni(NO_3)_2{\cdot}6H_2O$.

2.3. Characterization Techniques

The obtained BNi0–BNi10 samples were subjected to powder diffraction phase analysis using a Shimadzu XRD diffractometer-6000 (Japan) with a cobalt tube and a 2-theta scanning range of 10–60° at a 2°/min scan rate. Product morphology was established via scanning electron microscopy with the Inspect F50 (The Netherlands) apparatus. An XPS system (Thermo K Alpha spectrometer, Thermo Fisher Scientific, Waltham, MA, USA) was used to conduct X-ray photoelectron spectroscopy to determine the surface chemistry and conduct elemental analysis of the samples. Finally, a thermogravimetric (TGA) analyzer (TG 209 F1 Libra, Netzsch, Selb, Germany) served to ascertain the mass loss (≈100 mg) resulting from heating each product from 40 to 600 °C, at 5 °C min^{-1} increments under a helium atmosphere. An FT-IR spectrometer (Perkin–Elmer system 2000) was used for recording FTIR spectra in the range of 4000–400 cm^{-1}.

3. Results and Discussion

3.1. Mineralogical and Microstructural Analysis

The XRD patterns produced by all samples, including standard brushite ($CaHPO_4{\cdot}2H_2O$) and hexaaquanickel(II) hydrogenphosphate monohydrate ($Ni(H_2O)_6{\cdot}HPO_4{\cdot}H_2O$) (HNiP), are presented in Figure 1. The qualitative mineralogical analysis confirmed that after mixing $NaH_2PO_4{\cdot}2H_2O$ and $Ca(NO_3)_2{\cdot}4H_2O$ solutions with a Ca:P molar ratio of 1:1 (Table 1), pure brushite (BNi0) is precipitated. Moreover, after nucleation, its crystals grow proportionately to the three major crystallographic planes, i.e., (020), (12-1), and (14-1). In addition, all peaks identified in the pattern produced by BNi0 are indicative of brushite's monoclinic structure [19,30], while the peak at 11.7° 2-theta suggests that crystal growth primarily progresses along the (020) plane [31].

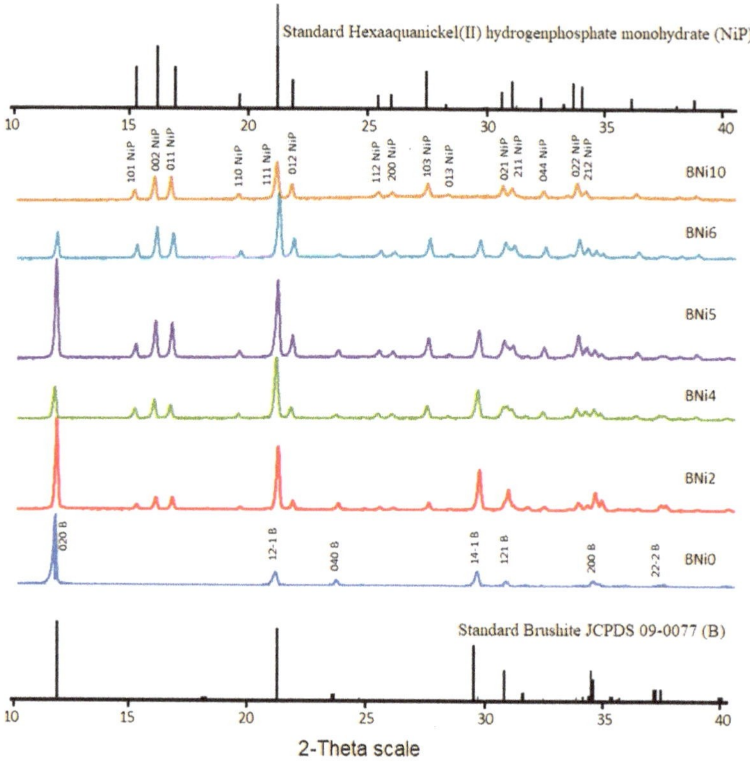

Figure 1. XRD patterns of the $Ca_xNi_{1-x}HPO_4{\cdot}nH_2O$ compounds synthesized under the conditions provided in Table 1.

In addition to the growth of XRD peaks corresponding to HNiP, all the main brushite peaks can be identified in the pattern, BNi2, and show higher intensity than in the BNi0 spectrum, especially for peaks associated with the (020), (121), and (141) planes [32,33].

The patterns produced by BNi2 to BNi6 (corresponding to the Ni/Ca molar ratio of 0.25 to 1.5), as expected, indicate that a greater proportion of Ca is substituted by Ni, while hexaaquanickel(II) hydrogenphosphate monohydrate ($Ni(H_2O)_6{\cdot}HPO_4{\cdot}H_2O$) (HNiP) also began to precipitate. The planes corresponding to $Ni(H_2O)_6{\cdot}HPO_4{\cdot}H_2O$—(101), (002), (011), (111) and (103)—are thus observed. The XRD patterns further show that as the Ni/Ca molar ratio increases, so does the HNiP peak intensity. Finally, the pattern produced by BNi10 (with the Ni/P ratio of 1.0) confirms the presence of pure HNiP with an orthorhombic crystal structure, as seen in Figure 1.

Figure 2 provides the SEM images of brushite (BNi0), along with those of biphasic compounds (BNi2, BNi4, and BNi6) based on different Ni/Ca molar ratios. The SEM image of pure HNiP ($Ni(H_2O)_6{\cdot}HPO_4{\cdot}H_2O$; BNi10) is presented in Figure 3.

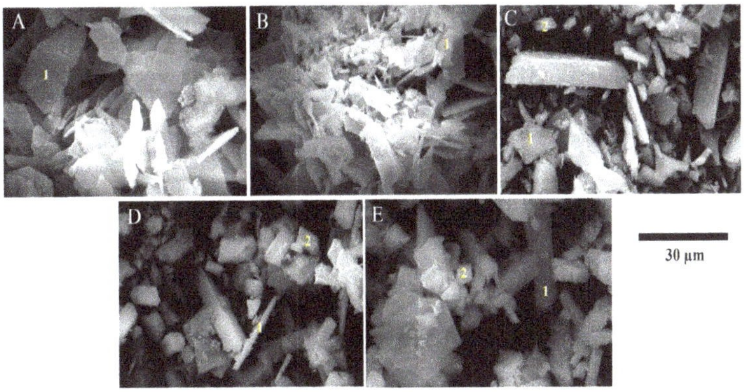

Figure 2. SEM images of $Ca_xNi_{1-x}HPO_4{\cdot}nH_2O$ compounds (labeled using product names provided in Table 1 and obtained with the same magnification): (**A**) BNi0; (**B**) BNi2; (**C**) BNi4; (**D**) BNi5, and (**E**) BNi6. (Point 1—brushite crystals, Point 2—hexaaquanickel(II) hydrogenphosphate monohydrate).

Figure 3. SEM images of hexaaquanickel(II) hydrogenphosphate monohydrate crystals ($Ni(H_2O)_6{\cdot}HPO_4{\cdot}H_2O$ (HNiP); product name BNi10).

As can be seen at Point 1 in Figure 2A (pertaining to BNi0 with the Ca/P molar ratio 1.0), the precipitation of pure brushite results in plate-like crystals, as expected for the solution pH used in this work [17,18]. The dimensions of these plate-like crystals are ≈500 nm × 15 μm × 30 μm, which is comparable to those obtained by other authors [10,34]. Moreover, flat-plate morphology is a typical crystal structure for precipitated brushite [35]. It is evident from Point 1 in Figure 2B that as Ca replacement with Ni intensifies due to the increase in the Ni/Ca molar ratio, the length of brushite crystals in the (020) direction decreases, ranging between ≈5 and ≈20 μm. Finally, as seen in Figure 2B–E—depicting a biphase compound—brushite crystals (Point 1) of different sizes are formed alongside HNiP crystals (Point 2).

BNi10 is depicted in Figure 3, which confirms that the full replacement of Ca by Ni results in a monophasic compound made up only of orthorhombic HNiP nanocrystals. Thus, the findings yielded by SEM analysis confirm those obtained via the XRD shown in Figure 1.

Rietveld refined unit cell parameters for brushite and HNiP materials are presented in Tables 2 and 3. As evidenced by the tabulated data, the values of unit cell parameters related to brushite increase as the Ni/Ca molar ratio increases from 0 to 1; hence, brushite is the dominant phase formed. Yet, when the Ni/Ca molar ratio exceeds 1, the brushite unit cell parameters decrease. Similarly, the parameters pertaining to the HNiP unit cell increase as the Ni/Ca molar ratio increases from 0.25 to 1 (Table 3). Still, the unit cell dimensions of pure HNiP are larger than those characterizing the standard unit cell.

Table 2. The unit cell parameters for brushite are derived from the XRD scans (Rietveld refinement).

Product ID	Brushite wt %	a (Å)	b (Å)	c (Å)	βo	V (Å3)
Standard Brushite	100	5.812	15.18	6.239	116.43	492.91
BNi0	100	5.8145	15.1693	6.2399	116.43	492.8455
BNi2	93	5.81143	15.17852	6.23839	116.43	492.7654
BNi4	25.5	5.8195	15.19959	6.24705	116.43	494.8206
BNi5	34.5	5.82386	15.21097	6.25173	116.43	495.9334
BNi6	14.6	5.81326	15.18328	6.24035	116.43	493.2301
BNi10	0	-	-	-	-	-

Table 3. The unit cell parameters for HNiP (product name BNi10) are derived from XRD scans (Rietveld refinement).

Product ID	Brushite wt %	a (Å)	b (Å)	c (Å)	V (Å3)
Standard HNiP	100	6.916	6.1032	11.1679	471.394
BNi0	0	-	-	-	-
BNi2	7	6.91997	6.09953	11.14389	470.3676
BNi4	74.5	6.93074	6.11028	11.17562	473.2737
BNi5	65.5	6.94044	6.11663	11.18034	474.629
BNi6	85.4	6.9211	6.10223	11.16156	471.3989
BNi10	100	6.9356	6.10915	11.17447	473.4692

The FT-IR spectrum of the $Ca_xNi_{1-x}HPO_4{\cdot}nH_2O$ compounds is shown in Figure 4. The broad absorption peak between 3500 and 2400 cm^{-1} is due to O-H stretching vibration [36]. The P-O-P asymmetric stretching vibration band was observed at 983 cm^{-1} as a result of P=O stretching vibrations; other bands at 654 and 568 cm^{-1} may be attributed to (H-O-)P=O for acid phosphates [36,37]. Another two peaks were observed at 3470 and 1640 cm^{-1}, indicating the existence of water, and their intensity decreased along with the decreasing trend in brushite contents in the samples (BNi2, BNi4, BNi5, and BNi6), whereas the BiN6 and BNi10 spectra reflected an absence of water [38]. On the other hand, the samples showed new peaks at 1645 and 1432 cm^{-1}, which were respectively indexed to O–H and P–O–bending vibrations, owing to the presence of Ni and its intensity, which reached a maximum in samples BNi6 and BNi10 [39,40].

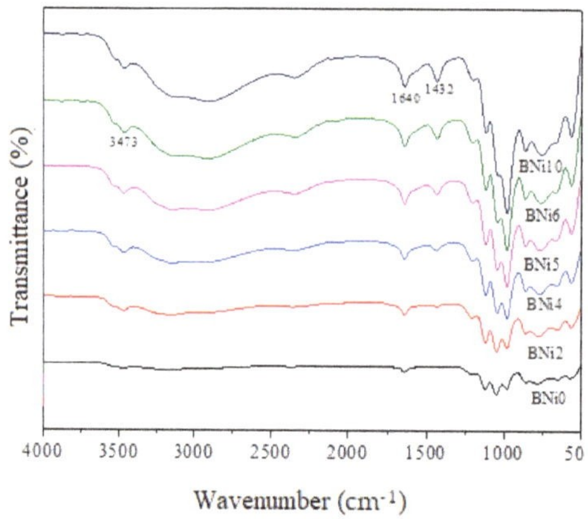

Figure 4. FT-IR spectrum of $Ca_xNi_{1-x}HPO_4{\cdot}nH_2O$ compounds.

3.2. Elemental and Chemical Composition of $Ca_xNi_{1-x}HPO_4{\cdot}nH_2O$ Compounds

The prepared samples were subjected to XPS analysis to appraise how the Ni/Ca ratio present in the starting solution affected the surface chemistry and chemical state of P, Ca, and Ni as the main constituents of the synthesized $Ca_xNi_{1-x}HPO_4{\cdot}nH_2O$ compounds; these findings are presented in Figure 5. As previously noted, the Ni/Ca ratio affects the degree of substitution of Ca with Ni in brushite as well as the precipitation of HNiP (when higher Ni/Ca ratios are used). The peaks corresponding to the Ni 2p orbital are visible in the XPS spectra of BNi2 to BNi10, and their intensity progressively increases, whereas the intensity of the Ca 2s and Ca 2p peaks decreases almost proportionally with the increase in the Ni/Ca ratio. Meanwhile, the intensity of the P 2s peaks remains nearly constant. This means that the intensity of P, Ca, and Ni peaks depend on the degree of Ca replacement with Ni as well as on the amount of the precipitated HNiP (for higher Ni/Ca ratios).

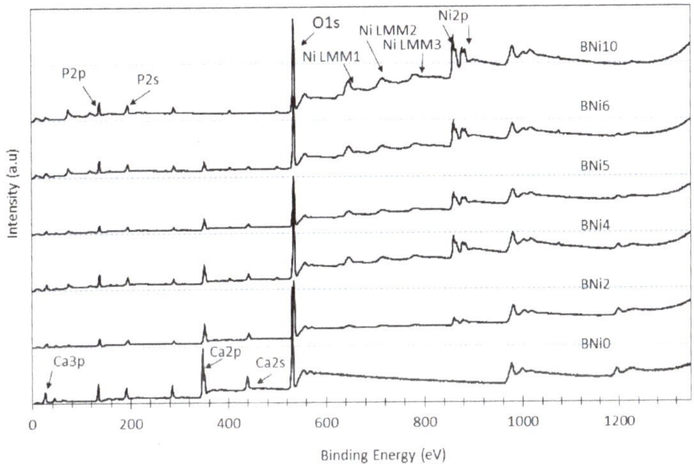

Figure 5. XPS spectra of $Ca_xNi_{1-x}HPO_4{\cdot}nH_2O$ compounds.

The effect of the Ni/Ca ratio on the binding energies of Ca 2s, P 2s, and Ni 2p peaks, which are shown in Figure 6A–C, respectively, evidence that increasing the Ni/Ca molar ratio from 0 to 0.25 (BNi2) results in an increase in the binding energies of the Ca 2s and P 2s peaks: from 438 to 442 eV and from 190 to 194 eV, respectively. In turn, the findings pertaining to sample BNi10 show that the binding energies of Ca 2s and P 2s remain relatively stable once nearly 100% of Ca is replaced by Ni.

The analyses also indicate that as the Ni/Ca ratio increases to 0.67 (BNi4), to 1 (BNi5), and further to 1.5 (BNi6), the intensity of the Ni 2p peak markedly increases—likewise to be expected, since the rate of replacement of Ca with Ni is accompanied by the precipitation of hexaaquanickel(II) hydrogenphosphate monohydrate. As confirmed by the XRD pattern shown in Figure 1, BNi4 is the first compound formed that contains HNiP crystals, while BNi10 comprises only pure hexaaquanickel(II) hydrogenphosphate monohydrate and thus produces Ni 2p and Ni Auger peaks of the highest intensity.

In sum, the XPS results reported above confirm that when the Ni/Ca ratio in the starting solution is low, Ca is partially replaced by Ni, but this is sufficient to alter the crystal structure of the $Ca_xNi_{1-x}HPO_4{\cdot}nH_2O$ compounds by increasing the binding energies of Ca 2s and P 2s peaks [41]. As the Ni concentration increases with the decrease in Ca concentration, supersaturation decreases (increases) with respect to brushite (hexaaquanickel(II) hydrogenphosphate monohydrate). Consequently, pure HNiP is obtained when no Ca is present in the system, as is the case for the BNi10 compound.

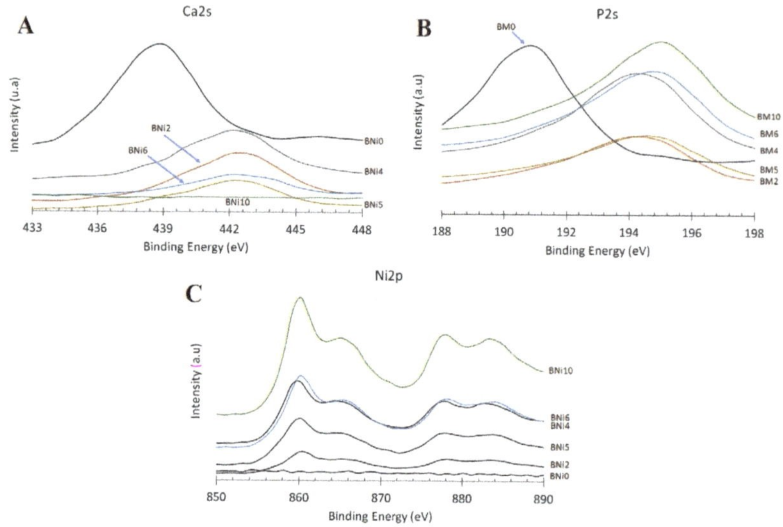

Figure 6. XPS analysis of the chemical state of (**A**) Ca 2s, (**B**) P 2s, and (**C**) Ni 2p orbitals in $Ca_xNi_{1-x}HPO_4{\cdot}nH_2O$ compounds.

3.3. *Thermogravimetric Analysis (TGA)*

The TGA results for all analyzed compounds (BNi0 to BNi10) are shown in Figure 7. Brushite has a crystal structure made up of compact sheets formed by parallel chains in which Ca ions are coordinated by six phosphate ions and two oxygen atoms belonging to the water molecules, making it a water-bearing phosphate [9]. Moreover, two sharp peaks reflect the mass loss as a result of heating from 80 to 220 °C. Brushite is characterized by two water molecules in its lattice and adsorbed water molecules on its surface [42,43]. Available evidence indicates that some of the chemically-bound water is released as brushite transforms to monetite ($CaHPO_4$) at ≈220 °C [44] and further to calcium pyrophosphate ($Ca_2P_2O_7$) as the temperature increases to ≈440 °C [8]. According to our results, when pure brushite (BNi0) is heated to 750 °C, approximately 21 wt % of its mass is lost [10], which is comparable to the theoretical mass loss of 20.93 wt % [45]. The BNi2−BNi6 samples with progressively greater Ni/Ca ratios lose more mass (27−35%), while a mass loss of 31.5 wt % is obtained for BNi10 containing solely hexaaquanickel(II) hydrogenphosphate monohydrate.

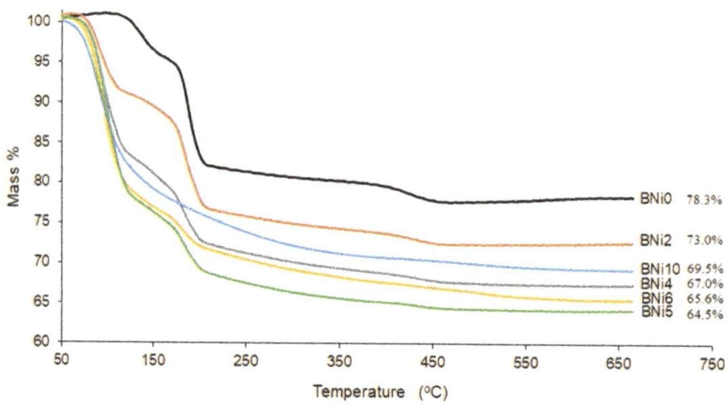

Figure 7. TG curves of $Ca_xNi_{1-x}HPO_4{\cdot}nH_2O$ compounds (product names Bni0–Bni10).

The brushite dehydration reaction is provided in Equation (2), while the formation of calcium pyrophosphate is described in Equation (3).

$$CaHPO_4{\cdot}2H_2O \rightarrow CaHPO_4 + 2H_2O \quad (2)$$

$$2CaHPO_4 \rightarrow Ca_2P_2O_7 + H_2O \quad (3)$$

The mass-loss rate for $Ca_xNi_{1-x}HPO_4{\cdot}nH_2O$ compounds as a function of temperature is shown in Figure 8. In Figure 8A–E, the dehydration peaks corresponding to the two water molecules of pure brushite (Bni0) are evident, and they are accompanied by the peaks characterizing the compounds obtained when the Ni/Ca ratio increased from 0.25 (BNi2) to 1.5 (BNi6). As indicated previously [24], partial replacement of Ca by Ni is initially accompanied by limited production of hexaaquanickel(II) hydrogenphosphate monohydrate. However, when the Ca is fully replaced by Ni in the starting solution (BNi10, Figure 8F), a primary zone of mass loss corresponding to HNiP is obtained at approximately 93 °C.

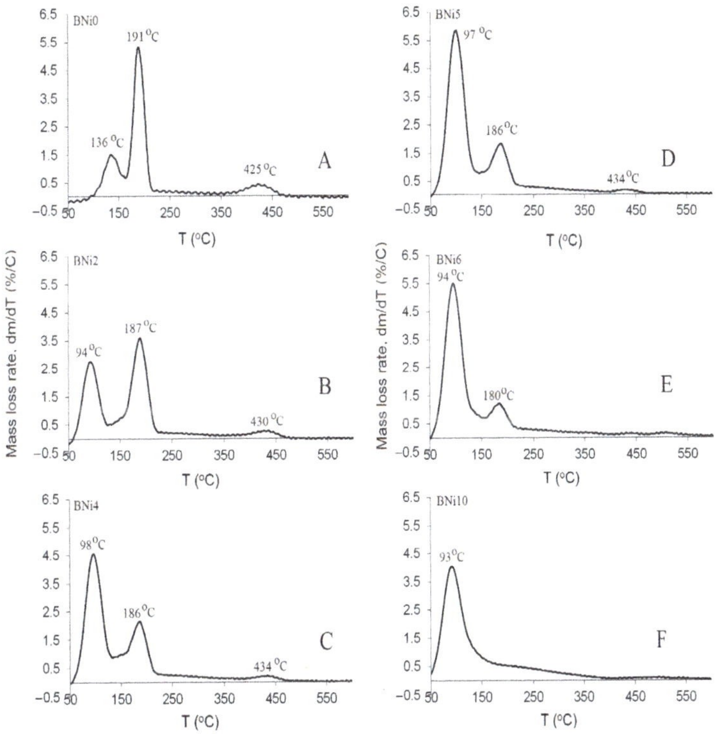

Figure 8. Differential TGA of diverse $Ca_xNi_{1-x}HPO_4{\cdot}nH_2O$ compounds: (**A**) BNi0; (**B**) BNi2; (**C**) BNi4; (**D**) BNi5; (**E**) BNi6, and (**F**) BNi10.

3.4. Phase Evolution during the Precipitation of $Ca_xNi_{1-x}HPO_4{\cdot}nH_2O$ Compounds

The results obtained in the present study indicate that the original shape of plate-like brushite crystals is preserved when the Ni/Ca ratio in the starting solution does not exceed 0.25 [39]. On the other hand, their size decreases due to the presence of Ni, as nickel inhibits crystal growth [20].

When the Ni/Ca ratio increases to 0.25 and even to 1.5 (BNi4–BNi6), HNiP with orthorhombic crystals starts to precipitate and eventually remains as the only phase when the Ni/Ca ratio increases further (with nano-scale crystals of approximately 500 nm size, as shown in Table 4).

Table 4. Phase evolution, crystal size, and structure as a function of the Ni/Ca molar ratio in solution.

Ni/Ca Ratio	Crystal Structure	Crystal Size (μm)	Compounds Formed
0 < 0.25	Monoclinic	~10	Brushite
$0.25 \leq x \leq 1.5$	Monoclinic/orthorhombic	~5 + 0.5	Brushite + HNiP
>1.5	Orthorhombic	~0.5	HNiP

4. Conclusions

In this study, the degree of replacement of Ca by Ni in brushite was investigated by examining a range of $Ca_xNi_{1-x}HPO_4{\cdot}nH_2O$ biomaterials that were subjected to XRD, SEM, XPS, FTIR, and TG analysis.

This study is consistent with previous studies [24], where the substitution of Ca^{2+} with Ni^{2+} by 20% or more leads to the formation of a mixture of compounds. The findings indicate that when the Ni/Ca molar ratio gradually increases to 1.5 (whereby the solution superstation with respect to Ni increases), both monoclinic Brushite and nanocrystals of orthorhombic HNiP precipitate. This study puts forth a novel procedure for HNiP synthesis (BNi10), utilizing the precipitation method. While brushite is considered a biomaterial precursor and a source of Ca ions, HNiP could be a promising compound as a source of Ni ions.

These insights might be valuable for the future synthesis of biomaterials with specific composition and properties, as they indicate that by controlling the Ni/Ca ratio in the starting solution, a variety of material compositions and morphologies can be attained. However, additional studies are required to assess the performance of the produced biomaterials, focusing specifically on their biological and mechanical aspects as well as physicochemical and antibacterial properties. The effect of doping with additional ions should also be investigated in depth, given that Ni may exhibit both beneficial and detrimental effects upon the biomaterials aimed for medical applications (such as kidney stone treatment and bone tissue engineering).

Author Contributions: Conceptualization, M.A. and K.I.; methodology, M.A.; validation, M.A., K.I. and A.S.A.; formal analysis, A.A.A. and A.E.; investigation, A.S.A. and M.A., resources, M.E.M.; data curation, A.E. and A.A.A.; writing—original draft preparation, M.A.; writing—review and editing, T.F.Q.; visualization, M.A.; supervision, M.A., M.E.M. and T.F.Q. All authors have read and agreed to the published version of the manuscript.

Funding: This research received no external funding.

Institutional Review Board Statement: Not applicable.

Informed Consent Statement: Not applicable.

Data Availability Statement: Not applicable.

Conflicts of Interest: The authors declare no conflict of interest.

References

1. Boanini, E.; Silingardi, F.; Gazzano, M.; Bigi, A. Synthesis and Hydrolysis of Brushite (DCPD): The Role of Ionic Substitution. *Cryst. Growth Des.* **2021**, *21*, 1689–1697. [CrossRef]
2. Radwan, N.H.; Nasr, M.; Ishak, R.A.; Abdeltawab, N.F.; Awad, G.A. Chitosan-calcium phosphate composite scaffolds for control of postoperative osteomyelitis: Fabrication, characterization, and in vitro–in vivo evaluation. *Carbohydr. Polym.* **2020**, *244*, 116482. [CrossRef] [PubMed]
3. Alshaaer, M.; Abdel-Fattah, E.; Saadeddin, I.; Al Battah, F.; Issa, K.I.; Saffarini, G. The effect of natural fibres template on the chemical and structural properties of Biphasic Calcium Phosphate scaffold. *Mater. Res. Express* **2020**, *7*, 065405. [CrossRef]

4. Cirillo, M.; Martelli, G.; Boanini, E.; Rubini, K.; Di Filippo, M.; Torricelli, P.; Pagani, S.; Fini, M.; Bigi, A.; Giacomini, D. Strontium substituted hydroxyapatite with β-lactam integrin agonists to enhance mesenchymal cells adhesion and to promote bone regeneration. *Colloids Surf. B Biointerfaces* **2021**, *200*, 111580. [CrossRef] [PubMed]
5. Khalifehzadeh, R.; Arami, H. Biodegradable calcium phosphate nanoparticles for cancer therapy. *Adv. Colloid Interface Sci.* **2020**, *279*, 102157. [CrossRef]
6. Shyong, Y.-J.; Chang, K.-C.; Lin, F.-H. Calcium phosphate particles stimulate exosome secretion from phagocytes for the enhancement of drug delivery. *Colloids Surf. B Biointerfaces* **2018**, *171*, 391–397. [CrossRef]
7. Liu, Y.; Ma, R.; Li, D.; Qi, C.; Han, L.; Chen, M.; Fu, F.; Yuan, J.; Li, G. Effects of calcium magnesium phosphate fertilizer, biochar and spent mushroom substrate on compost maturity and gaseous emissions during pig manure composting. *J. Environ. Manag.* **2020**, *267*, 110649. [CrossRef]
8. Revelo, C.F.; Colorado, H.A. A green composite material of calcium phosphate cement matrix with additions of car tire waste particles. *Int. J. Appl. Ceram. Technol.* **2021**, *18*, 182–191. [CrossRef]
9. Huertas, C.F.R.; Vieira, C.M.F.; Colorado, H.A. A hybrid composite for structural applications made of rubber waste tires and calcium phosphate cement. *Int. J. Appl. Ceram. Technol.* **2021**, *18*, 1342–1353. [CrossRef]
10. Boehm, A.V.; Meininger, S.; Gbureck, U.; Müller, F.A. Self-healing capacity of fiber-reinforced calcium phosphate cements. *Sci. Rep.* **2020**, *10*, 9430. [CrossRef]
11. Das, K.K.; Das, S.N.; Dhundasi, S.A. Nickel, its adverse health effects & oxidative stress. *Indian J. Med. Res.* **2008**, *128*, 412–425. [PubMed]
12. Priya, B.A.; Senthilguru, K.; Agarwal, T.; Narayana, S.N.G.H.; Giri, S.; Pramanik, K.; Pal, K.; Banerjee, I. Nickel doped nanohydroxyapatite: Vascular endothelial growth factor inducing biomaterial for bone tissue engineering. *RSC Adv.* **2015**, *5*, 72515–72528. [CrossRef]
13. Salnikow, K.; An, W.G.; Melillo, G.; Blagosklonny, M.V.; Costa, M. Nickel-induced transformation shifts the balance between HIF-1 and p53 transcription factors. *Carcinogenesis* **1999**, *20*, 1819–1823. [CrossRef] [PubMed]
14. Civjan, S.; Huget, E.F.; DeSimon, L.B. Potential Applications of Certain Nickel-Titanium (Nitinol) Alloys. *J. Dent. Res.* **1975**, *54*, 89–96. [CrossRef]
15. Fini, M.; Aldini, N.N.; Torricelli, P.; Giavaresi, G.; Borsari, V.; Lenger, H.; Bernauer, J.; Giardino, R.; Chiesa, R.; Cigada, A. A new austenitic stainless steel with negligible nickel content: An in vitro and in vivo comparative investigation. *Biomaterials* **2003**, *24*, 4929–4939. [CrossRef]
16. Sayahi, M.; Santos, J.; El-Feki, H.; Charvillat, C.; Bosc, F.; Karacan, I.; Milthorpe, B.; Drouet, C. Brushite (Ca,M)HPO_4, $2H_2O$ doping with bioactive ions (M = Mg^{2+}, Sr^{2+}, Zn^{2+}, Cu^{2+}, and Ag^{+}): A new path to functional biomaterials? *Mater. Today Chem.* **2020**, *16*, 100230. [CrossRef]
17. Alkhraisat, M.H.; Rueda, C.; Cabarcos, E.L. Strontium Ions Substitution in Brushite Crystals: The Role of Strontium Chloride. *J. Funct. Biomater.* **2011**, *2*, 31–38. [CrossRef]
18. Xue, Z.; Wang, Z.; Sun, A.; Huang, J.; Wu, W.; Chen, M.; Hao, X.; Huang, Z.; Lin, X.; Weng, S. Rapid construction of polyetheretherketone (PEEK) biological implants incorporated with brushite ($CaHPO_4{\cdot}2H_2O$) and antibiotics for anti-infection and enhanced osseointegration. *Mater. Sci. Eng. C* **2020**, *111*, 110782. [CrossRef]
19. Kim, Y.; Lee, S.Y.; Roh, Y.; Lee, J.; Kim, J.; Lee, Y.; Bang, J.; Lee, Y.J. Optimizing Calcium Phosphates by the Control of pH and Temperature via Wet Precipitation. *J. Nanosci. Nanotechnol.* **2015**, *15*, 10008–10016. [CrossRef]
20. Alshaaer, M.; Al-Kafawein, J.; Qazaq, A.; Issa, K.; Saffarini, G. Effects of magnetite incorporation in a chemically bonded phosphate ceramic. *J. Phys. Chem. Solids* **2022**, *162*, 110531. [CrossRef]
21. Luo, J.; Engqvist, H.; Persson, C. A ready-to-use acidic, brushite-forming calcium phosphate cement. *Acta Biomater.* **2018**, *81*, 304–314. [CrossRef] [PubMed]
22. Mert, I.; Mandel, S.; Tas, A.C. Do cell culture solutions transform brushite ($CaHPO_4$ $2H_2O$) to octacalium phosphate ($Ca_8(HPO_4)_2(PO_4)4\ 5H_2O$)? In *Advances in Bioceramics and Porous Ceramics IV*; Narayan, R., Colombo, P., Eds.; John Wiley & Sons, Inc.: Hoboken, NJ, USA, 2011; pp. 79–94.
23. Hurle, K.; Oliveira, J.; Reis, R.; Pina, S.; Goetz-Neunhoeffer, F. Ion-doped Brushite Cements for Bone Regeneration. *Acta Biomater.* **2021**, *123*, 51–71. [CrossRef] [PubMed]
24. Guerra-López, J.R.; Güida, J.A.; Ramos, M.A.; Punte, G. The influence of nickel on brushite structure stabilization. *J. Mol. Struct.* **2017**, *1137*, 720–724. [CrossRef]
25. Guerra-López, J.; Güida, J.; Bianchi, A.; Punte, G. Influence of carbonate and nickel(II) concentration on the synthesis of calcium phosphates. *J. Solid State Chem.* **2018**, *267*, 98–105. [CrossRef]
26. Patil, S.B.; Jena, A.; Bhargava, P. Influence of Ethanol Amount During Washing on Deagglomeration of Co-Precipitated Calcined Nanocrystalline 3YSZ Powders. *Int. J. Appl. Ceram. Technol.* **2012**, *10*, E247–E257. [CrossRef]
27. Piva, R.H.; Piva, D.H.; Pierri, J.; Montedo, O.R.K.; Morelli, M.R. Azeotropic distillation, ethanol washing, and freeze drying on coprecipitated gels for production of high surface area 3Y–TZP and 8YSZ powders: A comparative study. *Ceram. Int.* **2015**, *41*, 14148–14156. [CrossRef]
28. Lu, B.-Q.; Willhammar, T.; Sun, B.-B.; Hedin, N.; Gale, J.D.; Gebauer, D. Introducing the crystalline phase of dicalcium phosphate monohydrate. *Nat. Commun.* **2020**, *11*, 1546. [CrossRef]

29. Alshaaer, M.; Afify, A.S.; Moustapha, M.E.; Hamad, N.; Hammouda, G.A.; Rocha, F. Effects of the full-scale substitution of strontium for calcium on the microstructure of brushite: $(Ca_xSr_{1-x})HPO_4.nH_2O$ system. *Clay Miner.* **2020**, *55*, 366–374. [CrossRef]
30. Nosrati, H.; Le, D.Q.S.; Emameh, R.Z.; Perez, M.C.; Bünger, C.E. Nucleation and growth of brushite crystals on the graphene sheets applicable in bone cement. *Bol. Soc. Esp. Cerám. Vidr.* **2022**, *61*, 27–34. [CrossRef]
31. Wu, F.; Wei, J.; Guo, H.; Chen, F.; Hong, H.; Liu, C. Self-setting bioactive calcium–magnesium phosphate cement with high strength and degradability for bone regeneration. *Acta Biomater.* **2008**, *4*, 1873–1884. [CrossRef]
32. Alshaaer, M.; Issa, K.; Alanazi, A.; Mallouh, S.; Afify, A.; Moustapha, M.; Komnitsas, K. Gradual Replacement of Ca^{2+} with Mg^{2+} Ions in Brushite for the Production of $Ca_{1-x}Mg_xHPO_4{\cdot}nH_2O$ Materials. *Minerals* **2021**, *11*, 284. [CrossRef]
33. Alshaaer, M.; Al-Kafawein, J.; Afify, A.S.; Hamad, N.; Saffarini, G.; Issa, K. Effect of Ca^{2+} Replacement with Cu^{2+} Ions in Brushite on the Phase Composition and Crystal Structure. *Minerals* **2021**, *11*, 1028. [CrossRef]
34. Girisken, G.; Tas, A.C. Development of biomineralization solutions to facilitate the transformation of brushite ($CaHPO_4{\cdot}2H_2O$) into octacalcium phosphate ($Ca_8(HPO_4)_2(PO_4)_4{\cdot}5H_2O$). In Proceedings of the 15th National Biomedical Engineering Meeting (BIYOMUT), Antalya, Turkey, 21–24 April 2010.
35. Tamimi, F.; Le Nihouannen, D.; Eimar, H.; Sheikh, Z.; Komarova, S.; Barralet, J.; Tamimi, F. The effect of autoclaving on the physical and biological properties of dicalcium phosphate dihydrate bioceramics: Brushite vs. monetite. *Acta Biomater.* **2012**, *8*, 3161–3169. [CrossRef] [PubMed]
36. Madhurambal, G.; Subha, R.; Mojumdar, S.C. Crystallization and thermal characterization of calcium hydrogen phosphate dihydrate crystals. *J. Therm. Anal.* **2009**, *96*, 73–76. [CrossRef]
37. Rajendran, K.; Keefe, C.D. Growth and characterization of calcium hydrogen phosphate dihydrate crystals from single diffusion gel technique. *Cryst. Res. Technol.* **2010**, *45*, 939–945. [CrossRef]
38. Eldrehmy, E.; Alghamdi, Y.; Amer, H.; Yassin, M.; Mostafa, S.; Moustapha, M.E.; Menazea, A. Hydroxyapatite-based bio-ceramic of ternary nanocomposites containing cuprous oxide/graphene oxide for biomedical applications. *Diam. Relat. Mater.* **2022**, *126*, 109121. [CrossRef]
39. Septiani, N.L.W.; Kaneti, Y.V.; Fathoni, K.B.; Wang, J.; Ide, Y.; Yuliarto, B.; Nugraha; Dipojono, H.K.; Nanjundan, A.K.; Golberg, D.; et al. Self-assembly of nickel phosphate-based nanotubes into two-dimensional crumpled sheet-like architectures for high-performance asymmetric supercapacitors. *Nano Energy* **2019**, *67*, 104270. [CrossRef]
40. Chen, Z.; Zhou, D.; Gao, T.; Shen, W.; Dong, X.; Naito, S.; Qin, L.; Huang, Y. Unusual adsorption and desorption behaviors of NO and CO on nanoporous nickel phosphate VSB-5: In situ FT-IR and TPD study. *Catal. Today* **2015**, *258*, 199–204. [CrossRef]
41. Huang, H.; Li, J.; Li, B.; Zhang, D.; Zhao, N.; Tang, S. Comparison of different K-struvite crystallization processes for simultaneous potassium and phosphate recovery from source-separated urine. *Sci. Total Environ.* **2019**, *651*, 787–795. [CrossRef]
42. Gashti, M.P.; Stir, M.; Hulliger, J. Growth of strontium hydrogen phosphate/gelatin composites: A biomimetic approach. *New J. Chem.* **2016**, *40*, 5495–5500. [CrossRef]
43. Tortet, L.; Gavarri, J.R.; Nihoul, G.; Dianoux, A. Study of Protonic Mobility in $CaHPO_4{\cdot}2H_2O$ (Brushite) and $CaHPO_4$(Monetite) by Infrared Spectroscopy and Neutron Scattering. *J. Solid State Chem.* **1997**, *132*, 6–16. [CrossRef]
44. Dosen, A.; Giese, R.F. Thermal decomposition of brushite, $CaHPO_4{\cdot}2H_2O$ to monetite $CaHPO_4$ and the formation of an amorphous phase. *Am. Mineral.* **2011**, *96*, 368–373. [CrossRef]
45. Frost, R.L.; Palmer, S.J. Thermal stability of the 'cave' mineral brushite $CaHPO_4{\cdot}2H_2O$—Mechanism of formation and decomposition. *Thermochim. Acta* **2011**, *521*, 14–17. [CrossRef]

crystals

Article

Solvent-Induced Cobalt(II) Cyanoguanidine Bromides: Syntheses, Crystal Structure, Optical and Magnetic Properties

Jing Zhang [1], Alex. J. Corkett [1], Jan van Leusen [1], Ulli Englert [1] and Richard Dronskowski [1,2,*]

1 Institute of Inorganic Chemistry, RWTH Aachen University, 52056 Aachen, Germany
2 Hoffmann Institute of Advanced Materials, Shenzhen Polytechnic 7098 Liuxian Blvd, Nanshan District, Shenzhen 518055, China
* Correspondence: drons@HAL9000.ac.rwth-aachen.de

Abstract: Two different phase-pure cobalt(II) cyanoguanidine bromide coordination compounds, $Co(C_2N_4H_4)_2(H_2O)_4{\cdot}2Br{\cdot}2H_2O$ (**1**) and $Co_3(C_2N_4H_4)_8(H_2O)_8{\cdot}6Br$ (**2**), were precipitated from aqueous and methanol solutions, respectively, and their structures were solved and refined from X-ray single-crystal data at 100 K. Both **1** and **2** crystallize in the triclinic system with space group $P\bar{1}$. The structure of **1** consists of two crystallographically distinct isolated CoO_4N_2 octahedral units plus bromide anions and crystal water molecules, whereas **2** is built from both isolated octahedra and discrete binuclear cluster units made from edge-sharing octahedra. Diffuse reflectance spectra and IR analysis then go on to highlight optical and vibrational differences between these two compounds. The magnetic susceptibility of **1** is consistent with either isolated or very weakly interacting Co^{2+} centers whereas the magnetic susceptibility of **2** evidences the potential weak antiferromagnetic exchange interactions that may arise from superexchange within the binuclear clusters.

Keywords: solvent induction; crystal structure; optical properties; magnetic properties

Citation: Zhang, J.; Corkett, A.J.; van Leusen, J.; Englert, U.; Dronskowski, R. Solvent-Induced Cobalt(II) Cyanoguanidine Bromides: Syntheses, Crystal Structure, Optical and Magnetic Properties. *Crystals* **2022**, *12*, 1377. https://doi.org/10.3390/cryst12101377

Academic Editors: Aivaras Kareiva, Loreta Tamasauskaite-Tamasiunaite and Aleksej Zarkov

Received: 18 August 2022
Accepted: 24 September 2022
Published: 28 September 2022

1. Introduction

Over the last two decades, a large number of cyanoguanidine coordination compounds have been reported, such as $Zn(C_2N_4H_4)_2Cl_2$ [1–3], $[Ag(C_2N_4H_4)_2]F$, $[Ag(C_2N_4H_4)_2](BF_4)$ [4], $Cu(C_2N_4H_4)_2Br_2{\cdot}2H_2O$ [5] and others [6–9]. As a nitrogenous ligand, cyanoguanidine was discovered by Beilstein and Geuther at the end of the 19th century [10], and since then it has attracted widespread attention because of its interesting structure and important role in medicinal and industrial applications [11–13]. Cyanoguanidine is known to exist in two planar tautomeric forms, one with the cyano group at the imine nitrogen (Figure 1a) [14] and the other at the amine nitrogen (Figure 1b) [15]. Furthermore, cyanoguanidine is not only capable of acting as a monodentate ligand bonded to the metal by the cyano nitrogen alone, but also as a bridging ligand when it coordinates to two metal centers via the terminal (cyano) and inner (imine) nitrogens.

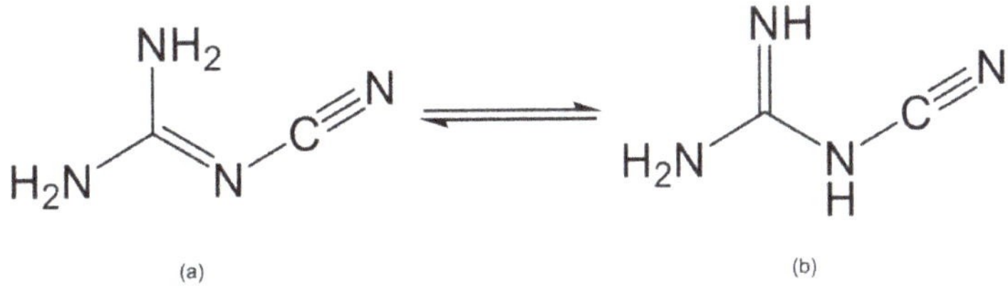

Figure 1. The two forms of cyanoguanidine with the cyano group at (**a**) imine (cyanoimine) nitrogen and (**b**) amine (cyanoamine) nitrogen.

As a late transition metal, divalent cobalt possesses an abundant coordination chemistry. In previous studies, many coordination compounds composed of cobalt and ligands of guanidine derivatives were found [16–18] but no coordination compound composed of cobalt and simple cyanoguanidine has been reported.

The structural motifs of coordination compounds are affected by several factors, not only the nature of the central metal atoms and the performance of ligands, but also the coordinated counter ions or small neutral molecules, the solvent system, temperature, pressure and so on [19,20]. Even if the same ligand reacts with the same metal salt in different solvents, the ligands, neutral molecules and the solvent often play an important role in the structural formation of the crystal such that different crystal structures will be obtained [21,22].

In order to investigate the combination of the coordination ability of cyanoguanidine with magnetic cobalt(II), we have designed two cobaltous bromide ($CoBr_2$) cyanoguanidine coordination complexes, $Co(C_2N_4H_4)_2(H_2O)_4{\cdot}2Br{\cdot}2H_2O$ (**1**) and $Co_3(C_2N_4H_4)_8(H_2O)_8{\cdot}6Br$ (**2**), in distilled water and methanol, respectively. Single-crystal X-ray diffraction revealed the structural differences of these two crystals, as also mirrored from diffuse reflectance and IR spectra in terms of optical differences. As is well-known, cobalt(II) compounds are particularly interesting as regards their magnetic properties [23–25], due to the $3d^7$ electron configuration of Co(II) and three unpaired electrons per ion ($S = 3/2$). The magnetic susceptibility of **1** was in agreement with either isolated or very weakly interacting Co^{2+} centers whereas the magnetic susceptibility of **2** evidenced the potential weak antiferromagnetic exchange interactions in terms of a superexchange model.

2. Materials and Methods

2.1. *Synthetic Procedures: All Reagents Were Commercially Bought (Alfa, Kandel, Rheinland-Pfalz, Germany) and Used without Further Purification*

1 $Co(C_2N_4H_4)_2(H_2O)_4{\cdot}2Br{\cdot}2H_2O$: stoichiometric 1:2 molar ratios of $CoBr_2$ (1 mmol) and $C_2N_4H_4$ (2 mmol), according to the empirical formula, were dissolved separately in 10 mL of distilled water. The solutions were then mixed in a 100 mL beaker and stirred for 30 minutes. After filtration, the clarified rose-red solution was left in a fume hood to evaporate naturally, yielding light-red crystals at the bottom of the beaker. After all of the solution was evaporated, the remaining crystals gave a 92% yield. (Figure S1).

2 $Co_3(C_2N_4H_4)_8(H_2O)_8{\cdot}6Br$: A similar method was used for synthesis; that is, $CoBr_2$ (1.5 mmol) and $C_2N_4H_4$ (4 mmol) with 3 to 8 ratios were dissolved separately in 10 mL of methanol. The solutions were then mixed in a 100 mL beaker and stirred for 30 minutes. After filtration, the clarified blue solution was left in a fume hood to evaporate naturally, resulting in dark-red crystals in the bottom of the beaker. The remaining crystals corresponded to a 90% yield (Figure S1). For illustration, in a methanol solvent, cyanoguanidine reacts with solvent molecules, displaying a clarified blue solution, and this reaction is reversible [26]. As the methanol solution evaporates, the reaction runs in reverse, so **2** is eventually formed. In contrast, the preparation of copper(II) and nickel(II) complexes under reflux conditions results in substituted cyanoguanidine ligands cnge-OCH_3 [12,27].

2.2. *PXRD Measurements*

The PXRD patterns of **1** and **2** were recorded on a STOE STADI-P powder diffractometer with a flat sample holder (Cu Kα_1, linear PSD, 2θ range 5–60°, with individual steps of 0.01°) at room temperature. The powder X-ray diffraction (PXRD) patterns match well with the simulated data generated from the single-crystal models, thereby demonstrating the purity of **1** (Figure S2) and **2** (Figure S3).

2.3. *CHN Measurements and Thermogravimetric Analysis (TGA)*

The CNH analysis for **1** and **2** was carried out using a CHN Rapid VarioEI analyzer from Heraeus. The TGA was performed on a STA1600 from Linseis, a combined system that measures HDSC and TGA simultaneously.

2.4. IR Spectra

The infrared spectra of the two samples were performed on a SHIMADZU IR Spirit QATR-S spectrometer in ATR mode using a diamond crystal.

2.5. UV-Vis Spectra

The UV-Vis diffuse reflectance spectra for **1** and **2** were recorded on a UV-2600 Shimadzu spectrophotometer in the range of 200–800 nm at room temperature, with barium sulfate pellets as background standard.

2.6. Single-Crystal Diffraction

The single-crystal X-ray diffraction data for **1** and **2** were collected on a Bruker D8 goniometer at 100 K (Oxford Cryostream 700), equipped with an Incoatec microsource (Mo Kα radiation, λ = 0.71073 Å, multilayer optics). The crystal structures were solved by intrinsic phasing as implemented in SHELXT, and SHELXL-2018 was used for full-matrix least-squares refinements on F^2. All non-hydrogen atoms were refined with anisotropic thermal parameters, and all hydrogen atoms were isotropically refined as riding on the parent atoms. In addition, the N−H, O−H bond lengths and H−N−H, H−O−H angles were restrained according to neutron-diffraction data [28,29]. All data reported in this paper were deposited in the Cambridge Crystallographic Data Centre (CCDC, numbers 2196005 for **1** and 2196006 for **2**) and can be obtained from there (12 Union Road, Cambridge CB2 1EZ, UK; fax: +44 1223 336033).

2.7. Magnetic Measurement

Magnetic data of **1** and **2** were collected on a Quantum Design MPMS-5XL SQUID magnetometer. A cylindrical PTFE capsule was used to immobilize the microcrystalline sample. Experimental data were collected in different sets, namely, at 2.0 K in the magnetic field scale 0.1–5.0 T and in the temperature range 2–290 K at 0.1 and 1.0 T. All experimental data were collected for the diamagnetic contributions of both sample holder and compound [$\chi_{m,dia}$ (10^{-4} cm^3 mol^{-1}) = −2.48 for **1** and −7.37 for **2**].

3. Results and Discussion

3.1. Stability and Structure

After slow evaporation of the solutions in a fume hood for several weeks, light-red (**1**) and dark-red (**2**) crystals were obtained (Figure S1, Supporting Information, SI). In the thermogravimetric analysis curve of **1** (Figure S4), one finds several steps for thermal decomposition. With increasing temperature, crystal water and coordinated water molecules (20.20%) are first lost between room temperature and 600 K. Then the ligands (35.54%) are sequentially removed (Table S1). As regards **2** depicted in Figure S5, the presence of a first step at 373 K is attributed to the loss of absorbed solvent [30]. The second step in the 400–600 K range corresponds to coordinated water molecules (9.51%). The final loss between 600–850 K indicates removal of ligands.

To reveal the structural differences, single-crystal X-ray diffraction data were collected at 100 K (Table S3). Both **1** and **2** crystallized in the triclinic system with space group $P\bar{1}$ (No. 2, Table S3) but differed structurally. For **1** ($Co(C_2N_4H_4)_2(H_2O)_4 \cdot 2Br \cdot 2H_2O$, in accordance with the CHN analysis (Table S4)), the cell parameters were a = 6.946(2) Å, b = 11.173(3) Å, c = 12.485(3) Å, α = 114.642(4)°, β = 98.284(4)°, γ = 100.028(4)°, Z = 2 and V = 841.5(3) Å^3. As shown in Figure 2, **1** was composed of two isolated Co-centered octahedra, bromide anions and additional crystal water molecules. The two crystallographically distinct octahedra were located at the centers of the (100) and (010) faces. In both octahedral units, the cobalt ion was coordinated by two N-containing cyanoguanidine ligands and four water molecules. It is also noteworthy that the two different octahedral units in every cell were both slightly distorted, as inferred from the divergence of the bond lengths. For example, the bond lengths for the Co1-centered octahedra were 2.046(2) Å for Co1−O1, while the Co1−O2 bond was 2.180(2) Å and Co1−N4 was 2.087(2) Å (Table S5).

A similar situation could be observed in the other isolated Co2-centered octahedral unit (Co2−O4 = 2.056(2) Å, Co2−O3 = 2.142(2) Å and Co2−N8 = 2.091(2) Å). Moreover, the isolated octahedra, crystal water molecules and bromide ions were connected by weak O−H···O, O−H···N, O−H···Br and N−H···Br hydrogen bonds (Figure S6).

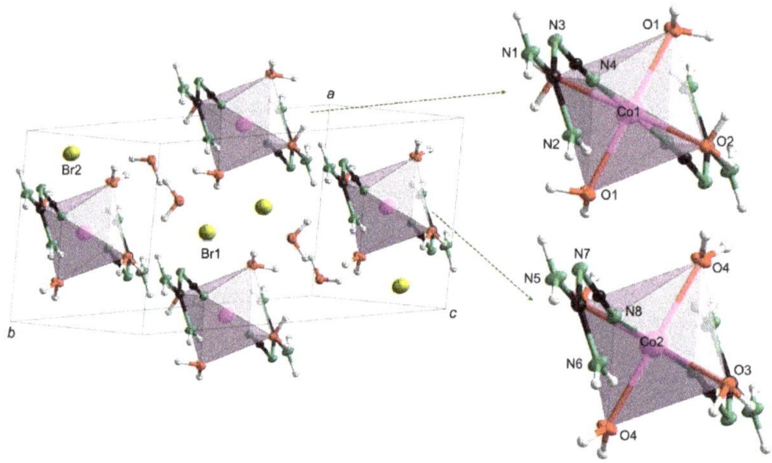

Figure 2. Crystal structure of **1** with Co in rose-red, O in red, N in green, C in black and H in white.

In contrast, **2** $Co_3(C_2N_4H_4)_8(H_2O)_8{\cdot}6Br$, in accordance with the CHN analysis (Table S6)) crystallized with a = 10.731(2) Å, b = 11.658(2) Å, c = 12.318(2) Å, α = 94.173(3)°, β = 114.733(3)°, γ = 108.593(3)°, Z = 1 and V = 1288.0(3) Å^3 (Table S3). In Figure 3, there are two types of structural moieties with distinct crystallographic configurations in the structure of **2**, namely, isolated octahedra and discrete binuclear clusters made from edge-sharing octahedra. For convenience, the binuclear cluster units and octahedra were labelled type A and type B, respectively. In type A, the framework consisted of edge-sharing octahedra, in which the adjacent central cobalt atoms are connected by two bridging oxygen atoms. In addition, every cobalt atom is coordinated to two ligands and two terminal oxygen water molecules. Interestingly, the edge-sharing octahedra could be considered almost regular, deduced from the statistically irrelevant divergence of the bond lengths (Co1−O2 = 2.073(4) Å, Co1−N4 = 2.064(6) Å, Co1−O1 = 2.048(5) Å and Co1−N5 = 2.062(6) Å, (Table S7)). In the isolated type B octahedra, the central atom, Co2, was coordinated to four ligands and two water molecules in a trans fashion along c. A modest distortion of the Co2 octahedra was also observed with distinct bond lengths of Co2−N12 = 2.118(6) Å and Co2−N13 = 2.072(6) Å. Moreover, the isolated octahedra, discrete binuclear clusters and bromide ions were bonded through weak O−H···O, O−H···N, O−H···Br and N−H···Br hydrogen bonds (Figure S7). As previously stated in the introduction, the presence of two hydrogen atoms at the terminal nitrogen atoms (**1**: N1, N2, N5, N6; **2**: N1, N2, N7, N8, N9, N10) in **1** and **2** prove these to be amino nitrogen atoms. This indicates that in both **1** and **2** the cyanoguanidine ligand favors the cyanoimine form. In addition, **1** and **2** differ from compounds that were structurally characterized before such as $Zn(C_2N_4H_4)_2Cl_2$ [1–3], $[Ag(C_2N_4H_4)_2]F$, $[Ag(C_2N_4H_4)_2](BF_4)$ [4], $Cu(C_2N_4H_4)_2Br_2{\cdot}2H_2O$ [5] and similar compounds [6–9]. In these earlier examples, the metal cations were coordinated by cyanoguanidine (cnge) and anions, whereas the cobalt cations in **1** and **2** were bonded to cnge and aqua ligands while the counter anions remained uncoordinated.

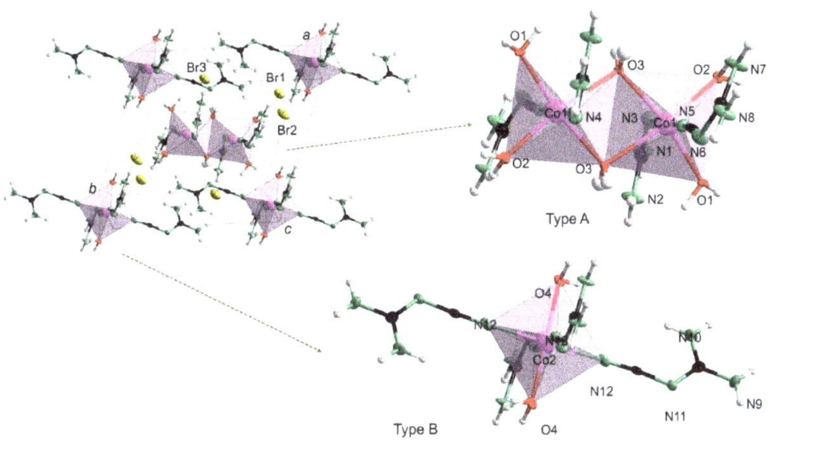

Figure 3. Crystal structure of **2** with Co in rose-red, O in red, N in green, C in black and H in white.

Compounds **1** and **2** are based on the same constituents but adopt different structures, in particular, with respect to their water content. The water capacity for **1** was 21.82%, much higher than for **2** with 9.78%. The reason is that compound **1** was synthesized in an aqueous solution, whereas compound **2** was obtained from methanol as a solvent. The presence of coordinated water in **2** presumably arose because the preparation was performed in ambient atmosphere and moisture was not excluded. The effects of solvent were obviously observed in the deviation of the coordination polyhedra from ideal octahedral symmetry. Notably, in **1**, the bond lengths subtended a wider range (Co1–O2 = 2.180(2) Å, Co1–N4 = 2.087(2) Å) than in **2** (Co2–O4 = 2.136(5) Å, Co2–N13 = 2.072(6) Å).

3.2. *UV–Vis and IR Optical Properties*

The UV-Vis absorption spectra for compounds **1** and **2** were measured on a UV-2600 Shimadzu spectrophotometer. As shown in Figure 4a, both spectra displayed two well-resolved bands, one at 266 nm, which was attributed to the $\pi\cdots\pi^*$ transitions of the cyanoguanidine ligand, and another one at 484 nm (**1**) and 505 nm (**2**) characteristic for *d-d* transitions in hexa-coordinate complexes [23,24], indicating that the ligands adopted an octahedral arrangement [31,32] around Co(II). In addition to that, Figure 4a reveals that the absorption peaks of **2** (505 nm) occured at a slightly higher wavelength than that in **1** (484 nm), coinciding with their divergent colors. Due to the presence of discrete binuclear clusters in **2**, there were four peaks in the range of 640 to 740 nm, whereas no peaks were observed in **1**.

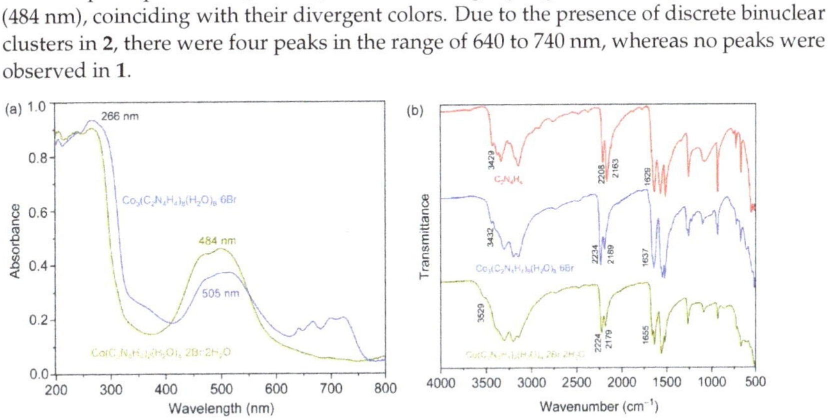

Figure 4. (**a**) Diffuse powder reflectance spectra for **1** (ochre line) and **2** (blue line); (**b**) IR spectrum of free cyanoguanidine (red line), **1** (ochre line) and **2** (blue line).

The IR spectra of both compounds (Figure 4b) showed an appreciable hypochromatic shift of the nitrile group (Table S8), compared to the free cyanoguanidine ligand, because of the coordination to the cobalt atoms [33]. Moreover, the separations of υ(C≡N) vibrations (45 cm^{-1}) were observed with a doublet character, which resulted from a Fermi resonance through the Sukhorukov and Finkel'shtein analyses [34,35]. The peaks at 1655 cm^{-1} (**1**) and 1637 cm^{-1} (**2**) were assigned to be υ(C=N) vibrations, while those at 3529 cm^{-1} and 3432 cm^{-1} were attributed to the υ(NH) vibrations [9,36,37].

3.3. Magnetic Properties

The magnetic data of **1** are shown in Figure 5 as $\chi_m T$ vs. T and M_m vs. B plots. At 290 K, the $\chi_m T$ value was 2.84 cm^{-1} K mol^{-1} at 0.1 T, and slightly smaller (2.81 cm^{-1} K mol^{-1}) at 1.0 T. These values were well within the range 2.3–3.4 cm^{-1} K mol^{-1} that is expected [38] for an isolated octahedral high-spin Co^{2+} center. Note that this range is above the spin-only value (1.876 cm^{-1} K mol^{-1}) due to relevant contributions of the orbital momentum that is typical of octahedral $3d^7$ metal centers. Upon cooling the compound, the values continuously decreased, significantly falling off at values below 100 K. At 2.0 K, $\chi_m T$ = 1.55 cm^{-1} K mol^{-1} at 0.1 T, and 1.31 cm^{-1} K mol^{-1} at 1.0 T. The observation regarding the decreasing values, in particular below 100 K, was due to the thermal depopulation of the energy states that originate from the $^4T_{1g}$ ground term, which is accordingly split due to the combined effects of spin-orbit coupling, electron–electron inter-repulsion and the ligand field. In addition to the energy splitting, the corresponding energy states are usually considerably mixed with respect to the m_J eigenstates. Below 10 K, $\chi_m T$ data at 0.1 and 1.0 T diverged due to the Zeeman effect, which becomes a relevant contribution at larger fields. At 2.0 K, the molar magnetization M_m rapidly grew with increasing fields up to about 2 T, and subsequently slightly increased up to 2.1 N_A μ_B at 5.0 T without reaching saturation (see inset in Figure 5). This value was also well below the saturation value of about 3.7 N_A μ_B (estimated from the $\chi_m T$ value at 290 K) due to the mixed nature of the ground state. Besides the latter, the cobalt centers were characterized by a medium magnetic anisotropy. Such anisotropy resulted in smaller values in powder measurements, since the data represent the mean value over randomly oriented crystallites.

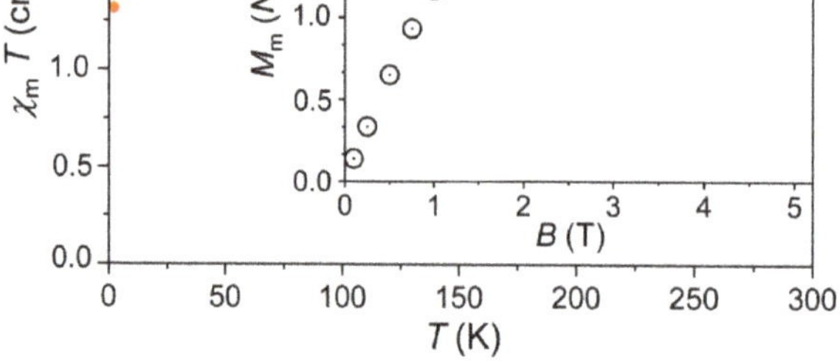

Figure 5. Temperature dependence of $\chi_m T$ of **1** at 0.1 (open circles) and 1.0 T (full circles); inset: molar magnetization M_m vs. applied magnetic field B at 2.0 K.

Therefore, the data are in agreement with either isolated or very weakly interacting Co^{2+} centers, and, thus, with the structural model.

For **2**, the $\chi_m T$ vs. T and M_m vs. B plots are shown in Figure 6. At 290 K, the value of $\chi_m T$ was 8.51 cm^{-1} K mol^{-1} at 0.1 and 1.0 T. This was in agreement with the range 6.9–10.2 cm^{-1} K mol^{-1} expected [38] for three non-interacting octahedral high-spin Co^{2+} centers. Upon decreasing temperature, $\chi_m T$ continuously decreased. Below 100 K, the values distinctly decreased, and dropped off below 5 K. At 2.0 K, $\chi_m T$ was 4.94 cm^{-1} K mol^{-1} at 0.1 T, and 4.20 cm^{-1} K mol^{-1} at 1.0 T. The shape of the curve, in particular, below 100 K, was—as for **1**—due to the thermal depopulation of the energy states originating from the $^4T_{1g}$ ground term. In addition to the Zeeman effect resulting in diverging $\chi_m T$ vs. T curves at 0.1 and 1.0 T below 10 K, the drop off observed below 5 K even at 0.1 T hints at antiferromagnetic exchange interactions. However, their magnitude was weak due to the rather small impact on the curves.

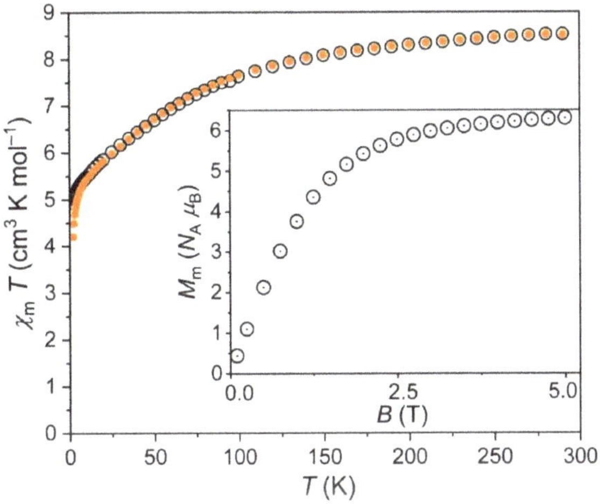

Figure 6. Same as before but for **2**.

The molar magnetization M_m at 2.0 K rapidly increased with increasing field up to about 2 T (see inset in Figure 6). At higher fields, M_m increased with a distinctly smaller rate reaching at 5.0 T a value of 6.3 N_A μ_B without being saturated. This value was well below the saturation value of about 11.1 N_A μ_B (estimated from the $\chi_m T$ value at 290 K) due to the reasons mentioned in the discussion of **1**.

Therefore, the data are in agreement with the structural information that shows in good approximation an isolated Co^{2+} center besides a dimeric unit of two Co^{2+} centers. In this unit, the two Co^{2+} ions were bridged by two oxygen atoms from water molecules forming Co–O–Co angles of 100.6° [39–41], which could explain the potential weak antiferromagnetic exchange interactions in terms of a superexchange model.

4. Conclusions

Two solvent-induced Co(II) cyanoguanidine bromides were synthesized and characterized. X-ray crystal structure determinations revealed different structural features, as **1** consisted of isolated octahedral units around Co with additional bromide anions and crystal water molecules whereas **2** contained both isolated octahedral units and discrete binuclear clusters. In the UV-vis absorption spectra, the absorption maximum of **2** (505 nm) was at lower energy than of **1** (484 nm), which was reflected in divergent crystal colors. The IR spectra displayed an appreciable hypochromatic shift of the C≡N nitrile groups because of the coordination to Co(II). Magnetic susceptibility measurements of **1** were consistent with the presence of isolated or very weakly interacting Co^{2+} centers, in accordance with the crystal structure, whereas **2** evidenced potential weak antiferromagnetic exchange interactions, which may have resulted due to superexchange between Co^{2+} centers via the bridging O in the binuclear clusters.

1. **Synopsis** Two solvent-induced Co(II) cyanoguanidine bromides (**1** and **2**) were synthesized in water and methanol, respectively. X-ray crystal structure, UV-vis absorption spectra and IR spectra analysis highlighted similarities and differences between these two compounds. Magnetic susceptibility measurements were consistent with the presence of cobalt(II) and evidence for weak antiferromagnetic interactions observed in **2**.
2. **TOC Figure**

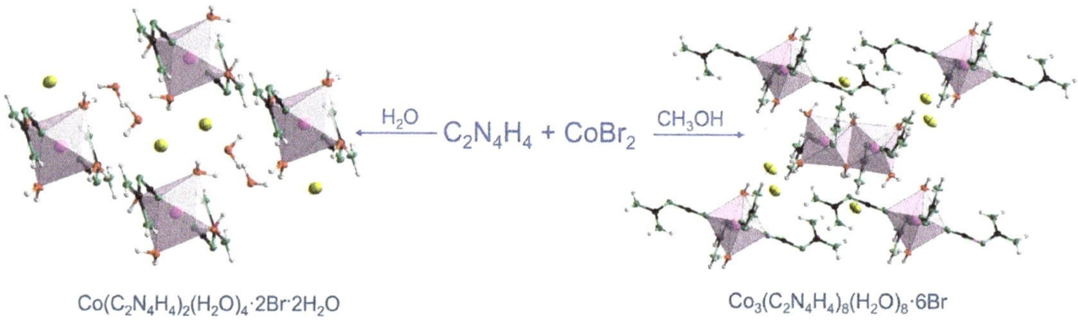

Supplementary Materials: The following supporting information can be downloaded at: https://www.mdpi.com/article/10.3390/cryst12101377/s1, Figure S1: The bulk crystals of **1** and **2**; Figure S2: Experimental and simulated powder X-ray diffraction patterns for **1**; Figure S3: Same as before but for **2**; Figure S4: TGA curve of **1**; Figure S5: Same as before but for **2**; Figure S6: Hydrogen-bonding interactions between O, N and Br atoms for **1**; Figure S7: Same as before but for **2**; Table S1: Thermal mass loss steps of **1**; Table S2; Same as before but for **2**; Table S3: Crystal data and structure refinement for **1** and **2**; Table S4: Elemental CHN analysis for **1**; Table S5: Bond lengths for **1**; Table S6: Elemental CHN analysis for **2**; Table S7: Bond lengths for **2**; Table S8: IR bands of cyanoguanidine, **1** and **2**.

Author Contributions: All authors contributed to this manuscript. Conceptualization, J.Z.; methodology, J.Z.; validation, J.Z.; formal analysis, J.Z., A.J.C., J.v.L. and U.E.; investigation, J.Z.; resources, R.D.; data curation, R.D., J.Z. and A.J.C.; writing—original draft preparation, J.Z. and J.v.L.; writing—review and editing, R.D., A.J.C. and J.Z.; visualization, J.Z. and A.J.C.; supervision, R.D.; project administration, R.D.; funding acquisition, R.D. All authors have read and agreed to the published version of the manuscript.

Funding: This research received funding from the China Scholarship Council.

Data Availability Statement: Not applicable.

Acknowledgments: We thank the China Scholarship Council for offering the stipend to Jing Zhang. We appreciate Tobias Storp collecting the single-crystal and powder X-ray diffraction data and Christina Houben measuring the magnetic susceptibility data. We thank Anne Frommelius for measuring the TGA data. Last but not least, we very much thank Simon Steinberg and Xiaoying Huang for help in the single-crystal structure analysis.

Conflicts of Interest: The authors declare no conflict of interest.

References

1. Pickardt, J.; Kühn, B. Crystal structure of dichlorobis(cyanoguanidine)zinc(II), $Zn(C_2N_4H_4)_2Cl_2$. *Z. Kristallogr. Cryst. Mater.* **1995**, *210*, 901. [CrossRef]
2. Fowkes, A.; Harrison, W.T.A. A monoclinic polymorph of dichlorobis-(cyanoguanidine)zinc(II). *Acta Cryst. E* **2005**, *61*, m2021–m2022. [CrossRef]
3. Ritche, L.K.; Harrison, W.T.A. catena-poly[[dichlorozinc(II)]-μ-cyanoguanidine]. *Acta Cryst. E* **2007**, *63*, m617–m618. [CrossRef]
4. Bessler, K.E.; de Sousa, A.T.; Deflon, V.M.; Niquet, E. Silver complexes with cyanoguanidine: Preparation and crystal structures of $[Ag(cgn)_2]F$ and $[Ag(cgn)_2][BF_4]$. *Z. Anorg. Allg. Chem.* **2003**, *629*, 1091–1095. [CrossRef]
5. Zhang, J.; Corkett, A.J.; van Leusen, J.; Nelson, R.; Dronskowski, R. $Cu(C_2N_4H_4)_2Br_2{\cdot}2H_2O$: An antiferromagnetic cyanoguanidine coordination compound and its characterization. *Z. Naturforsch. B* **2022**, *77*, 411–417. [CrossRef]

6. Dronskowski, R.; Liu, X.-H. Bis(cyanoguanidine)silver(I) nitrate–cyanoguanidine (1/1). *Acta Cryst. C* **2003**, *59*, m243–m245. [CrossRef] [PubMed]
7. Qiao, X.; Corkett, A.J.; Luo, D.; Dronskowski, R. Silver cyanoguanidine nitrate hydrate: $Ag(C_2N_4H_4)NO_3{\cdot}1/2H_2O$, a cyanoguanidine compound coordinating by an inner nitrogen atom. *Inorganics* **2020**, *8*, 64. [CrossRef]
8. Pickardt, J.; Kühn, B. Metallkomplexe mit Cyanoguanidin als Liganden: Kristallstrukturen von $[Cd(cnge)_2F_2]{\cdot}3H_2O$, $Cd(cnge)_2Br_2$ und $Hg(cnge)Cl_2$ (enge = Cyanoguanidin). *Z. Naturforsch. B* **1996**, *51*, 1701–1706. [CrossRef]
9. Irzoqi, A.A.; Salih, M.M. Synthesis and characterization complexes of Ni(II) that contain cyanoguanidine and phosphines Ligands. *Tikrit J. Pure Sci.* **2018**, *23*, 93–101. [CrossRef]
10. Beilstein, F.; Geuther, A. Ueber das Natriumamid. *Liebigs Ann. Chem.* **1858**, *108*, 88–102. [CrossRef]
11. Barrio, J.; Volokh, M.; Shalom, M. Polymeric carbon nitrides and related metal-free materials for energy and environmental applications. *J. Mater. Chem. A* **2020**, *8*, 11075–11116. [CrossRef]
12. Williams, P.A.; Ferrer, E.G.; Baeza, N.; Piro, O.E.; Castellano, E.E.; Baran, E.J. Transition metal promoted addition of methanol to cyanoguanidine. Molecular structure and properties of the generated Copper (II) and Nickel (II) complexes. *Z. Anorg. Allg. Chem.* **2005**, *631*, 1502–1506. [CrossRef]
13. Kose, M.; Duman, S.E.; McKee, V.; Akyol, I.; Kurtoglu, M. Hydrogen bond directed 1D to 3D structures of square-planar Ni(II) complexes and their antimicrobial studies. *Inorg. Chim. Acta* **2017**, *462*, 281–288. [CrossRef]
14. Bamberger, E.E. Ueber Dicyandiamid(I). *Ber. Dtsch. Chem. Ges.* **1883**, *16*, 1074–1078. [CrossRef]
15. Pohl, F. Zur Kenntnis des Dicyandiamids. *J. Prakt. Chem.* **1908**, *77*, 533–548. [CrossRef]
16. Bourosh, P.; Bologa, O.; Deseatnic-Ciloci, A.; Tiurina, J.; Bulhac, I. Synthesis, structure, and biological properties of mixed cobalt (III) dioximates with guanidine derivatives. *Russ. J. Coord. Chem.* **2017**, *43*, 591–599. [CrossRef]
17. Aremu, J.A.; Durosinmi, L.M.; Oluyemi, E.A.; Ojo, I.A.O. Synthesis and Characterization of Guanidine derivatives of Benzothiazole and their Cobalt(II), Nickel(II), Zinc(II), Copper(II) and Iron(II) Complexes. *IOSR J. Appl. Chem.* **2018**, *11*, 53–71.
18. Užarević, K.; Halasz, I.; Friščic, T. Real-time and in situ monitoring of mechanochemical reactions: A new playground for all chemists. *J. Phys. Chem. Lett.* **2015**, *6*, 4129–4140. [CrossRef]
19. Desiraju, G.R. Crystal gazing: Structure prediction and polymorphism. *Science* **1997**, *278*, 404–405. [CrossRef]
20. Nimax, P.R.; Reimann, D.; Sünkel, K. Solvent effects on the crystal structure of silver pentacyanocyclopentadienide: Supramolecular isomerism and solvent coordination. *Dalton Trans.* **2018**, *47*, 8476–8482. [CrossRef] [PubMed]
21. Dong, W.K.; He, X.N.; Yan, H.B.; Lv, Z.W.; Chen, X.; Zhao, C.Y.; Tang, X.L. Synthesis, structural characterization and solvent effect of copper(II) complexes with a variational multidentate Salen-type ligand with bisoxime groups. *Polyhedron* **2009**, *28*, 1419–1428. [CrossRef]
22. Bi, W.Y.; Lü, X.Q.; Chai, W.L.; Jin, W.J.; Song, J.R.; Wong, W.K. Synthesis, structure and near-infrared (NIR) luminescence of three solvent-induced pseudo-polymorphic complexes from a bimetallic Zn–Nd Schiff-base molecular unit. *Inorg. Chem. Commun.* **2008**, *11*, 1316–1319. [CrossRef]
23. Hchicha, K.; Korb, M.; Kliuikov, A.; Čižmár, E.; Naïli, H. A cobalt(II)-based semiconductor complex with two-channel slow magnetic relaxation. *J. Magn. Magn. Mater.* **2021**, *536*, 168140. [CrossRef]
24. Brader, M.L.; Kaarsholm, N.C.; Harnung, S.E.; Dunn, M.F. Ligand perturbation effects on a pseudotetrahedral Co(II)(His)3-Ligand Site: A Magnetic circular dichroism study of the Co(II)-substituted insulin hexamer. *J. Biol. Chem.* **1997**, *272*, 1088–1094. [CrossRef] [PubMed]
25. Rueff, J.M.; Masciocchi, N.; Rabu, P.; Sironi, A.; Skoulios, A. Structure and magnetism of a polycrystalline transition metal soap–$Co^{II}[OOC\,(CH_2)_{10}COO](H_2O)_2$. *Eur. J. Inorg. Chem.* **2001**, *11*, 2843–2848. [CrossRef]
26. Baker, W.A., Jr.; Daniels, M. Dicyandiamide-copper (II) complexes and the metal ion catalysed addition of alcohols to a nitrile. *J. Inorg. Nucl. Chem.* **1963**, *25*, 1194–1196. [CrossRef]
27. Quinzani, O.V.; Tarulli, S.; Marcos, C.; Granda, S.G.; Baran, E.J. Crystal structure, spectroscopic and thermal behaviour of bis (saccharinato) tetrakis (pyridine) nickel (II) dipyridine. *Z. Anorg. Allg. Chem.* **1999**, *625*, 1848–1852. [CrossRef]
28. Sawinski, P.K.; Meven, M.; Englert, U.; Dronskowski, R. Single-Crystal Neutron Diffraction Study on Guanidine, CN_3H_5. *Cryst. Growth Des.* **2013**, *13*, 1730–1735. [CrossRef]
29. Ceccarelli, C.; Jeffrey, G.A.; Taylor, R. A survey of OH$\cdots$O hydrogen bond geometries determined by neutron diffraction. *J. Mol. Struct.* **1981**, *70*, 255–271. [CrossRef]
30. Gašparovič, L.; Koreňová, Z.; Jelemenský, L'. Kinetic study of wood chips decomposition by TGA. *Chem. Pap.* **2010**, *64*, 174–181. [CrossRef]
31. Lever, A.B.P. Inorganic Electronic Spectroscopy. In *Studies in Physical and Theoretical Chemistry*, 2nd ed.; Elsevier: New York, NY, USA, 1984; Volume 33, pp. 480–490.
32. Papadopoulos, C.; Cristóvão, B.; Ferenc, W.; Hatzidimitriou, A.; Ciprioti, S.V.; Risoluti, R.; Lalia-Kantouri, M. Thermoanalytical, magnetic and structural investigation of neutral Co(II) complexes with 2,20-dipyridylamine and salicylaldehydes. *J. Therm. Anal. Calorim.* **2016**, *123*, 717–729. [CrossRef]
33. Witschard, G.; Griffin, C.E. Infrared absorption characteristics of alkyl and aryl substituted phosphonium salts. *Spectrochim. Acta* **1963**, *19*, 1905–1910. [CrossRef]
34. Sukhorukov, B.I.; Finkel'shtein, A.I. Optical Investigation of the Molecular Structure of Cyanamide and its Derivatives. 1. The Molecular Structure of Dicyandiamide. *Opt. Spektrosk.* **1959**, *6*, 637–641.

35. Sheludyakova, L.A.; Sobolev, E.V.; Kozhevina, L.I. Concerning the nature of the doublet at 2200 cm^{-1} in the vibrational spectra of cyanoguanidine. *J. Appl. Spectrosc.* **1991**, *55*, 661–665. [CrossRef]
36. Nakamoto, K. *Infrared Spectra of Inorganic and Coordination Compounds*; Wiley: New York, NY, USA, 1963.
37. Begley, M.J.; Hubberstey, P.; Walton, P.H. Bis(μ-2-cyanoguanidine)-bis ((2-cyanoguanidine) copper (I)), a planar dimeric cation containing co-ordinatively unsaturated copper (I). *J. Chem. Soc. Dalton Trans.* **1995**, *6*, 957–962. [CrossRef]
38. Lueken, H.; Magnetochemie, B. Teubner Stuttgart. *Leipzig* **1999**. [CrossRef]
39. Kanamori, J. Superexchange interaction and symmetry properties of electron orbitals. *J. Phys. Chem. Solids* **1959**, *10*, 87–98. [CrossRef]
40. van Leeuwen, G.C.M. Investigation of the angular dependency of Co-O-Co superexchange by means of magnetic susceptibility measurements in dilute systems. *Rec. Trav. Chim. Pays-Bas* **1973**, *92*, 1249–1266. [CrossRef]
41. Weihe, H.; Güdel, H.U. Angular and distance dependence of the magnetic properties of oxo-bridged iron (III) dimers. *JACS* **1997**, *119*, 6539–6543. [CrossRef]

crystals

MDPI

Article

Wet Chemical Synthesis and Characterization of Au Coatings on Meso- and Macroporous Si for Molecular Analysis by SERS Spectroscopy

Siarhei Zavatski [1,2], Anatoli I. Popov [3,4,*], Andrey Chemenev [1], Alma Dauletbekova [4] and Hanna Bandarenka [1,2,*]

1 Applied Plasmonics Laboratory, Belarusian State University of Informatics and Radioelectronics, 220013 Minsk, Belarus
2 Institute of Advanced Materials and Technologies, National Research University of Electronic Technology, 124498 Zelenograd, Russia
3 Institute of Solid State Physics, University of Latvia, LV-1063 Riga, Latvia
4 Department of Technical Physics, L.N. Gumilyov Eurasian National University, Astana 010008, Kazakhstan
* Correspondence: popov@latnet.lv (A.I.P.); h.bandarenka@bsuir.by (H.B.)

Abstract: Porous silicon (PS) is a promising material for nanostructure fabrication providing a precise control over its size, shape, and spatial distribution. This makes it an excellent candidate for constructing highly sensitive, reproducible, and low-cost platforms for surface enhanced Raman scattering (SERS) spectroscopy. In this work, we connect the PS structural parameters with the morphology of the gold nanostructures fabricated on its surface, placing the emphasis on the SERS response. Two different types of PS are considered here, namely meso- and macro-PS. The former is prepared by Si electrochemical etching, applying three different current densities: 100 mA cm^{-2}, 60 mA cm^{-2}, and 30 mA cm^{-2}, while the technological parameters for the latter are selected to mimic metal nanovoids' (Me NVs) configuration. The gold-coated PS surfaces are produced via an electroless chemical deposition method for different time periods. By performing comprehensive structural, morphological, and optical characterization, we show the importance of the size and density of the PS pore openings, which govern the Au growth kinetics. The results presented in this work assure a simple yet flexible approach for the fabrication of large-area plasmonic gold nanostructures, which are not only suitable for advanced SERS spectroscopy studies but can also serve for a wider range of plasmonic applications.

Keywords: porous silicon; gold; fabrication; SERS; XRD

Citation: Zavatski, S.; Popov, A.I.; Chemenev, A.; Dauletbekova, A.; Bandarenka, H. Wet Chemical Synthesis and Characterization of Au Coatings on Meso- and Macroporous Si for Molecular Analysis by SERS Spectroscopy. *Crystals* **2022**, *12*, 1656. https://doi.org/10.3390/cryst12111656

Academic Editor: George D. Verros

Received: 30 October 2022
Accepted: 14 November 2022
Published: 17 November 2022

1. Introduction

Nowadays, the rapid development of biomedicine has been increasingly faced with a demand for the detection, identification, and structural study of different organic analytes in trace amounts. Several sensing techniques, such as X-ray crystallography [1], nuclear magnetic resonance [2,3], and circular dichroism spectroscopy [4], have been applied to meet these needs. However, their routine application outside the framework of laboratory usage remains challenging: the low speed, high cost, complexity, and lack of complete databases [5] are major limitations for their practical implementation.

On the other hand, vibrational spectroscopy is a well-established molecular characterization technique [6], which does not require complex sample preparation and is widely used to study organic substances, e.g., in medicine [7], biology [8,9], the food industry [10], and pharmaceutics [11–13], to name a few. Utilizing the latest advancements of the Surface Enhanced Raman Scattering (SERS) spectroscopy technique, it is also possible to perform the detailed structure assignment of intricate substances, such as peptides, in small concentrations [14]. This is because of the extremely high sensitivity of the SERS technique along with the ability to provide molecule fingerprints [15,16]. In addition, the analysis procedure with SERS spectroscopy is currently straightforward, making it a solid candidate for future sensing techniques in a wide range of applications.

The SERS effect is associated with a local electromagnetic field enhancement in the vicinity of the metallic nanostructure as the result of the resonance light interaction with localized surface plasmons (LSP) [17–19]. The magnitude of this enhancement is defined by the material choice (usually one of the coinage metals), size, and shape of the nanostructure. Hence, the research directed towards maximizing the local electromagnetic fields of a nanoscale platform is of particular importance for SERS sensing. From a practical point of view, however, such a platform must also demonstrate remarkable spot-to-spot and sample-to-sample SERS signal reproducibility, preserve the stability of the signal intensity over the long storage time, and offer inexpensive large-scale production [20]. In attempts to meet these strict requirements, several types of SERS substrates have been already developed, including colloidal solutions of metallic nanoparticles [16,21] and solid-supported arrays of metallic nanostructures with different shapes [22–25]. Although colloids usually exhibit high enhancement factors [16,26], solid metallic nanostructures are more convenient for utilization and demonstrate better signal reproducibility. Nevertheless, existing commercially available solid SERS platforms are produced via multistage nanoengineering processes, which inevitably increases their fabrication cost [27–31]. Therefore, exploring other possible SERS platform fabrication strategies remains a crucial task for ensuring broader SERS spectroscopy implementation in practice.

One such promising strategy is the utilization of matrices based on porous silicon (PS) to create the metal nanostructures on their surfaces [32,33]. PS represents a net of the periodically altered pore channels formed in Si. It has recently enabled an accurate control over the morphology of deposited silver layers, leading to a considerable improvement in SERS signal reproducibility [34,35]. The extraordinary detection limit down to the attomolar concentration of lactoferrin protein adsorbed on silvered meso-PS covered with graphene has been also demonstrated [9]. The macroporous PS on the other hand has facilitated the formation of such sophisticated structures as nanothorns, nanospongy films, dendrites, and many others [36–38], reaching the prominent 10^{-11}–10^{-15} M detection limit range. Owing to the PS capability to precisely control the nanostructures' morphology, its utilization for SERS has attracted extensive research during the past decade. The major advantages and unveiled limitations of metal-coated PS SERS platforms have been comprehensively discussed in several recent review articles [39,40].

Even though many studies have highlighted the effectiveness of PS-based SERS platforms, most of them are associated with silver nanostructures, which display several critical drawbacks, while utilizing SERS, especially when studying biomolecules. One familiar issue that the LSP resonance band for the silver nanoparticles lies in the range of 370–500 nm, depending on the size of the nanoparticles. Hence, to achieve the optimum resonance conditions and thus the maximum SERS response, the LSP energy must closely match with the excitation laser energy. Therefore, the usage of high-energy (small wavelength) laser excitations is usually required in experiments. This excitation is destructive for the majority of organic analytes. Moreover, considering that silver itself exhibits antibacterial activity [41,42], the resulting SERS spectra intensities may be dramatically decreased. The solution for these is relatively simple: silver can be replaced with gold, but this has rarely been performed for the metalized PS platforms.

In this work, we therefore propose to replace the silver coverage with a more stable and biocompatible gold layer to develop an effective SERS platform based on PS, and we also fill the existing gap in the studies of gold deposition mechanisms on different types of PS.

2. Materials and Methods

2.1. PS Substrate Fabrication

The PS was fabricated by the electrochemical anodic etching of a monocrystalline silicon wafer with (100) surface crystallographic orientation, according to the procedure described elsewhere [43]. Briefly, an n^+- and p^--Si wafer was used to grow meso- and macro-PS, respectively. Just before anodization, the native silicon dioxide was removed from a silicon sample by etching in a diluted hydrofluoric acid (4.5%) for 60 s. After that, the Si sample was placed in a horizontal electrolytic cell made of polytetrafluoroethylene. An aqueous–alcoholic solution made of HF (45%), deionized water, and isopropanol at a volume ratio of 1:3:1 was used to prepare the meso-PS. Dimethylsulfoxide (DMSO) mixed with HF (45%) at the volume ratio of 10:46 was used for the macro-PS fabrication. The prepared solutions were gently added to the electrolytic cell. The meso-PS samples were fabricated at 100 mA cm^{-2}, 60 mA cm^{-2}, and 30 mA cm^{-2} current densities applied for 20 s, 22 s, and 20 s, respectively. The macro-PS samples were produced at an 8 mA cm^{-2} current density applied for 4 min. All freshly prepared PS samples were rinsed with deionized water and air dried.

2.2. Gold Deposition and Samples' Characterization

Gold was deposited by the immersion of a PS sample in an aqueous solution of 0.01 M $KAu(CN)_2$ and 0.15 M HF for different time periods according to the slightly changed procedure reported before [44]. Subsequently, freshly prepared samples were carefully washed with deionized water and air dried.

The organic dye rhodamine 6G (R6G) was used as an analyte for the SERS study. The dye molecules were adsorbed on the gold-coated PS sample by dipping it in a 10^{-6} M aqueous solution for 2 h. Each substrate was then thoroughly rinsed with deionized water for 30 s to remove the excess analyte molecules.

The morphology of the gold-coated PS was studied by scanning electron microscopy (SEM) with the help of an FESEM Hitachi−S4800 microscope. The crystallinity of the samples was studied with X-ray diffraction (XRD) using Cu K_α radiation (X-ray wavelength $\lambda = 0.15406$ nm). The SEM images were analyzed using ImageJ software. Reflectance spectra measurements were performed in the 350–900 nm wavelength range with the help of a Shimadzu UV-3600 equipped with an integrating sphere module (ISR-2200). SERS spectra were acquired with the help of a WITec Scanning Near-field Optical Microscope alpha300 S equipped with a 532 nm laser diode and an 60x air objective (0.9 NA). The spectra were acquired with 1 mW laser power for 1 s sample surface exposition.

3. Results and Discussion

3.1. Morphology of the Au-Coated Meso-PS Substrates

The electroless (galvanic, corrosion) deposition of the Au films on monocrystalline Si surfaces followed the oxidation–reduction (redox) mechanism, which assumes the reduction of Au^{3+} cations to atomic Au^0 by electron (e^-) exchange with Si. It is known that the redox potential for Au^{3+}/Au^0 is positive in the presence of Si atoms [45]. Hence, the Au deposition proceeded spontaneously as the result of the electron transfer from semiconductor valence band to metal cations:

$$Au^{3+} + 3e^- \rightarrow Au^0. \quad (1)$$

As the result of this process, the Si was simultaneously oxidized, since the above reduction reaction can be also viewed in terms of the holes (h^+) "injection" into the valence band of Si. This injection in the presence of H_2O induced silicon dioxide formation:

$$Si + H_2O + 4h^+ \rightarrow SiO_2 + 4H^+. \quad (2)$$

On a large timescale, the gold deposition is limited by the passivation of the Si surface with the oxide, reducing the process rate. However, with the presence of HF in the solution, the SiO_2 can be dissolved, producing new charge carriers and water soluble SiF_6^{2-} species:

$$SiO_2 + 6HF \rightarrow SiF_6^{2-} + 2H^+ + 2H_2O. \quad (3)$$

In these conditions, the Au electroless deposition on the Si is not limited by SiO_2 growth, and new Au nuclei form on the already adsorbed metal clusters and the available Si surface. Notably, the former is possible due to the galvanic cell formation between the noble metals and Si [46].

Considering a porous instead of monocrystalline Si surface, an important role for Au deposition is the presence of Si-H_x groups, which are usually generated via the hydration of uncompensated Si bonds produced upon electrochemical Si etching. These hydrated bonds are very reactive and can be easily oxidized, releasing additional electrons, and thus accelerating Au deposition:

$$2Si–H_{x(surface)} + H_2O \rightarrow SiO_{2(surface)} + 2H^+ + 4e^-. \quad (4)$$

Therefore, porous Si substrates can act not only as shape-forming platforms for plasmonic nanostructures due to the diversity of their surface configurations (meso-, macro-, and microporous surfaces of different porosity), but they also can control the redox reaction kinetics, which is an intriguing approach for producing even more sophisticated nanostructures, such as noble metal dendrites [47] or nanothorns [37].

Figure 1 shows the SEM images for the meso-PS templates prepared by different electrochemical etching regimes. Figure 1a reports a meso-PS etched at 30 mA cm^{-2} for 20 s, while Figure 1b,c show samples etched at 60 mA cm^{-2} and 100 mA cm^{-2} for 22 s and 20 s, respectively. Figure 1 confirms the PS formation with round pore openings of different sizes and densities. In all cases, a 1.3 μm thick PS layer was produced.

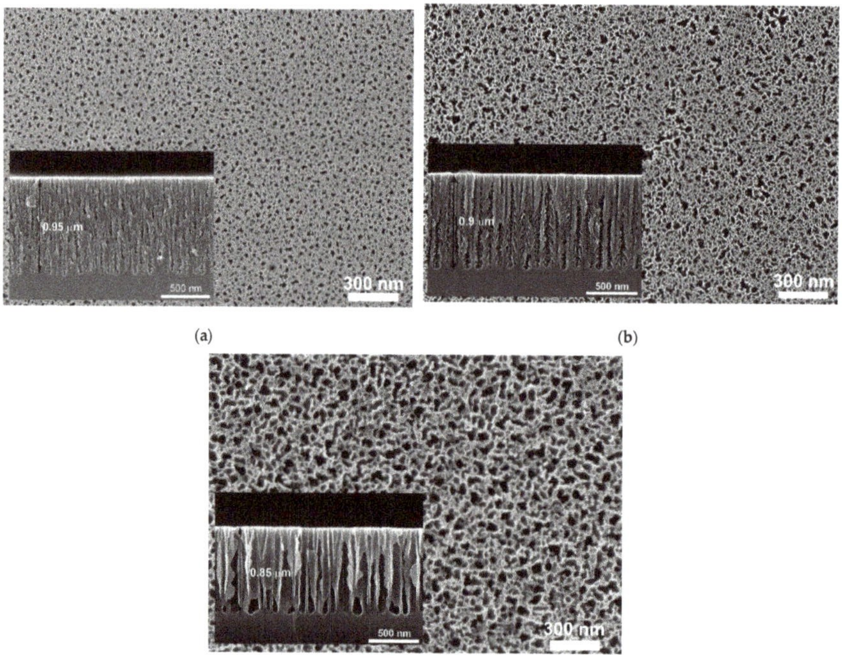

(a) (b) (c)

Figure 1. SEM images for a meso-PS sample after etching at a (**a**) 30 mA cm^{-2}, (**b**) 60 mA cm^{-2}, and (**c**) 100 mA cm^{-2} current density. The top views are presented in the background, while the cross sections are in the insets.

The quantitative characterization of the PS was carried out with the calculation of the mean pore opening diameter, pore density, and porosity. The mean pore diameter was calculated by averaging a Feret diameter distribution [48]. The pore density was calculated by dividing the number of pore openings by the visible surface area. The porosity of the meso-PS samples was estimated by gravimetric [49] and SEM image assessment methods. Both approaches resulted in the same porosities. When the applied etching current density was 30 mA cm^{-2}, we obtained an average pore diameter of 19 nm $\pm$ 8 nm, a pore density of 5.3 $\times$ 10^{10} cm^{-2}, and 49% porosity. For the sample etched at 60 mA cm^{-2}, we obtained an average pore diameter of 28 nm $\pm$ 14 nm, a pore density of 3.1 $\times$ 10^{10} cm^{-2}, and 59% porosity. For the highest etching current density of 100 mA cm^{-2}, we obtained an average pore diameter of 44 nm $\pm$ 32 nm, a pore density of 2.8 $\times$ 10^{10} cm^{-2}, and 72% porosity. These data are in good agreement with the well-studied mechanism of PS formation [50,51] and reported experimental observations [52,53], albeit the density of the pores slightly decreased for the more porous samples. Based on the obtained results, we can attribute the fabricated samples to the mesoporous material [54].

The fabricated meso-PS matrices were utilized for the following gold deposition step. Figure 2 represents the SEM images for different Au-coated meso-PS substrates. The first row shows the samples obtained after 30 min of Au deposition, while the second row represents the samples after 70 min of Au deposition. Each column in the figure corresponds to the gold nanostructure on the meso-PS etched at different current densities: 30 mA cm^{-2}, 60 mA cm^{-2}, and 100 mA cm^{-2}. The increase in the Au deposition time led to the gold layer densification, producing a void-free continuous film along with a second Au layer nucleation (observe the material excess on top of the formed layer in Figure 2a,d). The same morphology evolution of a gold layer was obtained when the porosity of the PS was reduced, while keeping the Au deposition time constant (Figure 2a–c). Notably, no significant layer alteration was revealed when the porosity of PS varied at higher Au deposition times (Figure 2d–f). The analysis of the SEM images indicated that the gold coverage level increased from ~83% to ~97% after 30 min and 70 min of Au deposition, respectively, and did not alter significantly for different PS matrixes. Although the produced gold coverage represented a rather quasicontinuous film, there was still a meaningful number of Au nanoparticles (NPs). Hence, we estimated their mean diameter and standard deviation, which were varied within the 22–60 nm range for PS matrices after 30 min Au deposition, while the 70 min Au deposition produced 31–46 nm NPs (Figure 3).

These results reveal the importance of the meso-PS matrix morphology for the accurate control over the Au layer fabrication process. The obtained Au layer was defined by a PS surface, which contained an array of densely packed voids on the silicon surface (Figure 1a–c). Their density primarily depended on the etching current density and could be characterized by the porosity of the PS. As a rule, the higher the porosity, the denser the void package on the surface, and the lower the "free" silicon area. Consequently, the gold deposition rate, which followed the redox reaction mechanism between the silicon atoms and gold cations, was dramatically reduced for the highly porous matrices. The number of seed nucleation centers, as well as the surface diffusion of the gold atoms, were restricted by the voids, preventing the growth of a continuous Au film. Therefore, the Au coverage appeared in the form of isolated islands, whose sizes strongly depended on the porosity of the parent meso-PS matrix. It should be emphasized, however, that the situation changed in the later stages of Au film growth (Figure 2d–f). In these stages, the produced Au nanoislands became new nucleation centers, leading to the film growth over the pore openings. Eventually, the Au nanoislands collapsed, and the quasicontinuous Au film on top of the porous surface was achieved. Continuing the Au deposition induced the new Au layer formation. Interestingly, the appearance of this layer started with NP nucleation, suggesting that the Au film formation followed the Stranski–Krastanov mechanism of thin films' growth [55–57].

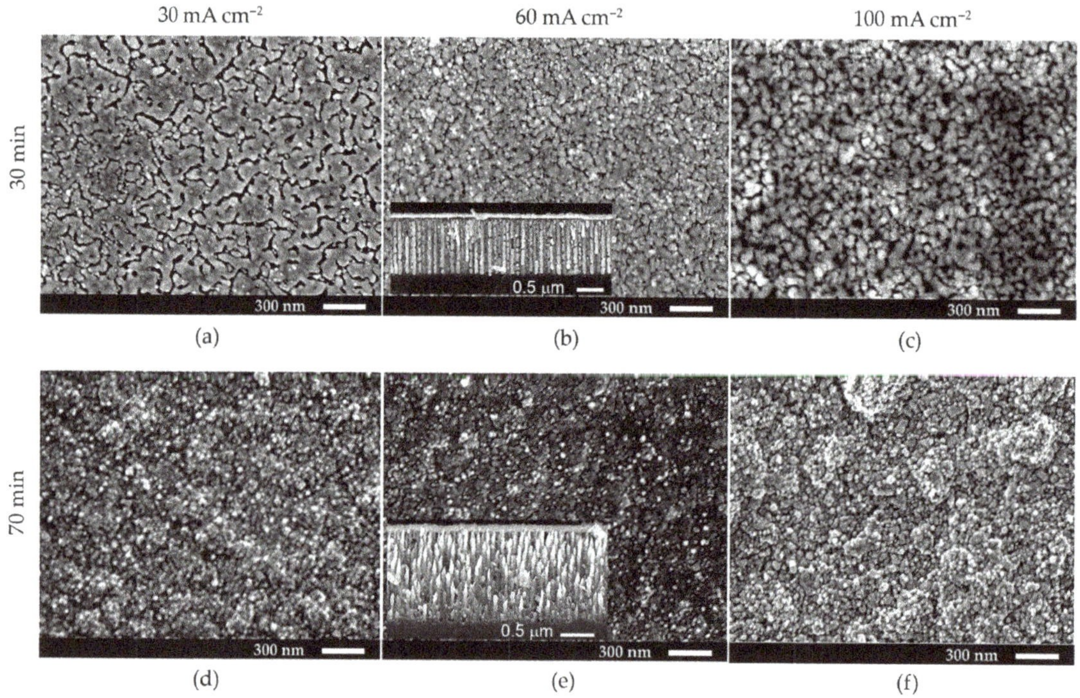

Figure 2. SEM images after Au deposition on the PS with a 49% (**a**,**d**), 59% (**b**,**e**), and 72% (**c**,**f**) porosity for 30 min (**a**–**c**) and 70 min (**d**–**f**). The insets in (**b**,**e**) depict the cross-section view for the meso-PS after the Au deposition.

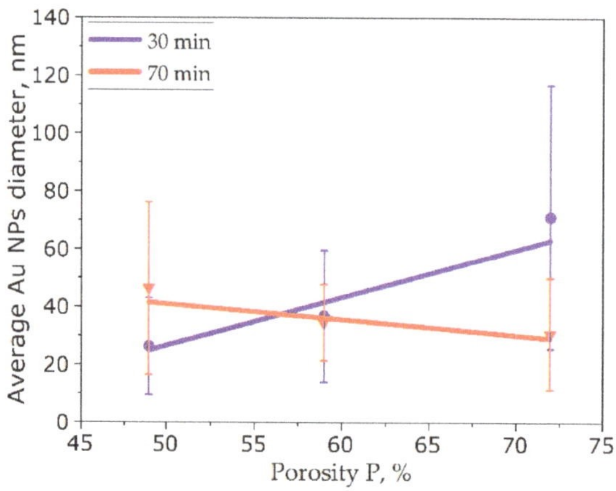

Figure 3. Evolution of the mean Au NP diameter as a function of the meso-PS porosity for different gold deposition times.

Figure 4 reports the XRD pattern for the Au-coated meso-PS after 30 min and 70 min of Au deposition on the PS etched at 60 mA cm^{-2}. As can be seen from Figure 4, both XRD patterns contained the characteristic peaks for the cubic cell Au crystals with orientations (111), (200), (220), and (311), confirming the polycrystalline nature of the Au coverage. Remarkably, the intensity of the (200) peak increased after 70 min of Au deposition, while the other peaks became less prominent, compare Figure 4a,b. These data indicate the gold coverage growth along with the deformation of a Au elementary crystal cell by a Si surface with (100) crystallographic orientation, leading to the monocrystallization of the Au coverage. The appearance of the Si (400) peak, however, suggested a weak uniformity of the Au layer for both samples. The same XRD patterns were observed for all the Au-coated meso-PS samples, albeit with a minor intensity variation of the characteristic peaks, depending on the porosity of the meso-PS and the Au deposition time.

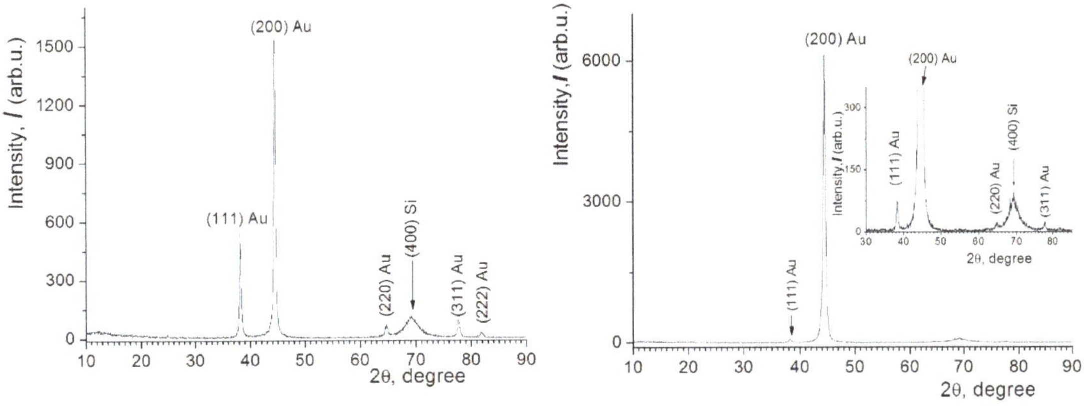

(a) (b)

Figure 4. XRD pattern after 30 min (**a**) and 70 min (**b**) of Au deposition on the meso-PS with 49% porosity.

3.2. Morphology of the Au-Coated Macro-PS Substrates

Figure 5 reports the SEM images for the macro-PS template and the corresponding distribution of the pore openings' diameter. The SEM images confirmed the 1.1 μm thick PS layer formation with round pore openings (Figure 5a). The pore diameters' distribution for the macro-PS was rather nonuniform, covering the 400–1000 nm range (Figure 5b). These sizes were about one order of magnitude larger compared to the meso-PS (Figure 1). We therefore attributed the obtained sample to the macroporous material [54]. These results were in good agreement with the established macro-PS formation models [51].

Figure 6 shows the SEM images for the Au-coated macro-PS substrates after 30–80 min of Au deposition. The morphology of the film altered dramatically at various Au deposition steps. An array of closely packed Au NPs was obtained after 30 min deposition, Figure 6a. These NPs coalesced after 40 min producing a quasicontinuous gold film, as shown in Figure 6b. The subsequent increase in the deposition time led to the film densification together with the new Au seeds' nucleation over the formed coverage, producing a multilayered Au film, as shown in Figure 6c–f. The analysis of the SEM images depicted in Figure 7 reported the average Au NP diameter after different deposition times. No noticeable dependence of NP size on the deposition time was observed. The average Au NP diameter for all samples varied from 50 nm to 60 nm. The statistical data for the samples after 30 min and 40 min of Au deposition is not reported due to the difficulties in distinguishing individual Au NPs.

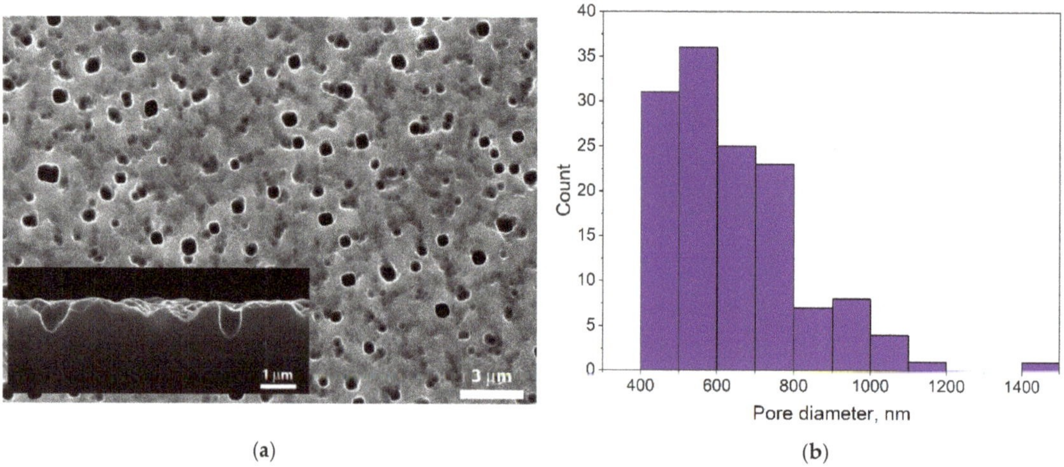

Figure 5. SEM images for a macro-PS sample (**a**) and the corresponding distribution of the pore opening diameters (**b**).

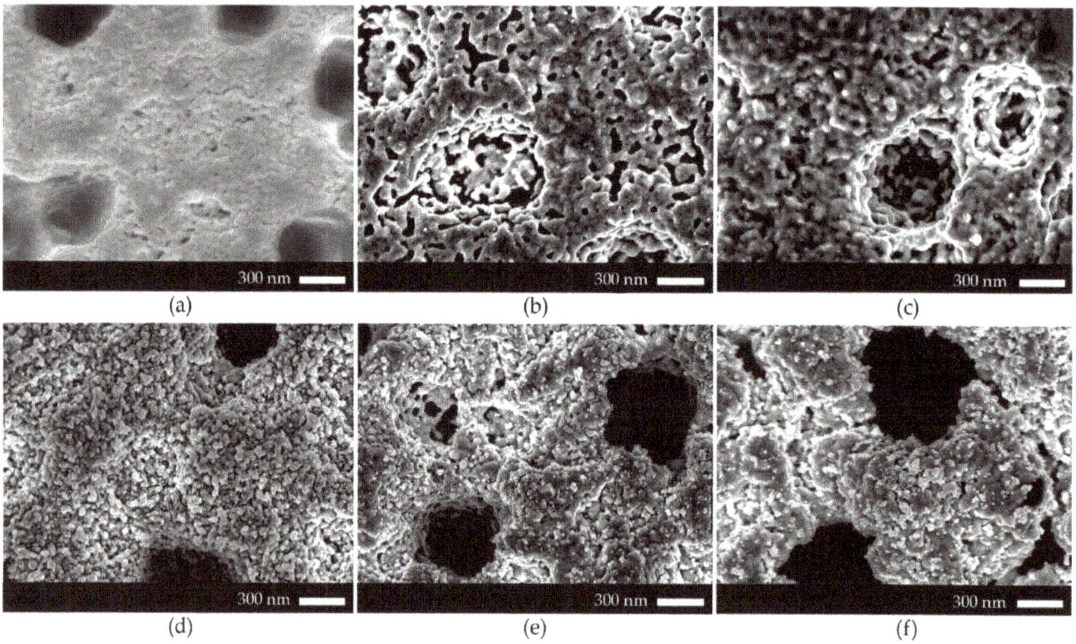

Figure 6. SEM images for the macro-PS after 30 min (**a**), 40 min (**b**), 50 min (**c**), 60 min (**d**), 70 min (**e**), and 80 min (**f**) of Au deposition.

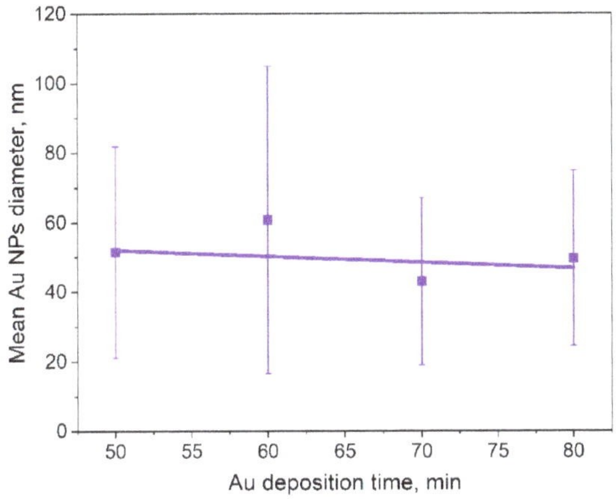

Figure 7. Mean Au NP diameter as a function of the gold deposition time on the macro-PS.

The observed kinetics of the Au film growth may be explained considering the presence of a thin microporous layer [58–60] over the entire macro-PS surface. This layer had a fractal-like geometry, which acted as the main source of the charge carriers for the redox reaction between the gold cations and the silicon wafer. This layer was also highly porous and composed of dense tiny voids. The growth of a continuous Au film on such a surface was even more confined by the voids compared with the meso-PS, leading to the formation of a dense Au NP layer in an early deposition stage (Figure 6a). As the redox reaction progressed with time, the thin microporous layer dissolved, leaving the bare macro-PS surface opened. At this growth stage, the number of Au seed nucleation centers increased, and the surface diffusion of gold atoms became possible. As the result, the Au NPs coalesced, and the quasicontinuous film was produced. For the rest of the deposition time, the gold layer grew similarly to the Au film on the meso-PS.

Figure 8 shows the XRD pattern for the Au-coated macro-PS after 50 min and 80 min of Au deposition. The patterns reported peaks associated with (111), (200), (220), and (311) Au crystal orientations, confirming the polycrystalline nature of the samples. Interestingly, in contrast to the XRD patterns for the Au-coated meso-PS (Figure 4a,b), no characteristic Si peak was observed in the XRD pattern for the gold-coated macro-PS, as shown in Figure 8a. This confirmed the introduced mechanism of Au film growth, where the coverage became uniform after the complete dissolution of the micro-PS layer. Apart from that, no noticeable difference between the Au-coated meso- and macro-PS samples was observed from the XRD; compare Figure 4a,b and Figure 8b.

We believe that the gold-coated macro-PS platform developed in this work can serve as an efficient nanoscale platform for SERS due to the strong plasmonic enhancement of the Raman signal. This can be better explained by closer inspection of the acquired SEM images in Figures 5 and 6. These images suggest that the obtained samples repeated the familiar nanoscale configuration, named metal nanovoids (NVs), which is one of the major candidates for mass utilization in SERS [61,62]. These NVs usually represent 500–2000 nm holes coated with various coinage metals and ensure the Raman signal enhancement by at least two basic optical effects: a localized plasmon mode with a strong rim component and multiple reflections of excitation light rays through the void. Moreover, the periodic morphology of the Me NVs guarantees a good spot-to-spot reproducibility of the SERS signal. The standard approach for Me NVs fabrication involves the deposition of a seed metallic layer, the coating with polymer spheres, the electrodeposition of a metal, and finally the dissolution of the polymer spheres. This procedure, however, is rather complex and limited by the small surface area covered with Me NVs. In contrast, our approach involved only two simple steps, electrochemical etching and chemical deposition, and can be extended up to wafer scale. It should be emphasized that the porosity and the thickness of the macro-PS in this work were optimized to meet the dimensions of the standard Me NV configurations, while the optimum Raman signal enhancement was determined by the variation in the Au deposition time in the 30–80 min range.

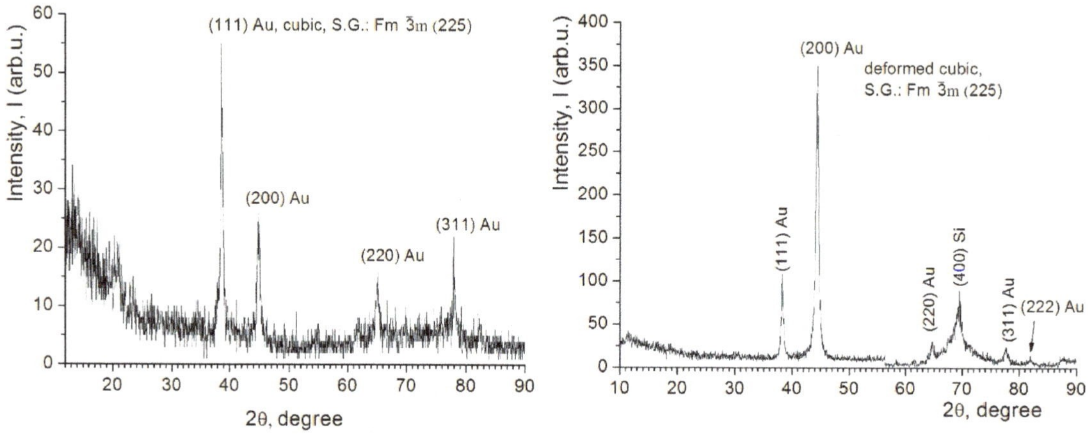

(**a**) (**b**)

Figure 8. XRD pattern for the macro-PS after 50 min (**a**) and 80 min (**b**) of Au deposition.

3.3. UV-Vis and SERS Measurements

Figure 9 depicts the reflectance spectra for the Au-coated meso-PS (Figure 9a) and macro-PS (Figure 9b) samples. These spectra clearly depict the reflectance minima around 480–495 nm for all samples, which could be associated with the (localized) surface plasmon resonance ((L)SPR). Considering that the highest Raman signal enhancement was achieved upon this (L)SPR condition, we therefore utilized the closest to this resonance and available for us laser wavelength of 532 nm. It should be emphasized that the (L)SPR observation for gold in the blue region of visible light is rather uncommon. It is known that isolated spherical Au nanoparticles demonstrate the LSPR band in the 520–580 nm wavelength region [63]. In some cases, however, it can shift towards the blue part of the spectrum, which is usually explained by the plasmons coupling between different nanoparticles upon shortening their spacing distance [64]. Therefore, our nanostructures were most likely to be affected by such a coupling effect, since they represented collections of densely packed Au nanoparticles comprising quasicontinuous films.

Figures 10 and 11 show the SERS spectra for 10^{-6} M R6G adsorbed on either the Au-coated meso- or macro-PS samples, respectively. To simplify the results representation and subsequent discussion, all the samples produced in this work and the corresponding fabrication regimes are summarized in Table 1. As can be seen from the figures, all SERS spectra exhibited the same characteristic Raman bands for R6G, namely 613 cm^{-1} (C-C-C ring in-plane bending); 772 cm^{-1} (C-H out of plane bending); 1185 cm^{-1} (C-H in-plane bending); 1311 cm^{-1} (C-O-C stretch); 1363 cm^{-1}, 1510 cm^{-1}, 1575 cm^{-1}, and 1650 cm^{-1} (aromatic C–C stretch). These results are in good agreement with the data reported elsewhere [65]. Figure 10 indicates that the intensity of the R6G SERS spectrum obtained for the best Au-coated macro-PS (sample 6) was two times higher than for the best Au-coated meso-PS (sample 2′, Figure 9). This confirms the stronger Raman signal enhancement provided by a configuration mimicking metal NVs compared with the usual metal NPs. The variation in the SERS spectra intensity between different nanostructures prepared on the same type of PS may be explained in terms of the stage of Au film growth. For example, the weak intensity of the R6G SERS spectra corresponded to the samples with a (quasi)continuous Au film (samples 2 and 5 in Figures 9 and 10, respectively; their morphology is highlighted in Figures 2a and 6b, respectively), while stronger SERS signals originated from the Au NPs (samples 2′ and 6 in Figure 9, respectively; their morphology is highlighted in Figures 2c and 6c, respectively). These results can be correlated with the number of so-called "hot spots", tiny gaps in a metal nanostructure where the localization of a giant electromagnetic field occurred. The number of these spots was greater for the nanostructures exhibiting nanoscale roughness compared with the smooth continuous surfaces, leading to a higher probability for a molecule to be located within it and contribute to the enhancement of Raman scattering. With only this in mind, the reasons for the weak SERS response for sample 4 (Figure 10) are rather unclear. One may expect comparable intensities of R6G SERS spectra for both samples 4 and 6 because of the similar nanostructure morphology (compare Figure 6a,c). However, the correct results interpretation also requires a particle size comparison. The calculated average particle diameter for sample 6 was 51 nm ± 30 nm, which was close to the optimum diameter of 60 nm for an isolated Au NP, demonstrating the maximum SERS enhancement [66,67]. On the other hand, the individual Au NPs for sample 4 could not be resolved from the SEM images, suggesting their sizes were far smaller than the known optimum. A similar analysis can be conducted for the other Au-coated samples prepared on the same type of PS matrixes. This means that the careful choice of the initial PS morphological parameters, as well as the Au deposition regimes, can help to obtain the optimum Au nanostructure with the corresponding NP size and spacing for maximum SERS enhancement.

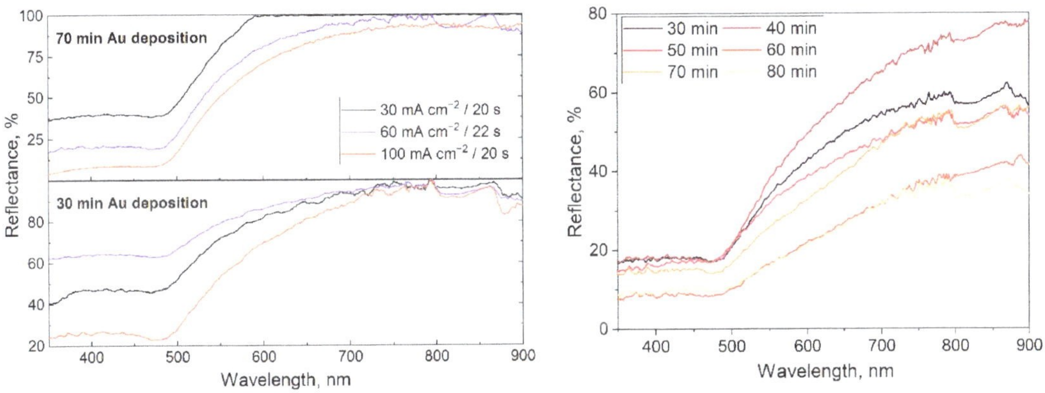

(a) (b)

Figure 9. Reflectance spectra for the meso-PS (**a**) and macro-PS (**b**) samples after Au deposition for different time periods.

Table 1. Fabrication regimes for the Au-coated meso- and macro-PS.

Sample	Etching Current Density for PS, mA cm^{-2}	Etching Time for PS, s	Au Deposition Time, min
1/1′	30	20	30/70
2/2′	60	22	30/70
3/3′	100	20	30/70
4	8	240	30
5	8	240	40
6	8	240	50
7	8	240	60
8	8	240	70
9	8	240	80

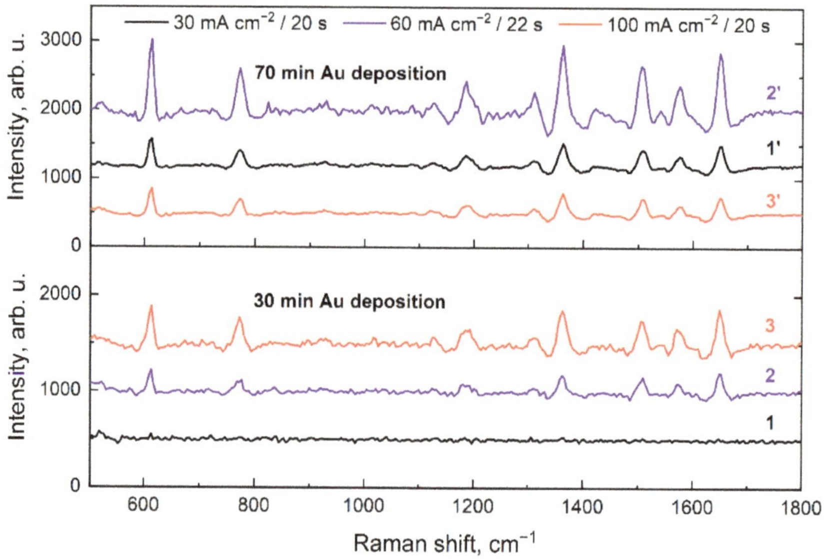

Figure 10. SERS spectra for 10^{-6} M R6G on the gold-coated meso-PS substrates.

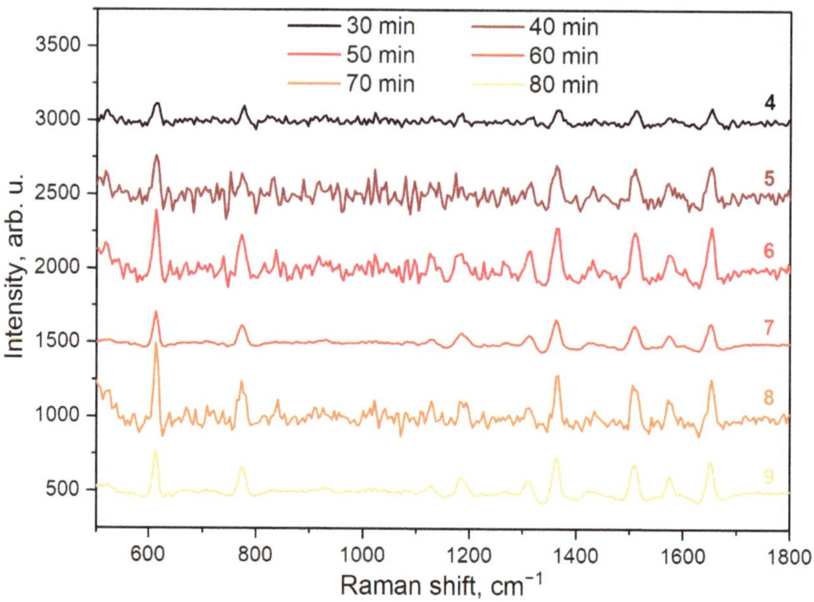

Figure 11. SERS spectra for 10^{-6} M R6G on the gold-coated macro-PS substrates.

4. Conclusions

In summary, we demonstrated that an effective SERS platform may be fabricated by Au electroless immersion deposition on meso- and macro-PS. SEM, XRD, and SERS analyses were carried out for sample characterization. In this work, a macro-PS template was fabricated to reproduce the Me NVs' structure [61,62], while the meso-PS matrixes were produced with three different porosities. SERS measurements were performed for the R6G molecules adsorbed on different Au-coated meso- and macro-PS samples at a

10^{-6} M concentration. It should be emphasized that the maximum SERS enhancement was achieved for the Me NVs' structures fabricated by a simple two-step procedure, while the current Me NVs' fabrication procedures involve at least four complex nanotechnology techniques, and the final structure is limited by a low surface area. We also demonstrated the correlation of the R6G SERS spectra intensity with the morphology and structural properties of Au coverage on the PS surface. Notably, the SERS response from the Au-coated PS samples may be tuned by the accurate choice of the initial PS morphological parameters, as well as the Au deposition regimes, which is explained by the formation of Au NPs with different sizes and spacings. Overall, both types of developed substrates demonstrated a sufficiently strong SERS response and show great promise for simple and versatile large-area plasmonic nanostructures fabrication.

Author Contributions: Author Contributions: conceptualization, S.Z., A.D., A.I.P. and H.B.; methodology, S.Z. and H.B.; software, S.Z.; validation, S.Z. and H.B.; formal analysis, S.Z.; investigation, S.Z. and A.C.; resources, H.B.; data curation, S.Z. and H.B.; writing—original draft preparation, S.Z.; writing—review and editing, S.Z., A.D., A.I.P. and H.B.; visualization, S.Z.; supervision, H.B.; project administration, H.B.; funding acquisition, H.B. All authors have read and agreed to the published version of the manuscript.

Funding: This research was funded by the Belarusian State Research Program "Photonics and Electronics for Innovations", tasks 1.8 (subprogram "Photonics and its application") and 2.2 (subprogram "Opto- and RF electronics"). Furthermore, S.Z., A.C. and H.B. thank the Russian Science Foundation (Project # 21-19-00761). In addition, the research of A.I.P. was partly supported by the RADON project (GA 872494) within the H2020-MSCA-RISE-2019 call and COST Action CA20129 "Multiscale Irradiation and Chemistry Driven Processes and Related Technologies" (MultIChem). A.I.P. thanks the Institute of Solid-State Physics, University of Latvia. ISSP UL as the Center of Excellence is supported through the Framework Program for European universities, Union Horizon 2020, H2020-WIDESPREAD-01–2016–2017-TeamingPhase2, under Grant Agreement No. 739508, CAMART2 project.

Data Availability Statement: The data presented in this study are available on request from the corresponding author.

Acknowledgments: The authors would like to thank Vitaly Bondarenko from the Belarusian State University of Informatics and Radioelectronics for the fruitful discussions on the mechanism of the gold coatings' formation on porous silicon.

Conflicts of Interest: The authors declare no conflict of interest.

References

1. Robertus, J. Principles of Protein X-Ray Crystallography, 3rd Edition By Jan Drenth (University of Groningen, The Netherlands). With a Major Contribution from Jeroen Mester (University of Lübeck, Germany). Springer Science + Business Media LLC: New York. 2007. Xiv + 3. *J. Am. Chem. Soc.* **2007**, *129*, 5782–5783. [CrossRef]
2. Bax, A.D.; Grzesiek, S. Methodological Advances in Protein NMR. *Acc. of Chem. Res.* **1993**, *26*, 131–138. [CrossRef]
3. Cavanagh, J.W.; Fairbrother, A.; Palmer, N., III; Skelton, M.R. *Protein NMR Spectroscopy: Principles and Practice*, 2nd ed.; Elsevier: Amsterdam, The Netherlands, 2007.
4. Kelly, S.M.; Jess, T.J.; Price, N.C. How to Study Proteins by Circular Dichroism. *Biochim. Biophys. Acta (BBA)-Proteins Proteom.* **2005**, *1751*, 119–139. [CrossRef]
5. Whitmore, L.; Miles, A.J.; Mavridis, L.; Janes, R.W.; Wallace, B.A. PCDDB: New Developments at the Protein Circular Dichroism Data Bank. *Nucleic Acids Res.* **2017**, *45*, D303–D307. [CrossRef] [PubMed]
6. Amendola, V.; Pilot, R.; Frasconi, M.; Maragò, O.M.; Iatì, M.A. Surface Plasmon Resonance in Gold Nanoparticles: A Review. *J. Phys. Condens. Matter* **2017**, *29*, 203002. [CrossRef]
7. Lane, L.A.; Qian, X.; Nie, S. SERS Nanoparticles in Medicine: From Label-Free Detection to Spectroscopic Tagging. *Chem. Rev.* **2015**, *115*, 10489–10529. [CrossRef]
8. Kneipp, J.; Kneipp, H.; Kneipp, K. SERS—A Single-Molecule and Nanoscale Tool for Bioanalytics. *Chem. Soc. Rev.* **2008**, *37*, 1052–1060. [CrossRef]
9. Zavatski, S.; Khinevich, N.; Girel, K.; Redko, S.; Kovalchuk, N.; Komissarov, I.; Lukashevich, V.; Semak, I.; Mamatkulov, K.; Vorobyeva, M.; et al. Surface Enhanced Raman Spectroscopy of Lactoferrin Adsorbed on Silvered Porous Silicon Covered with Graphene. *Biosensors* **2019**, *9*, 34. [CrossRef]

10. Xie, X.; Pu, H.; Sun, D.W. Recent Advances in Nanofabrication Techniques for SERS Substrates and Their Applications in Food Safety Analysis. *Crit. Rev. Food Sci. Nutr.* **2018**, *58*, 2800–2813. [CrossRef]
11. Burtsev, V.; Erzina, M.; Guselnikova, O.; Miliutina, E.; Kalachyova, Y.; Svorcik, V.; Lyutakov, O. Detection of Trace Amounts of Insoluble Pharmaceuticals in Water by Extraction and SERS Measurements in a Microfluidic Flow Regime. *Analyst* **2021**, *146*, 3686–3696. [CrossRef]
12. Cailletaud, J.; De Bleye, C.; Dumont, E.; Sacré, P.Y.; Netchacovitch, L.; Gut, Y.; Boiret, M.; Ginot, Y.M.; Hubert, P.; Ziemons, E. Critical Review of Surface-Enhanced Raman Spectroscopy Applications in the Pharmaceutical Field. *J. Pharm. Biomed. Anal.* **2018**, *147*, 458–472. [CrossRef] [PubMed]
13. Khinevich, N.; Zavatski, S.; Bandarenka, H.; Belyatsky, V.; Galyuk, E.; Ryneiskaya, O. Study of Diluted Meldonium Solutions by Surface Enhanced Raman Scattering Spectroscopy. *Int. J. Nanosci.* **2019**, *18*, 1940054. [CrossRef]
14. Khrustalev, V.V.; Khrustaleva, T.A.; Kahanouskaya, E.Y.; Rudnichenko, Y.A.; Bandarenka, H.V.; Arutyunyan, A.M.; Girel, K.V.; Khinevich, N.V.; Ksenofontov, A.L.; Kordyukova, L.V. The Alpha Helix 1 from the First Conserved Region of HIV1 Gp120 Is Reconstructed in the Short NQ21 Peptide. *Arch. Biochem. Biophys.* **2018**, *638*, 66–75. [CrossRef]
15. Kneipp, K.; Kneipp, H.; Bohr, H.G. Single-Molecule SERS Spectroscopy. In *Surface-Enhanced Raman Scatt*; Springer: Berlin/Heidelberg, Germany, 2006; pp. 261–277. [CrossRef]
16. Nie, S.; Emory, S.R. Probing Single Molecules and Single Nanoparticles by Surface-Enhanced Raman Scattering. *Science* **1997**, *275*, 1102–1106. [CrossRef] [PubMed]
17. Stiles, P.L.; Dieringer, J.A.; Shah, N.C.; Van Duyne, R.P. Surface-Enhanced Raman Spectroscopy. *Annu. Rev. Anal. Chem.* **2008**, *1*, 601–626. [CrossRef] [PubMed]
18. Langer, J.; Jimenez de Aberasturi, D.; Aizpurua, J.; Alvarez-Puebla, R.A.; Auguié, B.; Baumberg, J.J.; Bazan, G.C.; Bell, S.E.J.; Boisen, A.; Brolo, A.G. Present and Future of Surface-Enhanced Raman Scattering. *ACS Nano* **2019**, *14*, 28–117. [CrossRef] [PubMed]
19. Ding, S.-Y.; Yi, J.; Li, J.-F.; Ren, B.; Wu, D.-Y.; Panneerselvam, R.; Tian, Z.-Q. Nanostructure-Based Plasmon-Enhanced Raman Spectroscopy for Surface Analysis of Materials. *Nat. Rev. Mater.* **2016**, *1*, 16021. [CrossRef]
20. Natan, M.J. Concluding Remarks Surface Enhanced Raman Scattering. *Faraday Discuss.* **2006**, *132*, 321–328. [CrossRef]
21. Corni, S.; Tomasi, J. Surface Enhanced Raman Scattering from a Single Molecule Adsorbed on a Metal Particle Aggregate: A Theoretical Study. *J. Chem. Phys.* **2002**, *116*, 1156–1164. [CrossRef]
22. Mosier-Boss, P.A. Review of SERS Substrates for Chemical Sensing. *Nanomaterials* **2017**, *7*, 142. [CrossRef]
23. Poston, P.E.; Harris, J.M. Stable, Dispersible Surface-Enhanced Raman Scattering Substrate Capable of Detecting Molecules Bound to Silica-Immobilized Ligands. *Appl. Spectrosc.* **2010**, *64*, 1238–1243. [CrossRef] [PubMed]
24. Feng, L.; Xu, Y.-L.; Fegadolli, W.S.; Lu, M.-H.; Oliveira, J.E.B.; Almeida, V.R.; Chen, Y.-F.; Scherer, A. Experimental Demonstration of a Unidirectional Reflectionless Parity-Time Metamaterial at Optical Frequencies. *Nat. Mater.* **2013**, *12*, 108–113. [CrossRef]
25. Dubkov, S.V.; Savitskiy, A.I.; Trifonov, A.Y.; Yeritsyan, G.S.; Shaman, Y.P.; Kitsyuk, E.P.; Tarasov, A.; Shtyka, O.; Ciesielski, R.; Gromov, D.G. SERS in Red Spectrum Region through Array of Ag–Cu Composite Nanoparticles Formed by Vacuum-Thermal Evaporation. *Opt. Mater. X* **2020**, *7*, 100055. [CrossRef]
26. Kneipp, K.; Kneipp, H.; Manoharan, R.; Hanlon, E.B.; Itzkan, I.; Dasari, R.R.; Feld, M.S. Extremely Large Enhancement Factors in Surface-Enhanced Raman Scattering for Molecules on Colloidal Gold Clusters. *Appl. Spectrosc.* **1998**, *52*, 1493–1497. [CrossRef]
27. Jeon, T.Y.; Jeon, H.C.; Yang, S.-M.; Kim, S.-H. Hierarchical Nanostructures Created by Interference of High-Order Diffraction Beams. *J. Mater. Chem. C* **2016**, *4*, 1088–1095. [CrossRef]
28. Radu, A.I.; Ryabchykov, O.; Bocklitz, T.W.; Huebner, U.; Weber, K.; Cialla-May, D.; Popp, J. Toward Food Analytics: Fast Estimation of Lycopene and β-Carotene Content in Tomatoes Based on Surface Enhanced Raman Spectroscopy (SERS). *Analyst* **2016**, *141*, 4447–4455. [CrossRef]
29. Kim, S.; Zhang, W.; Cunningham, B.T. Coupling Discrete Metal Nanoparticles to Photonic Crystal Surface Resonant Modes and Application to Raman Spectroscopy. *Opt. Express* **2010**, *18*, 4300–4309. [CrossRef]
30. Gopalakrishnan, A.; Chirumamilla, M.; De Angelis, F.; Toma, A.; Zaccaria, R.P.; Krahne, R. Bimetallic 3D Nanostar Dimers in Ring Cavities: Recyclable and Robust Surface-Enhanced Raman Scattering Substrates for Signal Detection from Few Molecules. *ACS Nano* **2014**, *8*, 7986–7994. [CrossRef]
31. Yao, X.; Jiang, S.; Luo, S.; Liu, B.-W.; Huang, T.-X.; Hu, S.; Zhu, J.; Wang, X.; Ren, B. Uniform Periodic Bowtie SERS Substrate with Narrow Nanogaps Obtained by Monitored Pulsed Electrodeposition. *ACS Appl. Mater. Interfaces* **2020**, *12*, 36505–36512. [CrossRef]
32. Lazarouk, S.; Bondarenko, V.; Pershukevich, P.; La Monica, S.; Maiello, G.; Ferrari, A. Visible Electroluminescence from Al-Porous Silicon Reverse Bias Diodes Formed on the Base of Degenerate N-Type Silicon. *MRS Online Proc. Libr.* **1994**, *358*, 659. [CrossRef]
33. Bandarenka, H.; Redko, S.; Smirnov, A.; Panarin, A.; Terekhov, S.; Nenzi, P.; Balucani, M.; Bondarenko, V. Nanostructures Formed by Displacement of Porous Silicon with Copper: From Nanoparticles to Porous Membranes. *Nanoscale Res. Lett.* **2012**, *7*, 477. [CrossRef] [PubMed]
34. Bandarenka, H.V.; Khinevich, N.V.; Burko, A.A.; Redko, S.V.; Zavatski, S.A.; Shapel, U.A.; Mamatkulov, K.Z.; Vorobyeva, M.Y.; Arzumanyan, G.M. 3D Silver Dendrites for Single-Molecule Imaging by Surface-Enhanced Raman Spectroscopy. *ChemNanoMat* **2021**, *7*, 141–149. [CrossRef]

35. Korotcenkov, G. *Porous Silicon: From Formation to Application: Biomedical and Sensor Applications*; CRC Press: Boca Raton, FL, USA, 2016; Volume 2, ISBN 1482264579.
36. Bandarenka, H.V.; Girel, K.V.; Bondarenko, V.P.; Khodasevich, I.A.; Panarin, A.Y.; Terekhov, S.N. Formation Regularities of Plasmonic Silver Nanostructures on Porous Silicon for Effective Surface-Enhanced Raman Scattering. *Nanoscale Res. Lett.* **2016**, *11*, 262. [CrossRef] [PubMed]
37. Khajehpour, K.J.; Williams, T.; Bourgeois, L.; Adeloju, S. Gold Nanothorns–Macroporous Silicon Hybrid Structure: A Simple and Ultrasensitive Platform for SERS. *Chem. Commun.* **2012**, *48*, 5349–5351. [CrossRef] [PubMed]
38. Bandarenka, H.; Dolgiy, A.; Chubenko, E.; Redko, S.; Girel, K.; Prischepa, S.L.; Panarin, A.; Terekhov, S.; Pilipenko, V.; Bondarenko, V. Nanostructured Metal Films Formed onto Porous Silicon Template. *J. Nano Res. Trans. Tech. Publ.* **2016**, *39*, 235–255. [CrossRef]
39. Khinevich, N.; Bandarenka, H.; Zavatski, S.; Girel, K.; Tamulevičienė, A.; Tamulevičius, T.; Tamulevičius, S. Porous Silicon—A Versatile Platform for Mass-Production of Ultrasensitive SERS-Active Substrates. *Microporous Mesoporous Mater.* **2021**, *323*, 111204. [CrossRef]
40. Bandarenka, H.V.; Girel, K.V.; Zavatski, S.A.; Panarin, A.; Terekhov, S.N. Progress in the Development of SERS-Active Substrates Based on Metal-Coated Porous Silicon. *Materials.* **2018**, *11*, 852. [CrossRef]
41. Chernousova, S.; Epple, M. Silver as Antibacterial Agent: Ion, Nanoparticle, and Metal. *Angew. Chem.-Int. Ed.* **2013**, *52*, 1636–1653. [CrossRef]
42. Franci, G.; Falanga, A.; Galdiero, S.; Palomba, L.; Rai, M.; Morelli, G.; Galdiero, M. Silver Nanoparticles as Potential Antibacterial Agents. *Molecules* **2015**, *20*, 8856–8874. [CrossRef]
43. Dolgiy, A.; Redko, S.V.; Bandarenka, H.; Prischepa, S.L.; Yanushkevich, K.; Nenzi, P.; Balucani, M.; Bondarenko, V. Electrochemical Deposition and Characterization of Ni in Mesoporous Silicon. *J. Electrochem. Soc.* **2012**, *159*, D623. [CrossRef]
44. Magagnin, L.; Maboudian, R.; Carraro, C. Gold Deposition by Galvanic Displacement on Semiconductor Surfaces: Effect of Substrate on Adhesion. *J. Phys. Chem. B* **2002**, *106*, 401–407. [CrossRef]
45. Kong, L.; Dasgupta, B.; Ren, Y.; Mohseni, P.K.; Hong, M.; Li, X.; Chim, W.K.; Chiam, S.Y. Evidences for Redox Reaction Driven Charge Transfer and Mass Transport in Metal-Assisted Chemical Etching of Silicon. *Sci. Rep.* **2016**, *6*, 36582. [CrossRef]
46. Wang, C.H.; Sun, D.C.; Xia, X.H. One-Step Formation of Nanostructured Gold Layers via a Galvanic Exchange Reaction for Surface Enhancement Raman Scattering. *Nanotechnology* **2006**, *17*, 651–657. [CrossRef]
47. Grevtsov, N.; Burko, A.; Redko, S.; Khinevich, N.; Zavatski, S.; Niauzorau, S.; Bandarenka, H. Silicon Nanowire Arrays Coated with Ag and Au Dendrites for Surface-Enhanced Raman Scattering. *MRS Adv.* **2020**, *5*, 2023–2032. [CrossRef]
48. Merkus, H.G. *Particle Size Measurements*, 1st ed.; Springer: Dordrecht, The Netherlands, 2009; ISBN 1402090161.
49. Herino, R.; Bomchil, G.; Barla, K.; Bertrand, C.; Ginoux, J.L. Porosity and Pore Size Distributions of Porous Silicon Layers. *J. Electrochem. Soc.* **1987**, *134*, 1994. [CrossRef]
50. Smith, R.L.; Collins, S.D. Porous Silicon Formation Mechanisms. *J. Appl. Phys.* **1992**, *71*, R1–R22. [CrossRef]
51. Korotcenkov, G. *Porous Silicon: From Formation to Application: Formation and Properties*; CRC Press: Boca Raton, FL, USA, 2016; Volume 1, ISBN 1482264552.
52. Nativ-Roth, E.; Rechav, K.; Porat, Z. Deposition of Gold and Silver on Porous Silicon and inside the Pores. *Thin Solid Films* **2016**, *603*, 88–96. [CrossRef]
53. Panarin, A.Y.; Terekhov, S.N.; Kholostov, K.I.; Bondarenko, V.P. SERS-Active Substrates Based on n-Type Porous Silicon. *Appl. Surf. Sci.* **2010**, *256*, 6969–6976. [CrossRef]
54. Sing, K.S.W.; Everett, D.H.; Haul, R.A.W.; Moscou, L.; Pierotti, R.A.; Rouquerol, J.; Siemieniewska, T. Reporting Physisorption Data for Gas/Solid Systems with Special Reference to the Determination of Surface Area and Porosity (Recommendations 1984). *Pure Appl. Chem.* **1985**, *57*, 603–619. [CrossRef]
55. Venables, J. *Introduction to Surface and Thin Film Processes*; Cambridge University Press: Cambridge, UK, 2000; ISBN 0521785006.
56. Pimpinelli, A.; Villain, J. *Physics of Crystal Growth*; Cambridge University Press: Cambridge, UK, 1998; ISBN 0521551986.
57. Oura, K.; Lifshits, V.G.; Saranin, A.A.; Zotov, A.V.; Katayama, M. *Surface Science: An Introduction*; Springer: Berlin/Heidelberg, Germany, 2013; ISBN 3662051796.
58. Ponomarev, E.A.; Lévy-Clément, C. Macropore Formation on P-Type Si in Fluoride Containing Organic Electrolytes. *Electrochem. Solid State Lett.* **1998**, *1*, 42. [CrossRef]
59. Wehrspohn, R.B.; Chazalviel, J.; Ozanam, F. Macropore Formation in Highly Resistive P-Type Crystalline Silicon. *J. Electrochem. Soc.* **1998**, *145*, 2958. [CrossRef]
60. Cattarin, S.; Pantano, E.; Decker, F. Investigation by Electrochemical and Deflectometric Techniques of Silicon Dissolution and Passivation in Alkali. *Electrochem. Commun.* **1999**, *1*, 483–487. [CrossRef]
61. Cole, R.M.; Mahajan, S.; Bartlett, P.N.; Baumberg, J.J. Engineering SERS via Absorption Control in Novel Hybrid Ni/Au Nanovoids. *Opt. Express* **2009**, *17*, 13298–13308. [CrossRef] [PubMed]
62. Kelf, T.A.; Sugawara, Y.; Cole, R.M.; Baumberg, J.J.; Abdelsalam, M.E.; Cintra, S.; Mahajan, S.; Russell, A.E.; Bartlett, P.N. Localized and Delocalized Plasmons in Metallic Nanovoids. *Phys. Rev. B* **2006**, *74*, 245415. [CrossRef]
63. Haiss, W.; Thanh, N.T.K.; Aveyard, J.; Fernig, D.G. Determination of Size and Concentration of Gold Nanoparticles from UV-Vis Spectra. *Anal. Chem.* **2007**, *79*, 4215–4221. [CrossRef]

64. Han, X.; Liu, Y.; Yin, Y. Colorimetric Stress Memory Sensor Based on Disassembly of Gold Nanoparticle Chains. *Nano Lett.* **2014**, *14*, 2466–2470. [CrossRef]
65. Dieringer, J.A.; Wustholz, K.L.; Masiello, D.J.; Camden, J.P.; Kleinman, S.L.; Schatz, G.C.; Van Duyne, R.P. Surface-Enhanced Raman Excitation Spectroscopy of a Single Rhodamine 6G Molecule. *J. Am. Chem. Soc.* **2009**, *131*, 849–854. [CrossRef]
66. Zeman, E.J.; Schatz, G.C. An Accurate Electromagnetic Theory Study of Surface Enhancement Factors for Silver, Gold, Copper, Lithium, Sodium, Aluminum, Gallium, Indium, Zinc, and Cadmium. *J. Phys. Chem.* **1987**, *91*, 634–643. [CrossRef]
67. Krug, J.T.; Wang, G.D.; Emory, S.R.; Nie, S. Efficient Raman Enhancement and Intermittent Light Emission Observed in Single Gold Nanocrystals. *J. Am. Chem. Soc.* **1999**, *121*, 9208–9214. [CrossRef]

crystals

MDPI

Article

Wet Synthesis of Graphene-Polypyrrole Nanocomposites via Graphite Intercalation Compounds

Gintarė Rimkutė [1], Gediminas Niaura [2], Rasa Pauliukaitė [3], Justina Gaidukevič [1] and Jurgis Barkauskas [1,*]

1 Institute of Chemistry, Faculty of Chemistry and Geosciences, Vilnius University, Naugarduko Str. 24, LT-03225 Vilnius, Lithuania
2 Department of Organic Chemistry, Center for Physical Sciences and Technology (FTMC), Sauletekio Ave. 3, LT-10257 Vilnius, Lithuania
3 Department of Nanoengineering, Center for Physical Sciences and Technology, Savanoriu Ave. 231, LT-02300 Vilnius, Lithuania
* Correspondence: jurgis.barkauskas@chf.vu.lt

Abstract: Graphene-polypyrrole (GP) nanocomposites were synthesized by a wet-way protocol using a graphite bisulfate (GBS) precursor. Consequently, GBS, a type of graphite intercalation compound, was prepared in the presence of concentrated sulfuric acid in the presence of a potassium periodate oxidizer. Three different types of graphite precursor with particle sizes of <50 μm, ≥150, ≤830 μm, and ≤2000 μm were used for this purpose. It was found that in the Raman spectra of GBS samples, the characteristic D band, which is caused by defects in the graphene layer, disappears. Therefore, the proposed synthesis protocol of GBS could be considered as a prospective intermediate stage in the preparation of graphene with low defect concentration. In contrast to alkali metal intercalation, the intercalation process involving anions with a relatively complex structure (e.g., HSO_4^-), which has been much less studied and requires further research. On the basis of the results obtained, structural models of graphite intercalation compounds as well as GP nanocomposites were discussed. The most relevant areas of application for GP nanocomposites, including energy storage and (bio)sensing, were considered. This work contributes to the development of cost-effective, scalable, and highly efficient intercalation methods, which still remain a significant challenge.

Keywords: graphite intercalation compounds; graphite bisulfate; exfoliated graphite; graphene-polypyrrole nanocomposites; structural analysis

Citation: Rimkutė, G.; Niaura, G.; Pauliukaitė, R.; Gaidukevič, J.; Barkauskas, J. Wet Synthesis of Graphene-Polypyrrole Nanocomposites via Graphite Intercalation Compounds. *Crystals* **2022**, *12*, 1793. https://doi.org/10.3390/cryst12121793

Academic Editors: Pingheng Tan, Aleksej Zarkov, Aivaras Kareiva and Loreta Tamasauskaite-Tamasiunaite

Received: 15 November 2022
Accepted: 7 December 2022
Published: 9 December 2022

1. Introduction

Graphene-based materials comprise a wide and diverse class. They originates from graphite, including graphene itself, graphene oxide, reduced graphene oxide, exfoliated graphene flakes, graphene nanoplatelets, and chemically functionalized versions of all these. In graphene technology, the preparation of graphene with low/controllable defect concentration is among the most important trends today. However, the production of graphene is primarily facilitated by high temperature processes [1–3]. The high temperature process is prone to defect formation. The solution to the problem could be the use of wet synthesis methods, though wet-chemical preparation of graphene of the same quality has been challenging up until now. In principle, if using a wet synthesis protocol, one could not only reduce the defect concentration but also introduce preferred functional groups at the same time. With chemical modification, graphene can become more functional, enabling more scalable applications in medicine, energy storage, nanotechnology, electronics, and other fields [4–9]. The conventional wet synthesis protocol of graphene includes intercalation, oxidation into graphene oxide, delamination, and subsequent reduction. A ruptured carbon lattice is usually the result of these consecutive operations [10]. Graphene quality can be improved by both reducing the number of sequential steps and understanding the underlying mechanisms. By eliminating oxidation and subsequent reduction from the preparation protocol of graphene, the number of sequential steps can be reduced. It has been reported that single-layer graphene flakes can be obtained from graphene intercalation compounds directly by spontaneous, mechanical, electrochemical, microwave, or liquid phase exfoliation [11–15]. The prevailing opinion is that graphene prepared in this way has the best ratio of quality, purity, production costs, scalability, and yield [16].

The graphite intercalation compound with sulfuric acid—graphite bisulfate—was one of the first graphite intercalation compounds synthesized by C. B. Brodie in 1855 [17]. Its formula is written as $C_{24}^{+} \cdot HSO_4^{-} \cdot 2\ H_2SO_4$; nevertheless, it represents the phase of maximal saturation, where a single layer of graphene is regularly altered with H_2SO_4 molecules and HSO_4^{-} ions (also known as the phase of stage 1). In the general case, the number of graphene layers, n, sandwiched between the two layers of intercalant is referred to as the stage number, and the corresponding GBS phase is called stage n GBS. Despite the fact that GBS was discovered long ago, it remains one of the least studied and poorly understood among graphite intercalation compounds. This is determined in large part by the fact that sulfuric acid does not spontaneously intercalate into graphite, as the reaction has a positive Gibbs free energy [18]. The Daumas–Herold model can be used to understand phase coexistence and stage transition in GBS [19]. Although the model explains the coexistence of GBS stages with $n > 1$, it does not always fit well with the experimental data [20]. The process of GBS formation is initiated by oxidation of graphite and leads to charge redistribution in the system. A surface charge transfer effect can control the redistribution of charges. Among low-dimensional materials, because of their high fraction of surface atoms, this phenomenon was originally demonstrated for conductive polymers [21]. Since graphene (as well as graphene intercalation compounds) is a representative 2D material, it interacts strongly with different dopants (halogens, alkali metals, oxygen-containing anions) [22]. A substantial change in its Fermi level and charge density is observed as a result. Surface charge transfer enables graphene and graphene intercalation compounds to be exploited for semiconducting 2D materials [23,24]. However, doping remains unclear in its mechanistic details despite its broad functional tunability and application possibilities. Due to the intertwined roles of oxygen, water, and substrates, this is particularly relevant for oxygen-mediated hole doping [25–28]. It should be noted that the formation of GBS is strongly dependent on the surface charge transfer effect; therefore, the role of the oxidant in the intercalation reaction is significant [29].

In this work, structural models of GBS compounds and carbon nanostructures derived from them were discussed. For this purpose, the GBS compounds were chemically synthesized using a KIO_4 oxidizer and three different graphite precursors with grain sizes of <50 μm, ≥150, ≤830 μm, and ≤2000 μm. The effect of precursor grain size and chemical constitution on defect concentration and phase constitution in prepared samples was evaluated. All samples obtained were annealed at 800 °C or microwave treated to obtain exfoliated graphite (EG), which was further modified with conductive polymer polypyrrole (PPy). The Raman spectra of GBS, EG, and composite samples with PPy were analysed to reveal the defect concentration. A relatively low defect concentration was found for EG samples, whereas the defect band completely disappeared in GBS samples. Thus, the proposed synthesis protocol could be considered as a prospective intermediate stage in the preparation of graphene with a low defect concentration. The applications of prepared samples in energy storage and biosensing devices were also considered.

2. Materials and Methods

2.1. Materials

All reagents were used without further purification. Extra pure fine graphite powder (<50 μm, ≥99.5%) was purchased from Merck KGaA (Darmstadt, Germany), synthetic graphite powder (≥150, ≤830 μm, 99.9%), natural graphite flakes (≤2000 μm, 99.9%), and Nafion (5%) were purchased from Alfa Aesar (Kandel, Germany), Na_2SO_4 (≥99%) was obtained from Eurochemicals, KIO_4 (99.8%), H_2SO_4 (98%) and pyrrole (≥98%) were purchased from Sigma-Aldrich (Darmstadt, Germany), $(NH_4)_2S_2O_8$ (98%) from Lach-

Ner (Neratovice, Czech Republic), and HCl (37%) and KBr (≥99.5%) from Carl Roth (Karlsruhe, Germany).

2.2. Synthesis of GBS Products

For GBS synthesis, three different graphite precursors with different particle sizes (<50 μm, ≥150, ≤830 μm and ≤2000 μm) denoted as Gr_1, Gr_2 and Gr_3, respectively, were used. In this work, the oxidizing mixture was prepared in a 50 mL Erlenmeyer flask with a ground glass joint by adding 0.1 g of a KIO_4 oxidizer to 15 mL of conc. H_2SO_4 at constant swirling. When the oxidizer dissolved, 0.25 g of graphite was slowly added to the mixture. The swirling procedure was continued for 1.5 h. Three synthesized samples were denoted as GBS_1, GBS_2, and GBS_3.

To achieve liquid phase exfoliation, the GBS_1, GBS_2, and GBS_3 samples were mixed with anhydrous Na_2SO_4. For this purpose, 2 mL of the GBS reaction mixture was added to 8 mL of H_2SO_4 and 5 g of Na_2SO_4. The mixture was left to stir for 3 days.

2.3. Thermal Treatment of GBS Samples

In order to obtain thermally exfoliated graphite (EG), 6 GBS samples (without and with liquid phase exfoliation) were annealed under thermal shock conditions. Samples were placed into ceramic combustion boats, which were inserted into a quartz glass tube. The tube was sealed and placed in a tubular furnace (SNOL 0.4/1250, Utena, Lithuania) preheated to 800 °C. The samples were heat treated under argon flow (80 mL·min^{-1}) until the sulfuric acid had stopped evaporating. Then, the tube was taken out of the furnace and cooled up to room temperature. The annealed products were washed with deionized water and dried at room temperature. The obtained EG products were denoted as EG_1, EG_2, EG_3, and EG_1_Na, EG_2_Na, EG_3_Na.

2.4. Microwave Treatment of GBS Samples

Furthermore, the GBS_1, GBS_2, and GBS_3 samples were subjected to direct microwave irradiation using a domestic microwave oven (Electrolux, 800 W EMS20405X, Stockholm, Sweden) for 1 min. During the treatment, the samples were placed in a capped weighting bottle with a ground glass joint. The obtained EG products were denoted as EG_1_MW, EG_2_MW, and EG_3_MW.

2.5. Synthesis of Graphene-Polypyrrole (GP) Nanocomposites

For the synthesis of GP nanocomposites by a wet-way protocol, EG_1, EG_2, and EG_3 samples were used. For this purpose, a 10% pyrrole solution in 1.0 M HCl, an EG sample, and $(NH_4)_2S_2O_8$ were mixed and ground in an agate mortar for 30 min. The ratio of pyrrole, EG, and $(NH_4)_2S_2O_8$ used was 1:2:20 by weight, respectively. The samples obtained were purified by dialysis in a tube and kept in a drying oven at a temperature of 40 °C until the water evaporated. Resultant samples were denoted as GP_1, GP_2, and GP_3.

2.6. Characterization

Optical microscopy images were acquired using an optical microscope (BX51 Olympus, Tokyo, Japan) at a magnification of 50x. Raman spectroscopy was performed using a Renishaw inVia spectrometer (Wotton-under-Edge, UK) equipped with an optical microscope at a magnification of 20×, a CCD camera, an objective numerical aperture of 0.40, and 1800 grooves mm^{-1} grating. The laser excitation wavelength was 532 nm, the beam was focussed to an area of 2 μm^2, and the integration time was 100 s. To analyse the changes in characteristic peaks, the Raman bands were fitted with a pseudo-Voigt function (a linear combination of Gaussian and Lorentzian functions) [30]. XRD measurements were performed in the 2θ range 10–45° with a Rigaku diffractometer (Miniflex II, Neu-Isenburg, Germany) (Cu K_α radiation with a graphite monochromator). The interplanar spacing d of the samples was calculated using Bragg's equation [31]. The crystallite size L was calculated by the Debye–Scherrer equation [32]. XPS analysis was performed using a Kratos analytical spectrometer (Axis Supra, Manchester, UK) equipped with a monochromatized Al K_α X-ray source (hν = 1486.69 eV, 15 kV, 25 mA). The spectra were acquired with an electron analyzer pass energy of 20 eV and calibrated using the C 1s peak at 284.6 eV. Core-level peaks were analysed using a nonlinear Shirley-type background subtraction, and the calculation of the elemental composition was performed on the basis of Scofield's relative sensitivity factors. SEM micrographs were obtained with the scanning electron microscope (Hitachi SU-70, Tokyo, Japan) at an accelerating voltage of 10.0 kV and magnification of 50,000×.

2.7. Electrochemical Measurements

Electrochemical analysis of the synthesized materials was performed using a glassy carbon electrode (GCE) (PalmSens BV, Houten, The Netherlands) modified with EG and GP samples. For this purpose, GCE was first polished with alumina powder (0.05 μm, Kemet, Kent, UK), rinsed with deionized water, ultrasonicated with deionized water, and then with ethanol. For the construction of the modified electrodes EG/GCE and GP/GCE, 1.0 mg of synthesized carbon material was dispersed in 1.0 mL of 0.05% Nafion (Alfa Aesar, Kandel, Germany) solution by sonicating the mixture in an ultrasonic bath for 30 min. In total, 10 μL of the suspension obtained was drop cast onto the polished surface of the GCE. The prepared electrodes were dried in air at room temperature.

Characterization of the electrodes was performed using cyclic voltammetry (CV). CV measurements were performed at room temperature using a CompactStat potentiostat/galvanostat (Ivium Technologies, Eindhoven, The Netherlands). All electrochemical measurements were run in the three-electrode system. A platinum wire, Ag/AgCl bare or EG/GCE, and GP/GCE (diameter of 3.0 mm) were used as the counter, reference, and working electrodes, respectively. CV curves were recorded in the potential range of −1.0 to +1.0 V (vs Ag/AgCl) and at a scan rate of 100 mV·s^{-1} in 0.1 M phosphate buffer solution (PBS) at a pH of 7.0.

3. Results

Precursors, or synthesized and heat-treated GBS products, were characterized using optical and SEM microscopy, Raman spectroscopy, and XPS and XRD analysis. GBS itself is not stable in air, therefore, SEM, XPS, and XRD measurements were only applied to those GBS products that were either subjected to the MW treatment or annealed at higher temperatures, i.e., samples in which H_2SO_4 was removed.

3.1. Optical Spectroscopy and SEM Characterization

The optical micrographs of the GBS samples are presented in Figure 1.

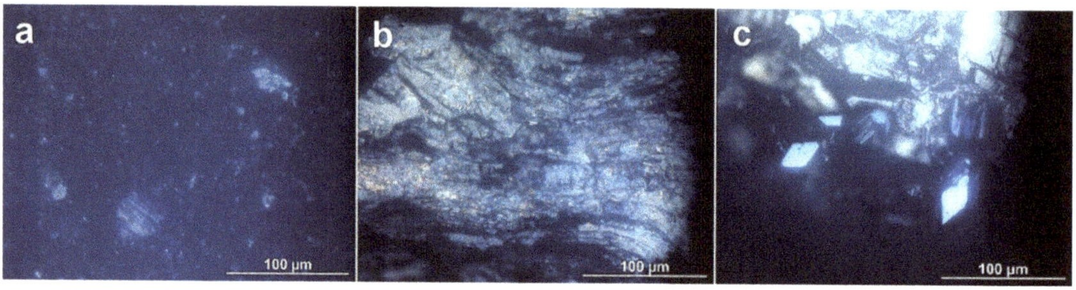

Figure 1. Optical micrographs of GBS_1 (**a**), GBS_2 (**b**), and GBS_3 (**c**).

It is known that stage 1 and stage 2 GBS phases are coloured. Stage 1 GBS is characterized by a deep blue colour, and stage 2 GBS particles are white, whereas mixed stage zones (they are also called intermediate phases) can be identified by their iridescent colours [33]. Optical microscopy pictures of GBS confirmed successful intercalation using all three graphite precursors (Figure 1). Both blue and white particles of various sizes as well as blurred iridescent areas (Figure 1b,c) can be observed. In GBS_1 (Figure 1a), the proportion of bluish particles is the highest. Optical micrographs of the GBS_2 and GBS_3 compounds (Figure 1b,c) show the presence of larger grains of both white and blue colours, as well as iridescent zones in smaller particles. The obtained results are in good agreement with those published by Dimiev et al. [33]. Thus, we can assume that the use of graphite precursor composed of smaller particles is preferable for obtaining a stage 1 GBS, where the intercalant is interposed between each graphene layer and can be represented by the formula $C_{24}{}^{+}{\cdot}HSO_4{}^{-}{\cdot}2H_2SO_4$ [34].

SEM micrographs were used to assess the structure of the EG and GP samples and compare it with that of pristine graphite (Figure 2). The graphite precursors (Figure 2a–c) can be characterized by the lamellar structure and flat surfaces with regular edges. The EG samples after heating also retain their likewise lamellar structure; however, the edges of the layers are less regular and more jagged (Figure 2d–i). When analysing the SEM micrographs, one can see an increase in morphological changes along the direction (Gr_1, Gr_2, Gr_3) < (EG_1, EG_2, EG_3) $\leq$ (EG_1_MW, EG_2_MW, EG_3_MW) < (EG_1_Na, EG_2_Na, EG_3_Na) < (GP_1, GP_2, GP_3), which also depend on the particle size. The phenomenon of partial delamination is typical for the EG and EG_MW samples (Figure 2d–i), whereas the splitting of the basal graphene planes can be seen in the micrographs of the EG_Na samples (Figure 2j–l). During the procedures of thermal shock and microwave treatment, the molecules of H_2SO_4 evaporate and expand the graphite structure. In the case of microwave treatment, the edges of the basal planes are more jagged, presumably due to the presence of oxygen in the reaction zone. The EG_Na samples are obtained using the liquid phase exfoliation protocol (see Section 2.2), when Na_2SO_4 is added to the reaction mixture. $NaHSO_4$ formed in the solution of H_2SO_4 is eutectic at room temperature and is suitable as an ionic liquid for exfoliation [35]. Yao et al. found $NaHSO_4$ as an efficient exfoliation agent for covalent organic frameworks, which are morphologically similar to graphene [36]. SEM micrographs (Figure 2j–l) show that the procedure of liquid exfoliation in the eutectic of the H_2SO_4–$NaHSO_4$ matrix is effective to reduce the particle size; however, the process of particle size reduction occurs not by exfoliation, but by splitting of the basal planes. The morphology of GP samples obtained by mechanochemical synthesis of GBS products with pyrrole is significantly different from the other samples of pristine graphite and other EG products (Figure 2m–o). An obvious layered structure is no longer visible. It is replaced by smaller particles of various sizes and individual sheets. Furthermore, in the micrographs of the GP samples round, regular-shaped particles can be seen, indicating the presence of the formed PPy. Similar morphology was reported by Wu and co-authors [37]. Therefore, it can be argued that the core–shell nanocomposite structure, where the EG nanoparticles are covered with a PPy shell, was obtained during the mechanochemical synthesis procedure. In our future research, we are planning to apply the TEM technique for more detailed examination of the composite structure.

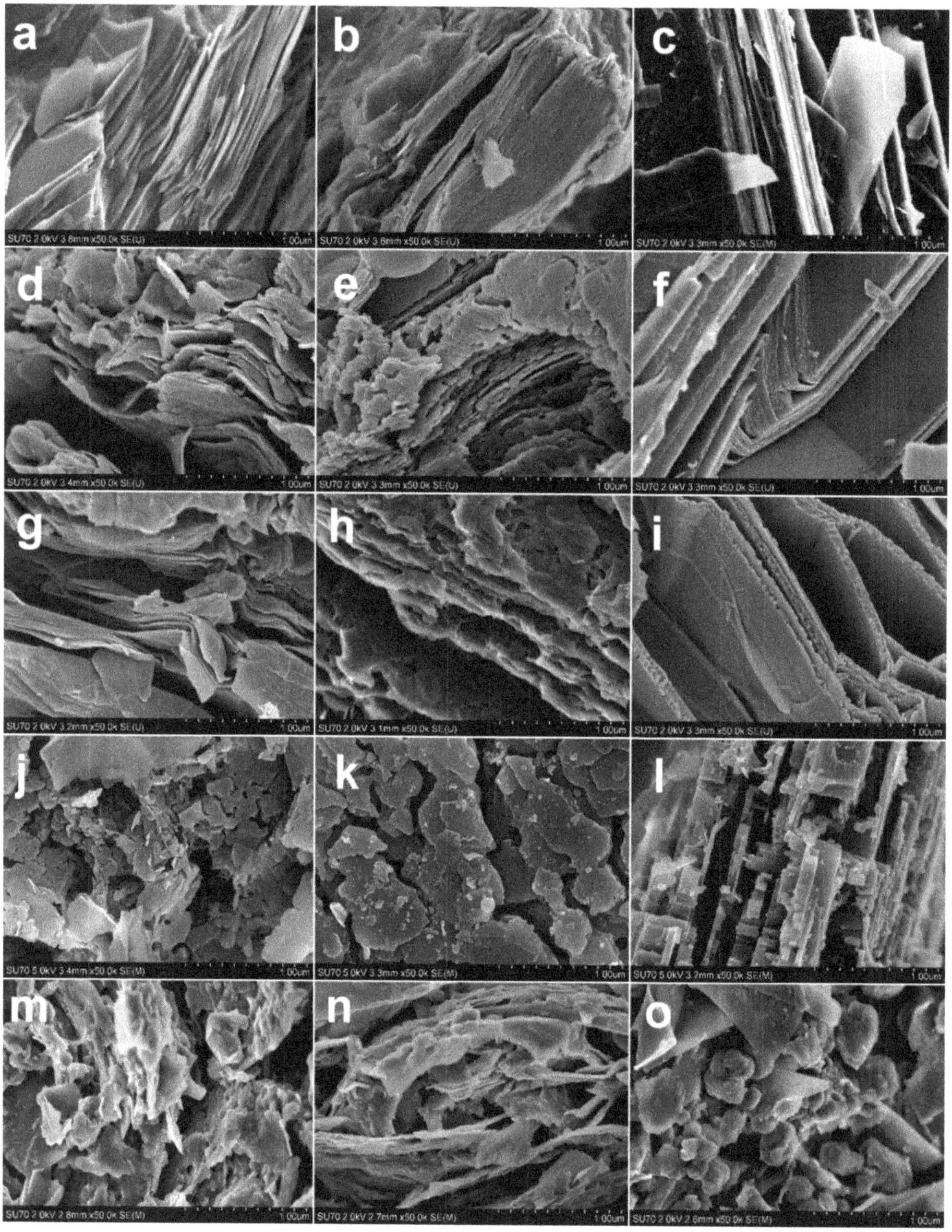

Figure 2. SEM micrographs of precursor graphite of different grain sizes, EG and GP products: Gr_1 (**a**), Gr_2 (**b**), Gr_3 (**c**), EG_1 (**d**), EG_2 (**e**), EG_3 (**f**), EG_1_MW (**g**), EG_2_MW (**h**), EG_3_MW (**i**), EG_1_Na (**j**), EG_1_Na (**k**), EG_1_Na (**l**), GP_1 (**m**), GP_2 (**n**), and GP_3 (**o**).

3.2. XRD Analysis

More detailed analysis of the structure of pristine graphite, EG, and nanocomposite GP samples was performed using the XRD technique. The obtained XRD patterns are shown in Figure 3. The values of the peak positions, the interplanar spacing d_{002} and the crystallite size L are presented in Table 1. Both pristine graphite as well as EG and nanocomposite samples exhibit a characteristic diffraction peak for the (002) plane. Therefore, we can conclude that original graphitic structure was maintained during the procedure of exfoliation or nanocomposite formation [38]. More precisely, the shift of the XRD peak and d_{002} values determined using Bragg's equation is presented in Table 1. The interplanar distance of all pristine graphite samples equals 0.336 nm; meanwhile, the crystallite size varies from 31 to 68 nm. The pristine graphite sample Gr_3 with the largest particle size (≤2000 μm) is characterized by larger crystallites, whereas the other two samples with smaller particle sizes (150–800 μm and <50 μm) are characterized by the crystallites of almost the same size. Thermal exfoliation of all GBS samples prepared from three different pristine graphite samples results in a slight increase in interplanar spacing. The most pronounced increase is for EG_1—the sample prepared from the pristine graphite with the smallest particle size. Meantime, for EG_3, the sample obtained from Gr_3 with the largest particle size, this increase is insignificant. This result is comparable to that previously obtained from optical microscopy; the amount of stage 1 GBS is found to be the highest in GBS_1, which is prepared from Gr_1. Therefore, the small particle size of Gr_1 is beneficial for the preparation of GBS stage 1, which in turn is the most suitable for efficient thermal exfoliation. In the case of EG_3, it is possible that part of pristine graphite did not exfoliate and retained the original graphitic structure [38]. Analysis of the XRD data shows a slight increase in d_{002} spacing equally for all three EG_MW samples. The crystallite size L is significantly reduced after MW exfoliation; it is the most significant in the case of EG_3_MW, which is obtained from Gr_3 with the largest particle size. Presumably, the disintegration of the basal plane of graphene should be more intense in the case of MW exfoliation in comparison with that of thermal exfoliation.

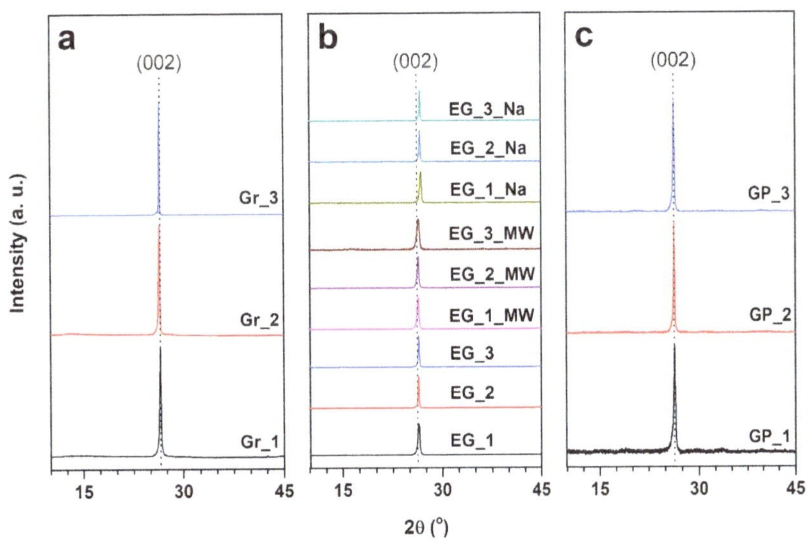

Figure 3. XRD patterns of pristine graphite (**a**), EG (**b**), and EG samples modified with PPy (**c**).

Table 1. Data obtained from XRD analysis (positions, interplanar spacing d_{002}, and crystallite size), and Raman spectra (I_D/I_G ratio).

Sample	2θ (deg)	d_{002} (nm)	L (nm)	I_D/I_G
Gr_1	26.54	0.336	31.29	0.297
Gr_2	26.52	0.336	33.80	0.104
Gr_3	26.50	0.336	62.87	0.028
EG_1	26.38	0.338	20.54	0.185
EG_2	26.45	0.337	40.21	0.114
EG_3	26.52	0.336	39.09	0.093
EG_1_MW	26.43	0.337	23.68	0.259
EG_2_MW	26.46	0.337	24.10	0.114
EG_3_MW	26.42	0.337	16.97	0.118
EG_1_Na	26.81	0.332	24.03	0.196
EG_2_Na	26.74	0.333	36.92	0.100
EG_3_Na	26.76	0.333	44.18	0.065
GP_1	26.41	0.337	26.35	0.580
GP_2	26.46	0.337	36.42	0.483
GP_3	26.45	0.337	32.32	0.504

As a conclusion, one can expect that it is better to use thermal shock than MW exfoliation to obtain graphene layers with less defect concentration. The liquid phase exfoliation in the melt of $NaHSO_4$ results in the contraction of the interplanar distance. Some authors (e.g., A.J. Cooper et al.) attribute this phenomenon to planes narrowing as a result of neighbouring planes separating [39]. Together with the results of the SEM micrographs (Figure 2j–l), one could suggest that the number of defects after liquid exfoliation in the melt of $NaHSO_4$ should be higher in comparison with that obtained after thermal exfoliation. For this reason, thermally exfoliated samples were further used for the preparation of nanocomposites with PPy. The procedure of mechanochemical synthesis affects the structure of GP samples. The interplanar distance d_{002} of all three GP samples converges to the value of 0.337 nm; this may be the effect of the PPy coverage on the EG particles. The crystallite size L either remains the same or is slightly reduced after the procedure of mechanochemical synthesis.

3.3. Raman Spectroscopy Analysis

Raman spectroscopy was used to assess the amount of stacked graphene layers, defectiveness of precursors and GBS products, and the precise staging indices of intercalated products [40–43]. Raman spectra in G (1500–1700 cm^{-1}) and 2D (2500–2900 cm^{-1}) band frequencies are provided in Figure 4. Continuous Raman spectra running from 1200 cm^{-1} to 3200 cm^{-1} are shown in the Supplementary Information file (Figure S1).

One can notice the presence of three characteristic bands in almost all pristine and synthesized samples: D, G and 2D (Figure S1). The D band at about 1350 cm^{-1} is caused by defects in the graphene layer (A_{1g} symmetry mode), the G band at about 1581 cm^{-1} is designated to the first order scattering of the E_{2g} mode that arises from the sp^2 bonded carbon, and the 2D mode at approximately 2709 cm^{-1} is an overtone of the D band. The 2D band is Raman-allowed and defects or disorders are not required to activate this mode [40,42–48]. In addition to the already mentioned bands, at 1620 cm^{-1}, a low intensity mode D′ can be seen as a shoulder at the high-frequency side of the G peak in the Raman spectra of some samples (Gr_1, EG_1, EG_MW_1, EG_1_Na, and EG_2_Na; see Figure 4a,c,d,e). This band can appear when accidentally spread impurities interact with extended graphene phonon modes [49]. Therefore, one can suggest that impurities present in the pristine graphite with the smallest particle size Gr_1 have a significant impact on defect formation and remain noticeable in exfoliated GBS products.

The Raman spectra of the synthesized GBS samples are shown in Figure 4b and Figure S1b. Compared with graphite, the shift of the G peak towards the higher wavenumbers can be seen. This shift, which indicates the process of graphite intercalation, occurs when the graphene layers are charged with intercalant molecules [33,50]. Another attribute observed in the Raman spectra of the GBS samples is the splitting of the G band. The splitting occurs when the charged graphite layers adjacent to the intercalated layers are differentiated from those uncharged next to the empty galleries [51]. Peaks G1 at about 1630 cm^{-1} and G2 at about 1619 cm^{-1} represent stage 1 and stage 2 GBS phases, respectively [20]. A comparison of the intensities of the G1 and G2 peaks, evaluated by the fit procedure, was used to quantify the different intercalation phases in the GBS samples [52]. The GBS_2 sample contains more than half of stage 2 compounds (62.47%), and GBS_3 was composed of the majority of stage 2 phases (84.42%). However, Raman analysis reveals that the GBS_1 sample contains an almost pure GBS phase of stage 1, which is characterized by an apparent peak at 1630 cm^{-1}. From that, one can assume that the particle size of pristine graphite is substantial in determining the constitution of GBS phases of various staging indices.

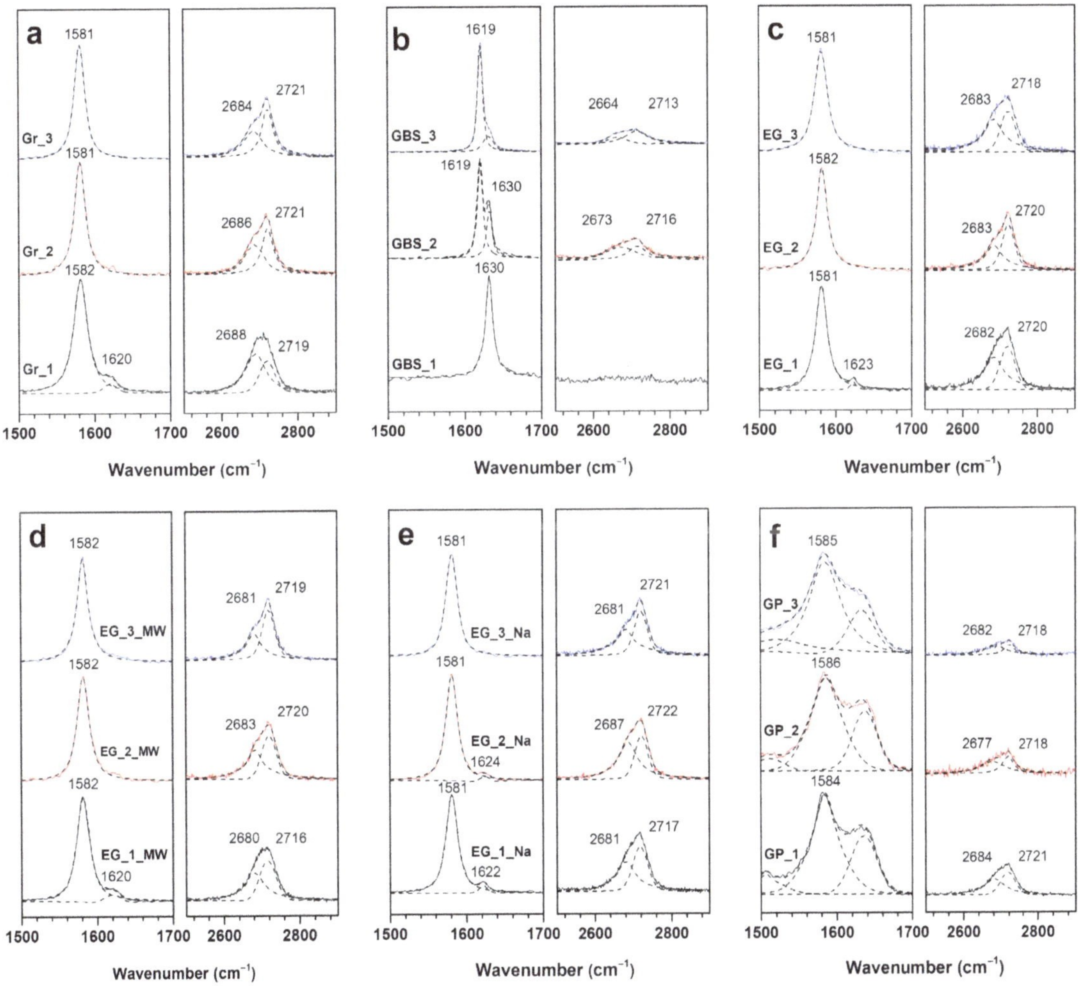

Figure 4. Raman spectra of precursors and GBS products in G (1500–1700 cm^{-1}) and 2D (2500–2900 cm^{-1}) band frequencies: pristine graphite of different grain sizes (**a**), GBS samples (**b**), EG samples (**c**), EG_MW samples (**d**), EG_Na samples (**e**), and EG samples modified with PPy (**f**).

Heated microwave treated samples (Figure 4c–e and Figure S1c–e) show the only one G peak at approximately 1581 cm^{-1}. Together with the XRD data, this supports the idea that the elements of the graphite structure are restored after the exfoliation procedure [53]. The Raman spectra of nanocomposite GP samples (Figure 4f and Figure S1f) are very different from those previously discussed, yet, they coincide with those reported by the other authors, where composites of graphitic materials with PPy were examined [54]. In the range of 900–1700 cm^{-1}, new appeared peaks can be seen. Although some peaks overlap, the existence of D, G, and 2D bands in the GP spectrum is observed at approximately 1352 cm^{-1}, 1585 cm^{-1}, and 2705 cm^{-1}, respectively. Furthermore, some broad peaks at approximately 1410 cm^{-1} (C–N stretching), 1056 cm^{-1}, and 987 cm^{-1} (C–H ring deformation vibration) are associated with the characteristics of PPy [55].

The I_D/I_G ratio (ratio of the intensity of the D peak and the G peak) is an important Raman spectra parameter, which is used as an indicator of the defect concentration in graphene-like materials. The values of I_D/I_G determined for pristine graphite, EG, and nanocomposite samples are presented in Table 1. Comparing the obtained results with the data of other authors, it can be concluded that the EG samples are characterized by a low concentration of defects [56]. Comparing pristine graphite samples, the lowest concentration of defects is found in Gr_3 with the largest particle and crystallite size. In general, a pronounced inverse correlation can be observed between the I_D/I_G ratio and crystallite size in each group of thermally exfoliated samples, whereas MW exfoliated and GP nanocomposite samples show no correlation between these two parameters. The MW exfoliated samples can be characterized by a lower crystallite size, whereas the GP samples—by a higher I_D/I_G ratio. Therefore, the MW exfoliation and mechanochemical synthesis can be considered as the procedures affecting the structure of the GBS products in the most significant results.

A characteristic feature emerges in the Raman spectra of the GBS samples—the absence of the D band (Figure S1b). However, the low-intensity 2D band, which is an overtone of the D band, remains in the Raman spectra of the GBS_2 and GBS_3 samples. Apparently, the use of KIO_4 oxidizer during the preparation of GBS reduces the defect concentration in a graphene layer. Therefore, the proposed synthesis protocol could be considered as a prospective intermediate stage in the preparation of graphene with a low defect concentration. The disappearance of the D band in graphite intercalation compounds is rather rarely mentioned by the other authors. Hardwick and co-authors observed this phenomenon during formation of the Li intercalation compound [57], and they attributed it either to the change in stacking order (from ABAB to AAAA) during the intercalant formation, or to solid interphase formation on the surface of crystallites. Dimiev et al. [20] also reported the phenomenon of D band disappearance during the synthesis of GBS. The main conclusion of their work was that the D band disappears in the Raman spectra of GBS consisting of pure stage 1 or stage 2, whereas in the transition stages this band appears. Considering our results (see Figure 4b), we can suggest that using a KIO_4 oxidizer, the pure GBS of stage 1 was synthesized from the Gr_1 precursor with the smallest particle size. Precursors Gr_2 and Gr_3, composed of larger graphite particles, yield the mixture of stage 1 and stage 2. Furthermore, we plan for a more detailed investigation of GBS samples and their Raman spectra, which would allow for a more detailed explanation of the phenomenon of disappearance of the D band.

The values of FWHM(2D) (full width at half maximum of the 2D band) can be used to identify the number of layers in graphene-like material [26,46]. It was determined that the FWHM(2D) value of a single layer graphene is approximately 30 cm^{-1}. When the number of stacked graphene layers increases, the values of FWHM(2D) also increases [20]. In addition, the shape of the 2D band can help to distinguish few-layer graphene from multilayer. Monolayer graphene has a single 2D component, bilayer graphene can be fitted into four components, and the 2D band of the multilayer graphene can be decomposed into two components. The calculated FWHM(2D) values of the synthesized samples (Figure 4) exceed 60 cm^{-1} and can be best fitted with two components, indicating that the materials obtained in this work have a multilayer structure. Different from the other samples, the 2D band is not visible in the Raman spectra of GBS_1. The other Raman data show that GBS_1 consists of a rather pure stage 1 phase. As the 2D peak is very sensitive to the stacking order of the graphite material along the c-axis, the decrease in the 2D band may occur due to the breaking of the mentioned stacking order [58]. The disappearance of the 2D band is an additional argument for the existing of AAAA stacking order in the GBS stage 1 phase.

3.4. XPS Analysis

Pristine graphite and EG and nanocomposite GP samples were analysed using XPS. Data about surface elemental composition of all analysed samples are summarised in Figure 5.

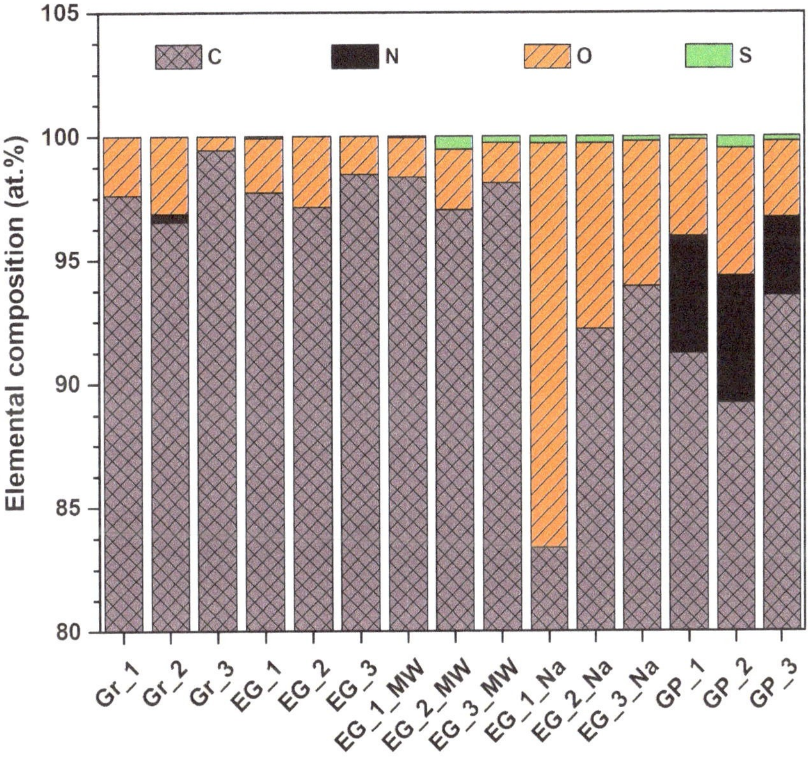

Figure 5. Surface elemental composition of pristine graphite and EG and nanocomposite samples determined by XPS analysis.

It should be noted that the difference in elemental composition is evident beginning from the pristine graphite samples although they are composed mostly of carbon but also contain traces of oxygen and nitrogen. The relatively high nitrogen content in the Gr_2 sample (0.33%) could be attributed to its synthetic nature. This precursor sample also contains the highest amount of oxygen. High resolution spectra in the C 1s region show that the oxygen is present in both graphite precursor samples as well as in GBS products in the form of functional groups attached to the edges of the graphene basal planes (see Figure S2). The signals of the C–O, C–O–C, C=O, and COOR groups are found in different proportions in all samples analysed. The intercalation procedure with subsequent both thermal and MW exfoliation, in principle, does not change the oxygen content in the analysed samples. On the contrary, the procedure of liquid phase exfoliation in the melt of $NaHSO_4$ results in a significant increase in the oxygen content. This increase is inversely proportional to the particle size. Presumably, it occurs due to the reaction between graphite particles and the remains of the KIO_4 oxidizer in GBS samples during the exfoliation procedure. Another result to be noted is the increase in the oxygen content in the nanocomposite GP samples. One can suggest that the reason for this increase is the presence of the $(NH_4)_2S_2O_8$ oxidizer during the mechanochemical reaction.

Some samples (EG_MW, EG_Na and GP) contain traces of sulfur. MW-treated samples can contain traces of H_2SO_4 not removed, whereas $NaHSO_4$-treated samples can contain traces of Na_2SO_4. The source of sulfur in nanocomposite GP samples should be the $(NH_4)_2S_2O_8$ oxidizer present in the mechanochemical reaction.

The nitrogen content is significantly increased in the EG nanocomposite samples, which is a consequence of PPy shell formation on the surface of the EG core. The highest concentration of N is found in GP_2 (5.12%), whereas the lowest—in GP_3 (3.15%). The most probable explanation of this should be the highest concentration of oxygen and nitrogen in the pristine Gr_2, which is beneficial for the formation of PPy.

The high-resolution spectra in the N 1s region of the nanocomposite GP samples are presented in Figure 6.

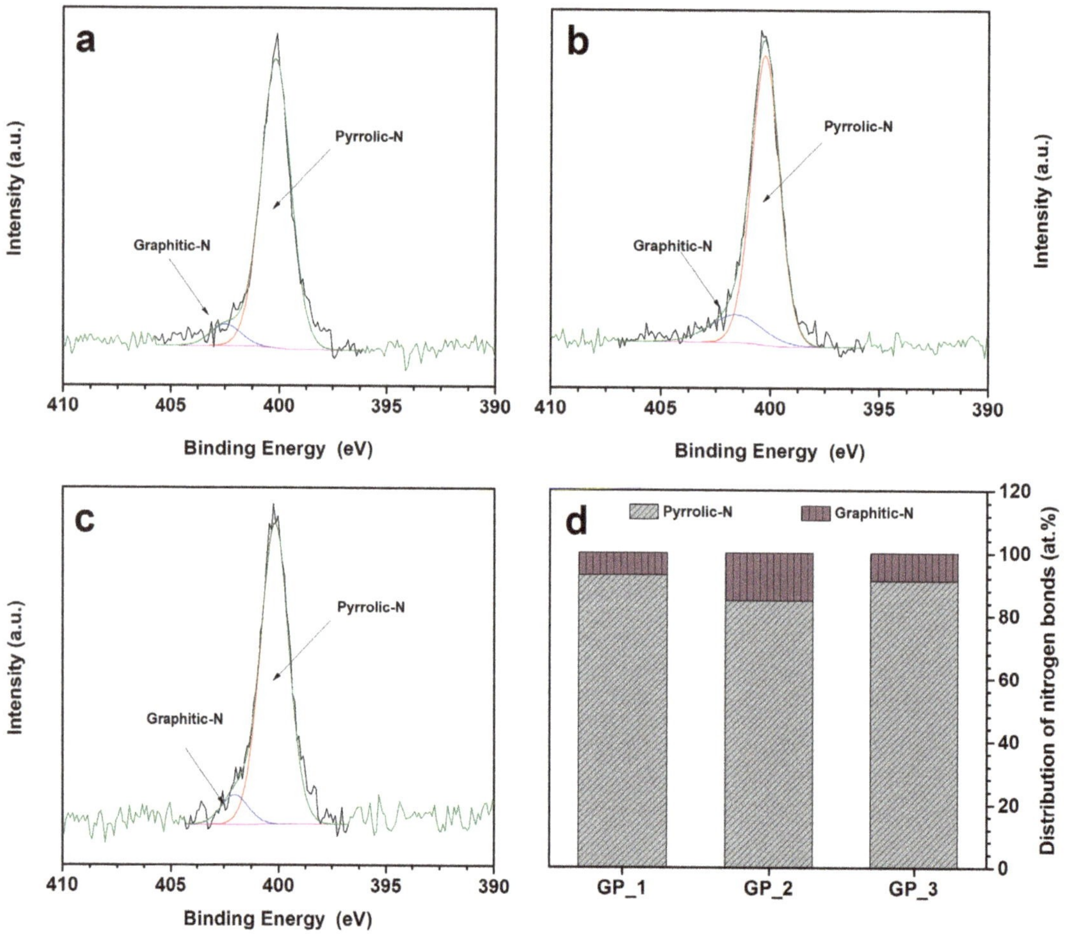

Figure 6. N 1s XPS spectra of nanocomposite GP samples: GP_1 (**a**), GP_2 (**b**), and GP_3 (**c**). Distribution of nitrogen bonds in GP samples (**d**).

Deconvolution of N 1s shows two nitrogen functional groups in the GP samples. The peak at around 400.0 eV is assigned to pyrrolic-N and the peak at around 402.0 eV belongs to graphitic-N. Pyrrolic-N refers to nitrogen atoms that are bonded to two carbon atoms and contribute to the π-system with two p-electrons. Graphitic-N refers to nitrogen atoms incorporated into graphene [59–61]. In this study, the pyrrolic-N has the most intense peak for all GP samples. It is evident that the main part of nitrogen in nanocomposite GP samples should be derived from PPy. Graphitic-N either may be found in pristine graphite samples (Gr_2) or can be formed near the graphitic core in the PPy shell. Huang et al. reported that graphitic-N contributes more effectively to catalytic activity [62]. Taking into account the data in Figure 6d, one can suggest that the GP_2 sample should provide the best electrocatalytic performance.

3.5. *Electrochemical Study*

The electrochemical study of synthesized EG, EG_MW, EG_Na, and nanocomposite GP samples was performed using CV. It is a standard technique for the characterization of the capacitive behaviour of cathode or anode materials [63]. CV at bare GCE is presented in Figure S3, whereas CVs obtained using modified GCEs are shown in Figure 7. A reduction of the oxy-species is observed at −0.8 V at bare GCE. GCEs modified with EG, EG_MW, and EG_Na samples showed a weak electrocatalytic response (Figure 7a–c), i.e., peak potential is observed at almost the same position as an unmodified electrode. Thereby, they are unsuitable as effective electrode materials. In Figure 7d, representing the CVs at nanocomposite GP samples, a cathodic peak at around –0.36 V and a new anodic peak at ca. –0.18 V are observed.

Therefore, the nanocomposite GP samples with a core–shell structure have much better electrocatalytic activity in comparison with those without a PPy shell. As seen, the most pronounced cathodic and anodic peaks are obtained using the GP_2 sample. As mentioned, this sample contains the highest amount of graphitic-N. The cathodic and anodic peaks appear due to the electron transfer associated with functional groups present on the surface and the incorporation of N atoms into graphene layers, since it could have provided the formation of active sites suitable for electrochemical events [64,65]. In addition, an increase in the electrochemical performance of GCE after modification with GP_1, GP_2, and GP_3 samples could be caused by good electrical conductivity acquired after the wet synthesis with the conducting polymer PPy. Thus, electrochemical investigations demonstrated that GP nanocomposites could be prospective for energy storage and/or biosensing.

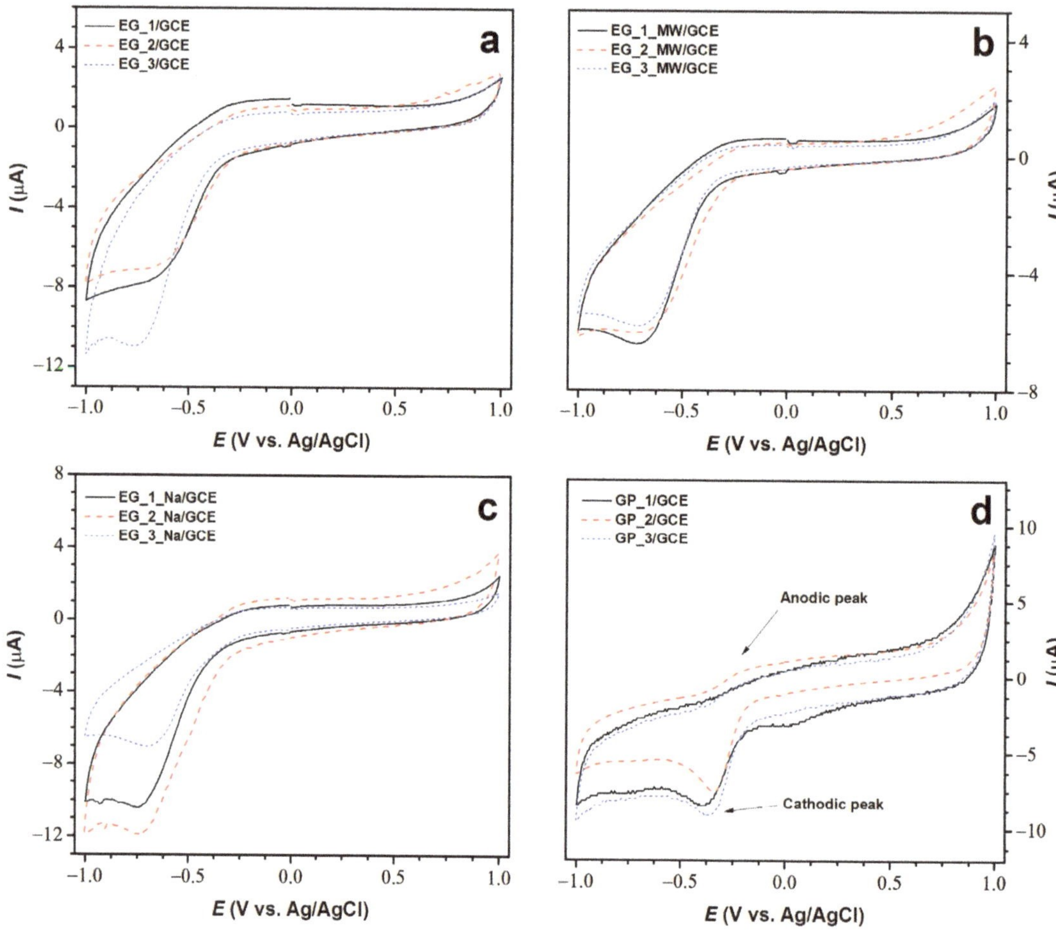

Figure 7. CVs of EG/GCE (**a**), EG_MW/GCE (**b**), EG_Na/GCE (**c**) and GP/GCE (**d**) samples.

4. Conclusions

The graphite bisulfate intercalation compound was prepared from three graphite precursors with different grain sizes using the KIO_4 oxidising agent to assist the intercalation. The intercalation compounds obtained were further annealed, subjected to microwave irradiation, or stirred in the $NaHSO_4$ melt to achieve the delamination/exfoliation effect. The heat-treated samples were used for the preparation of nanocomposites with polypyrrole by a wet synthesis protocol via a mechanochemical reaction. It was found that the particle size of the graphite precursor is substantial in determining the constitution of GBS phases of various staging indices; the smaller particles are beneficial for the formation of the stage 1 phase, whereas the larger particles result in the formation of the mixture of stage 1 and stage 2. The characteristic D band, which is caused by defects in the graphene layer, disappears in all Raman spectra of the GBS samples. Apparently, the use of the KIO_4 oxidizer during the preparation of GBS reduces the defect concentration in a graphene layer. Thus, the proposed synthesis protocol could be considered as a prospective intermediate stage in the preparation of graphene with a low defect concentration. Therefore, this work contributes to the development of cost-effective, scalable, and highly efficient intercalation methods, which remain a significant challenge. Furthermore, the obtained results suggest that there was a structural change during the GBS formation reaction; the stacking order should change from ABAB to AAAA. In addition to Raman spectroscopy, SEM, XRD, and XPS data demonstrate that the exfoliation of GBS using thermal shock, microwave irradiation, or liquid phase delamination leads to a more defective structure than that of pristine graphite. It was established that the nanocomposites prepared from exfoliated graphene and polypyrrole have a core–shell structure, where the particle core is composed of exfoliated graphene covered by a polypyrrole shell. The electrochemical investigation of exfoliated samples and nanocomposites showed that the nanocomposite samples exhibit cathodic and anodic peaks and electrocatalytic properties. Among the samples of pristine graphite, the graphite containing the highest amount of graphitic-N has the highest electrocatalytic efficiency. Therefore, synthesized nanocomposites could be prospective materials for energy storage and/or biosensing.

Supplementary Materials: The following supporting information can be downloaded at: https://www.mdpi.com/article/10.3390/cryst12121793/s1, Figure S1. Raman spectra of graphite (a), GBS products (b), EG products (c), EG_MW products (d), EG_Na products (e), and EG samples modified with PPy (f). Figure S2. C 1s XPS spectra of pristine graphite, EG, and nanocomposite GP samples. Figure S3. CV at bare GCE.

Author Contributions: Conceptualization, J.G. and J.B.; methodology, J.G., R.P. and J.B.; software, G.R. and J.G.; validation, J.G., G.N. and R.P.; formal analysis, G.R. and G.N.; investigation, G.R. and J.G.; data curation, J.G. and J.B.; writing—original draft preparation, G.R.; writing—review and editing, J.G. and J.B.; visualization, G.R. and J.G.; supervision, J.G. and J.B. All authors have read and agreed to the published version of the manuscript.

Funding: This research received no external funding.

Data Availability Statement: Data are contained within the article.

Conflicts of Interest: The authors declare no conflict of interest.

References

1. Garcia-Hernandez, M.; Coleman, J. Materials science of graphene: A flagship perspective. *2D Mater.* **2016**, *3*, 010401. [CrossRef]
2. Lee, J.H.; Lee, E.K.; Joo, W.J.; Jang, Y.; Kim, B.S.; Lim, J.Y.; Choi, S.H.; Ahn, S.J.; Ahn, J.R.; Park, M.H.; et al. Wafer-scale growth of single-crystal monolayer graphene on reusable hydrogen-terminated germanium. *Science* **2014**, *344*, 286–289. [CrossRef] [PubMed]
3. Hao, Y.; Bharathi, M.S.; Wang, L.; Liu, Y.; Chen, H.; Nie, S.; Wang, X.; Chou, H.; Tan, C.; Fallahazad, B.; et al. The role of surface oxygen in the growth of large single-crystal graphene on copper. *Science* **2013**, *342*, 720–723. [CrossRef] [PubMed]
4. Eigler, S.; Hirsch, A. Chemistry with graphene and graphene oxide challenges for synthetic chemists. *Angew. Chem. Int. Ed.* **2014**, *53*, 7720–7738. [CrossRef] [PubMed]
5. Georgakilas, V.; Otyepka, M.; Bourlinos, A.B.; Chandra, V.; Kim, N.; Kemp, K.C.; Hobza, P.; Zboril, R.; Kim, K.S. Functionalization of graphene: Covalent and non-covalent approaches, derivatives and applications. *Chem. Rev.* **2012**, *112*, 6156–6214. [CrossRef]
6. Mao, H.Y.; Laurent, S.; Chen, W.; Akhavan, O.; Imani, M.; Ashkarran, A.A.; Mahmoudi, M. Graphene: Promises, facts, opportunities, and challenges in nanomedicine. *Chem. Rev.* **2013**, *113*, 3407–3424. [CrossRef]
7. Dreyer, D.R.; Todd, A.D.; Bielawski, C.W. Harnessing the chemistry of graphene oxide. *Chem. Soc. Rev.* **2014**, *43*, 5288–5301. [CrossRef]
8. Dreyer, D.R.; Park, S.; Bielawski, C.W.; Ruoff, R.S. The chemistry of graphene oxide. *Chem. Soc. Rev.* **2010**, *39*, 228–240. [CrossRef]
9. Criado, A.; Melchionna, M.; Marchesan, S.; Prato, M. The covalent functionalization of graphene on substrates. *Angew. Chem. Int. Ed.* **2015**, *54*, 0734–10750. [CrossRef]
10. Eigler, S.; Dotzer, C.; Hirsch, A. Visualization of defect densities in reduced graphene oxide. *Carbon* **2012**, *50*, 3666–3673. [CrossRef]
11. Du, W.; Geng, H.; Yang, Y.; Zhang, Y.; Rui, X.; Li, C.C. Pristine graphene for advanced electrochemical energy applications. *J. Power Sources* **2019**, *437*, 226899. [CrossRef]
12. Kovtyukhova, N.I.; Wang, Y.; Berkdemir, A.; Cruz-Silva, R.; Terrones, M.; Crespi, V.H.; Mallouk, T.E. Non-oxidative intercalation and exfoliation of graphite by Brønsted acids. *Nat. Chem.* **2014**, *6*, 957–963. [CrossRef]
13. Aghamohammadi, H.; Eslami-Farsani, R.; Torabian, M.; Amousa, N. Recent advances in one-pot functionalization of graphene using electrochemical exfoliation of graphite: A review study. *Synth. Met.* **2020**, *269*, 116549. [CrossRef]
14. Ruse, E.; Larboni, M.; Lavi, A.; Pyrikov, M.; Leibovitch, Y.; Ohayon-Lavi, A.; Vradman, L.; Regev, O. Molten salt in-situ exfoliation of graphite to graphene nanoplatelets applied for energy storage. *Carbon* **2021**, *176*, 168–177. [CrossRef]
15. Du, W.; Jiang, X.; Zhu, L. From graphite to graphene: Direct liquid-phase exfoliation of graphite to produce single- and few-layered pristine graphene. *J. Mater. Chem. A* **2013**, *1*, 10592. [CrossRef]

16. Xu, Y.; Cao, H.; Xue, Y.; Li, B.; Cai, W. Liquid-phase exfoliation of graphene: An overview on exfoliation media, techniques, and challenges. *Nanomaterials* **2018**, *8*, 942. [CrossRef]
17. Brodie, B. Note sur un nouveau procédé pour la purification et la désagrégation du graphite. *Ann. Chim. Phys.* **1855**, *45*, 351–448.
18. Aronson, S.; Frishberg, C.; Frankl, G. Thermodynamic properties of the graphite-bisulfate lamellar compounds. *Carbon* **1971**, *9*, 715–723. [CrossRef]
19. Daumas, N.; Herold, A. Notes des membres et correspontants et notes présentées ou transmises par leurs soins. *C. R. Acad. Sci. Ser. C* **1969**, *268*, 373–375.
20. Dimiev, A.M.; Ceriotti, G.; Behabtu, N.; Zakhidov, D.; Pasquali, M.; Saito, R.; Tour, J.M. Direct real-time monitoring of stage transitions in graphite intercalation compounds. *ACS Nano* **2013**, *7*, 2773–2780. [CrossRef]
21. Chiang, C.K.; Fincher, C., Jr.; Park, Y.W.; Heeger, A.J.; Shirakawa, H.; Louis, E.J.; Gau, S.C.; MacDiarmid, A.G. Electrical conductivity in doped polyacetylene. *Phys. Rev. Lett.* **1977**, *39*, 1098. [CrossRef]
22. Ohta, T.; Bostwick, A.; Seyller, T.; Horn, K.; Rotenberg, E. Controlling the electronic structure of bilayer graphene. *Science* **2006**, *313*, 951–954. [CrossRef] [PubMed]
23. Mouri, S.; Miyauchi, Y.; Matsuda, K. Tunable photoluminescence of monolayer MoS2 via chemical doping. *Nano Lett.* **2013**, *13*, 5944–5948. [CrossRef] [PubMed]
24. Tongay, S.; Zhou, J.; Ataca, C.; Liu, J.; Kang, J.S.; Matthews, T.S.; You, L.; Li, J.; Grossman, J.C.; Wu, J. Broad-range modulation of light emission in two-dimensional semiconductors by molecular physisorption gating. *Nano Lett.* **2013**, *13*, 2831–2836. [CrossRef]
25. Novoselov, K.S.; Geim, A.K.; Morozov, S.V.; Jiang, D.; Zhang, Y.; Dubonos, S.V.; Grigorieva, I.V.; Firsov, A.A. Electric field effect in atomically thin carbon films. *Science* **2004**, *306*, 666–669. [CrossRef]
26. Ryu, S.; Liu, L.; Berciaud, S.; Yu, Y.-J.; Liu, H.; Kim, P.; Flynn, G.W.; Brus, L.E. Atmospheric oxygen binding and hole doping in deformed graphene on a SiO_2 substrate. *Nano Lett.* **2010**, *10*, 4944–4951. [CrossRef]
27. Park, K.; Kang, H.; Koo, S.; Lee, D.; Ryu, S. Redox-governed charge doping dictated by interfacial diffusion in two-dimensional materials. *Nat. Commun.* **2019**, *10*, 4931. [CrossRef]
28. Kang, H.; Ryu, S. Optical imaging of chemically and geometrically controlled interfacial diffusion and redox in 2D van der Waals space. *J. Phys. Chem. C* **2021**, *125*, 16819–16826. [CrossRef]
29. Puech, P.; Hu, T.; Sapelkin, A.; Gerber, I.; Tishkova, V.; Pavlenko, E.; Levine, B.; Flahaut, E.; Bacsa, W. Charge transfer between carbon nanotubes and sulfuric acid as determined by Raman spectroscopy. *Phys. Rev. B* **2012**, *85*, 205412. [CrossRef]
30. Lünsdorf, N.K. Raman spectroscopy of dispersed vitrinite—Methodical aspects and correlation with reflectance. *Int. J. Coal Geol.* **2016**, *153*, 75–86. [CrossRef]
31. Zaaba, N.I.; Foo, K.L.; Hashim, U.; Tan, S.J.; Liu, W.W.; Voon, C.H. Synthesis of graphene oxide using modified hummers method: Solvent influence. *Procedia Eng.* **2017**, *184*, 469–477. [CrossRef]
32. Ivanov, A.V.; Maksimova, N.V.; Kamaev, A.O.; Malakho, A.P.; Avdeev, V.V. Influence of intercalation and exfoliation conditions on macrostructure and microstructure of exfoliated graphite. *Mater. Lett.* **2018**, *8*, 403–406. [CrossRef]
33. Dimiev, A.M.; Shukhina, K.; Behabtu, N.; Pasquali, M.; Tour, J.M. Stage transitions in graphite intercalation compounds: Role of the graphite structure. *J. Phys. Chem. C* **2019**, *123*, 19246–19253. [CrossRef]
34. Bonaccorso, F.; Lombardo, A.; Hasan, T.; Sun, Z.; Colombo, L.; Ferrari, A.C. Production and processing of graphene and 2D crystals. *Mater. Today* **2012**, *15*, 564–589. [CrossRef]
35. Freire, M.G.; Cláudio, A.F.M.; Araújo, J.M.M.; Coutinho, J.A.P.; Marrucho, I.M.; Canongia Lopes, J.N.; Rebelo, L.P.N. Aqueous biphasic systems: A boost brought about by using ionic liquids. *Chem. Soc. Rev.* **2012**, *41*, 4966–4995. [CrossRef]
36. Yao, J.; Liu, C.; Liu, X.; Guo, J.; Zhang, S.; Zheng, J.; Li, S. Azobenzene-assisted exfoliation of 2D covalent organic frameworks into large-area, few-layer nanosheets for high flux and selective molecular separation membrane. *J. Membr. Sci.* **2020**, *601*, 117864. [CrossRef]
37. Wu, W.; Yang, L.; Chen, S.; Shao, Y.; Jing, L.; Zhao, G.; Wei, H. Core–shell nanospherical polypyrrole/graphene oxide composites for high performance supercapacitors. *RSC Adv.* **2015**, *5*, 91645–91653. [CrossRef]
38. Hou, B.; Sun, H.J.; Peng, T.J.; Zhang, X.Y.; Ren, Y.Z. Rapid preparation of expanded graphite at low temperature. *New Carbon Mater.* **2020**, *35*, 262–268. [CrossRef]
39. Cooper, A.J.; Wilson, N.R.; Kinloch, I.A.; Dryfe, R.A.W. Single stage electrochemical exfoliation method for the production of few-layer graphene via intercalation of tetraalkylammonium cations. *Carbon* **2014**, *66*, 340–350. [CrossRef]
40. Trusovas, R.; Ratautas, K.; Račiukaitis, G.; Niaura, G. Graphene layer formation in pinewood by nanosecond and picosecond laser irradiation. *Appl. Surf. Sci.* **2019**, *471*, 154–161. [CrossRef]
41. Seiler, S.; Halbig, C.E.; Grote, F.; Rietsch, P.; Börrnert, F.; Kaiser, U.; Meyer, B.; Eigler, S. Effect of friction on oxidative graphite intercalation and high-quality graphene formation. *Nat. Commun.* **2018**, *9*, 836. [CrossRef] [PubMed]
42. Ferrari, A.C.; Robertson, J. Raman spectroscopy of amorphous, nanostructured, diamond-like carbon, and nanodiamond. *Philos. Trans. R. Soc.* **2004**, *362*, 2477–2512. [CrossRef] [PubMed]
43. ChacónChac, J.C.; Wirtz, L.; Pichler, T. Raman spectroscopy of graphite intercalation compounds: Charge transfer, strain, and electron-phonon coupling in graphene layers. *Phys. Status Solidi B* **2014**, *251*, 2337–2355. [CrossRef]
44. Gurzęda, B.; Buchwald, T.; Krawczyk, P. Thermal exfoliation of electrochemically synthesized graphite intercalation compound with perrhenic acid. *J. Solid State Electrochem.* **2020**, *24*, 1363–1370. [CrossRef]
45. Eigler, S. Graphite sulphate—A precursor to graphene. *Chem. Commun.* **2015**, *51*, 3162–3165. [CrossRef]

46. Pei, S.; Wei, Q.; Huang, K.; Cheng, H.M.; Ren, W. Green synthesis of graphene oxide by seconds timescale water electrolytic oxidation. *Nat. Commun.* **2018**, *9*, 145. [CrossRef]
47. Trusovas, R.; Račiukaitis, G.; Niaura, G.; Barkauskas, J.; Valušis, G.; Pauliukaite, R. Recent advances in laser utilization in the chemical modification of graphene oxide and its applications. *Adv. Opt. Mater.* **2016**, *4*, 37–65. [CrossRef]
48. Ferrari, A.C. Raman spectroscopy of graphene and graphite: Disorder, electron-phonon coupling, doping and nonadiabatic effects. *Solid State Commun.* **2007**, *143*, 47–57. [CrossRef]
49. Douda, J.; González Vargas, C.R.; Basiuk, E.V.; Díaz Cano, A.I.; Fuentes García, J.A.; Hernández Contreras, X.A. Optical properties of amine-functionalized graphene oxide. *Appl. Nanosci.* **2019**, *9*, 567–578. [CrossRef]
50. Zhou, X.; Liu, Q.; Jiang, C.; Ji, B.; Ji, X.; Tang, Y.; Cheng, H.-M. Strategies towards low-cost dual-ion batteries with high performance. *Angew. Chem. Int. Ed.* **2020**, *59*, 3802–3832. [CrossRef]
51. Zou, J.; Sole, C.; Drewett, N.E.; Velický, M.; Hardwick, L.J. In situ study of li intercalation into highly crystalline graphitic flakes of varying thicknesses. *J. Phys. Chem. Lett.* **2016**, *7*, 4291–4296. [CrossRef]
52. Salvatore, M.; Carotenuto, G.; de Nicola, S.; Camerlingo, C.; Ambrogi, V.; Carfagna, C. Synthesis and characterization of highly intercalated graphite bisulfate. *Nanoscale Res. Lett.* **2017**, *12*, 167. [CrossRef]
53. Chung, D.D.L. A review of exfoliated graphite. *J. Mater. Sci.* **2015**, *51*, 554–568. [CrossRef]
54. Okan, B.S.; Yürüm, A.; Gorgülü, N.; Gürsel, S.A.; Yürüm, Y. Polypyrrole coated thermally exfoliated graphite nanoplatelets and the effect of oxygen surface groups on the interaction of platinum catalysts with graphene-based nanocomposites. *Ind. Eng. Chem. Res.* **2011**, *50*, 12562–12571. [CrossRef]
55. Kulandaivalu, S.; Suhaimi, N.; Sulaiman, Y. Unveiling high specific energy supercapacitor from layer-by-layer assembled polypyrrole/graphene oxide | polypyrrole/manganese oxide electrode material. *Sci. Rep.* **2019**, *9*, 4884. [CrossRef]
56. Shams, S.S.; Zhang, R.; Zhu, J. Graphene synthesis: A Review. *Mater. Sci. Pol.* **2015**, *33*, 566–578. [CrossRef]
57. Hardwick, L.J.; Buqa, H.; Novák, P. Graphite surface disorder detection using in situ Raman microscopy. *Solid State Ion.* **2006**, *177*, 2801–2806. [CrossRef]
58. Krishnamoorthy, K.; Veerapandian, M.; Yun, K.; Kim, S.J. The chemical and structural analysis of graphene oxide with different degrees of oxidation. *Carbon* **2013**, *53*, 38–49. [CrossRef]
59. Shao, Y.; Zhang, S.; Engelhard, M.H.; Li, G.; Shao, G.; Wang, Y.; Liu, J.; Aksay, I.A.; Lin, Y. Nitrogen-doped graphene and its electrochemical applications. *J. Mater. Chem.* **2010**, *20*, 7491–7496. [CrossRef]
60. Lai, L.; Potts, J.R.; Zhan, D.; Wang, L.; Poh, C.K.; Tang, C.; Gong, H.; Shen, Z.; Lin, J.; Ruoff, R.S. Exploration of the active center structure of nitrogen-doped graphene-based catalysts for oxygen reduction reaction. *Energy Environ. Sci.* **2012**, *5*, 7936–7942. [CrossRef]
61. Lee, Y.H.; Chang, K.H.; Hu, C.C. Differentiate the pseudocapacitance and double-layer capacitance contributions for nitrogen-doped reduced graphene oxide in acidic and alkaline electrolytes. *J. Power Sources* **2013**, *227*, 300–308. [CrossRef]
62. Huang, Z.; Zhou, H.; Yang, W.; Fu, C.; Chen, L.; Kuang, Y. Three-dimensional hierarchical porous nitrogen and sulfur-codoped graphene nanosheets for oxygen reduction in both alkaline and acidic media. *ChemCatChem* **2017**, *9*, 987–996. [CrossRef]
63. Zhang, Y.; Ma, M.; Yang, J.; Huang, W.; Dong, X. Graphene-based three-dimensional hierarchical sandwich-type architecture for high performance supercapacitors. *RSC Adv.* **2014**, *4*, 8466–8471. [CrossRef]
64. Gaidukevic, J.; Aukstakojyte, R.; Barkauskas, J.; Niaura, G.; Murauskas, T.; Pauliukaite, R. A novel electrochemical sensor based on thermally reduced graphene oxide for the sensitive determination of dopamine. *Appl. Surf. Sci.* **2022**, *592*, 153257. [CrossRef]
65. Oliveira, L.S.; Alba, J.F.G.; Silva, V.L.; Ribeiro, R.T.; Falcão, E.H.L.; Navarro, M. The effect of surface functional groups on the performance of graphite powders used as electrodes. *J. Electroanal. Chem.* **2018**, *818*, 106–113. [CrossRef]

crystals

MDPI

Article

Hydrothermal Synthesis of Well-Defined Red-Emitting Eu-Doped $GdPO_4$ Nanophosphors and Investigation of Their Morphology and Optical Properties

Egle Ezerskyte, Aleksej Zarkov, Vaidas Klimkevicius * and Arturas Katelnikovas *

Institute of Chemistry, Faculty of Chemistry and Geosciences, Vilnius University, Naugarduko 24, LT-03225 Vilnius, Lithuania

* Correspondence: vaidas.klimkevicius@chf.vu.lt (V.K.); arturas.katelnikovas@chf.vu.lt (A.K.)

Abstract: Rare-earth-doped $GdPO_4$ nanoparticles have recently attracted much scientific interest due to the simultaneous optical and magnetic properties of these materials and their possible application in bio-imaging. Herein, we report the hydrothermal synthesis of $GdPO_4$:Eu^{3+} nanoparticles by varying different synthesis parameters: pH, <Gd>:<P> molar ratio, and Eu^{3+} concentration. It turned out that the Eu^{3+} content in the synthesized nanoparticles had little effect on particle shape and morphology. The synthesis media pH, however, has showed a pronounced impact on particle size and distribution, i.e., the nanoparticle length can be adjusted from hundreds to tens of nanometers by changing the pH from 2 to 11, respectively. Increasing the <Gd>:<P> molar ratio resulted in a decrease in nanoparticle length and an increase in its width. The temperature-dependent measurements in the 77–500 K range revealed that the $GdPO_4$:50%Eu^{3+} sample maintains half of its emission intensity, even at room temperature ($TQ_{1/2}$ = 291 ± 19 K).

Keywords: hydrothermal synthesis; morphology control; rare-earth phosphates; red emission; rod-like nanoparticles

Citation: Ezerskyte, E.; Zarkov, A.; Klimkevicius, V.; Katelnikovas, A. Hydrothermal Synthesis of Well-Defined Red-Emitting Eu-Doped $GdPO_4$ Nanophosphors and Investigation of Their Morphology and Optical Properties. *Crystals* **2023**, *13*, 174. https://doi.org/10.3390/cryst13020174

Academic Editor: Zhi Lin

Received: 17 December 2022
Revised: 12 January 2023
Accepted: 14 January 2023
Published: 19 January 2023

1. Introduction

Inorganic nanosized materials, doped with various lanthanide ions, have distinctive chemical and optical properties; thus, they are applied in multiple fields, including catalysis [1,2], temperature sensing [3], magnetic resonance imaging [4,5], biomedicine [6,7], anti-counterfeiting [8,9], dye-sensitized solar cells [10], etc. Such luminescent nanomaterials should possess the following characteristics if they are to be applied practically, i.e., particles should be non-toxic to humans and the environment [11], stable in colloidal suspensions [12], possess the desired morphology and narrow particle size distribution (PSD) [13], possess strong absorption and efficient emission, etc. [14,15]. The inorganic nanosized phosphors are usually synthesized via sol–gel, precipitation, or other wet chemistry routes [16]. Unfortunately, many of these synthesis methods often yield agglomerated or even bulk materials (especially if post-synthesis annealing is performed); therefore, additional milling or crashing is required to obtain nanosized particles. The thermal decomposition or hot-injection synthesis methods are more suitable for nanoparticle preparation, ensuring the reproducibility and narrow PSD of various inorganic nanophosphors [17]. However, such synthesis approaches ignore the principles of green chemistry since they depend on environmentally hazardous precursors and solvents. These methods also lead to the formation of hydrophobic nanoparticles, narrowing down their potential fields of application (e.g., biomedicine) [18,19]. In order to apply such nanoparticles in the biomedical field, additional surface modification is essential to transfer such hydrophobic nanoparticles into the aqueous colloids. There are numerous approaches reporting stabilization of optically active nanoparticles in aqueous dispersions: utilization of surfactants, such as TWEEN or SPAN [20], exchange of ligands using citric acid [21], usage of low molecular weight functional phosphates [22], usage of phosphoryl-PEG derivatives [23], or commercially available polymers, e.g., PVP [24]. Meanwhile, the hydrothermal synthesis method allows researchers to obtaining nanophoshors with various morphologies (both nano- and microrods [25], wires [26], prisms [13], cubes [27], spheres [28], etc.) by altering the synthesis parameters. This method meets the fundamentals of green chemistry since no volatile or toxic materials are released into the environment during and after the synthesis process. Moreover, the luminescent phosphors, synthesized via the hydrothermal method, are hydrophilic, indicating that the solid form of these phosphors can be effortlessly redispersed into an aqueous media. This, in turn, heightens the odds of these materials finding practical use in the biomedical field [12]. Therefore, the hydrothermal synthesis method is considered as a high-ranking chemical engineering tool for synthesizing novel luminescent, electronic, magnetic, and catalytic materials [29]. Recently, the $REPO_4$ nanoparticles, possessing a rhabdophane crystal structure, have gained much scientific interest, especially in the field of luminescent materials [30,31]. Among all rare-earth orthophosphates, $GdPO_4$ shows the most exceptional properties. Gd^{3+} stands out from other trivalent lanthanide ions by having seven unpaired electrons in the $4f$ orbital ($[Xe]4f^7$ electronic configuration) and demonstrating magnetic properties. Materials containing trivalent gadolinium ions are widely used as MRI agents, host lattices for fluorescent lamp phosphors, X-ray intensifying screens, scintillators for X-ray tomography, etc. [32–35]. In the meantime, luminophores doped with trivalent europium are utilized in red fluorescent lamps, LEDs, and as bio-imaging or anti-counterfeit pigments [36–38]. Eu^{3+} possesses six electrons in $4f$ orbital (adopts $[Xe]4f^6$ electronic configuration). When excited, europium(III) ions emit red to reddish-orange light caused by $^5D_0 \rightarrow {}^7F_{0\text{–}4}$ optical transitions [36]. Moreover, due to the spin-forbidden nature of emission transitions, Eu^{3+} possesses longer photoluminescence (PL) lifetimes (in the order from several to a few hundred milliseconds [36,39]) if compared to other rare-earth ions emitting in the red spectral region (for instance, Er^{3+} ($^4F_{9/2} \rightarrow {}^4I_{15/2}$ transition at about 650 nm) $\tau \approx 550$ µs; Tm^{3+} ($^3H_4 \rightarrow {}^3H_6$ transition at about 790 nm) $\tau \approx 700$ µs; Ho^{3+} ($^5F_5 \rightarrow {}^5I_8$ transition at about 660 nm) $\tau \approx 1$ ms [40]). This is an exceptionally advantageous feature, since long lifetimes allow researchers to avoid an undesirable protein autofluorescence (of a short lifetime) by employing time-resolved detection methods. Thus, Eu^{3+}-doped luminescent nanoparticles could be easily detected in biological tissues. Ergo, doping the $GdPO_4$ host lattice with Eu^{3+} could extend the application fields of such unique materials. The combination of both trivalent Gd and Eu ions empowers the creation of dual-modal opto-magnetic inorganic nanoprobes for theranostics. Several interesting papers were published recently regarding the hydrothermal synthesis of $GdPO_4$ nanoparticles; however, most of them yielded long (more than 500 nm) nanorods or nanowires. Such materials are unsuitable for biological applications, requiring sizes < 100 nm [41]. Zhang et al., for instance, reported the hydrothermal synthesis of $GdPO_4$ nanowires 20–200 nm in width and 1−3 µm in length [42]. Interestingly, the monoclinic ($P2_1/n$, #14) phase was obtained even after performing the synthesis in water, which was assigned to a high synthesis temperature (240 °C). Hernandes et al. [43], on the other hand, performed the hydrothermal synthesis of $GdPO_4:Eu^{3+}$ nanowires at 160 °C using glycerol as a co-solvent. In this case, nanowires possessing trigonal structure were obtained with width in tens of nanometers and length in hundreds of nanometers (reaction media pH = 1.6). The authors also showed that increasing pH to 12 yields irregularly shaped nanoparticles which were tens of nanometers in diameter. Yan et al., in turn, demonstrated that the $GdPO_4:Eu^{3+}$ nanoparticle size and shape could be controlled by selecting the appropriate solvent or mixture of solvents [44]. They showed that 50−500 nm and 20−50 nm spheres of $GdPO_4:Eu^{3+}$ can be obtained using dimethylaniline (DMA) and N-methyl-2pyrrolidone as a solvent. On the other hand, rod-like particles were obtained when water was used as a solvent. However, the synthesis duration was three days, which is not in favor of practical applications. It is also interesting to note that the $GdPO_4$ nanowires, obtained by the hydrothermal synthesis method, can be converted into magnetic $GdPO_4$ aerogel, as recently reported by Janulevius et al. [26]. There are only a few techniques that were developed to yield smaller lanthanide orthophosphate nanoparticles in the form of nanospheres [13], nanorods [13,45], or less uniform elongated nanoparticles (length in the range from 100 to 200 nm) [46,47], hexagons (ca. 15 nm) [48], and nanocubes (ca. 75 nm) [49]. Besides the hydrothermal synthesis, the $GdPO_4$ nanoparticles can also be prepared by co-precipitation or sol–gel methods. For instance, Di et al. reported the synthesis of $GdPO_4$ by the aqueous co-precipitation method (80 °C for 12 h), yielding nanowires from 30 to 100 nm in diameter and from several hundred nanometers to several micrometers in length [50]. Unfortunately, such particles are too large for bio-applications. Huang et al., in turn, reported the co-precipitation synthesis of $GdPO_4$ nanorods employing water/alcohol as synthesis media [51]. Surprisingly, the urchin-like structures were obtained at the beginning, consisting of ca. 120 nm needle-shaped particles radiating from the center. However, after several minutes, the urchin-like structure started to collapse, and $GdPO_4{\cdot}H_2O$ hydrogel was formed in the end. The synthesis of urchin-like $GdPO_4{:}Eu^{3+}$ hollow spheres was also reported by Xu et al. The two-step procedure first involved the co-precipitation synthesis of $Gd(OH)CO_3{:}Eu^{3+}$ colloidal spheres followed by hydrothermal synthesis at 180 °C for 24 h, yielding relatively large (ca. 250 nm in diameter) hollow spheres [52]. In 2018, Rosas Camacho et al. reported the sol–gel synthesis of $GdPO_4$ phosphors doped with lanthanide ions [53]. Unfortunately, in order to obtain single-phase materials and eliminate the organic reagents, the annealing step at 1000 °C was performed, resulting in highly agglomerated particles. It should also be noted that Kumar et al. reported the Pechini-type sol–gel synthesis of $GdPO_4$ nanowires (20−50 nm in diameter and from several hundreds of nanometers to several micrometers in length) [54]. However, a mixture of hexagonal and monoclinic phases was obtained. Subsequently, the monoclinic phase was obtained after annealing at 1000 °C for two hours, but the particle size increased even further.

In this study, the manipulation of $GdPO_4$ nanoparticle morphology via the different synthesis parameters, such as the pH of the reaction media, and the initial molar ratio of gadolinium to phosphorus (<Gd>:<P>) were reported. The evolution of the emission and excitation spectra, as well as the average PL lifetime values as a function of Eu^{3+} concentration in the $GdPO_4$ host lattice, was investigated in detail. The temperature-dependent photoluminescence properties of the sample exhibiting the highest PL intensity ($GdPO_4$:50% Eu^{3+}) were also examined and presented in this study.

2. Materials and Methods

Materials used were as follows: Gd_2O_3 (99.99%, Tailorlux, Münster, Germany), Eu_2O_3 (99.99%, Tailorlux, Münster, Germany), $NH_4H_2PO_4$ (≥99%, Carl Roth, Karlsruhe, Germany), tartaric acid (99.99%, Eurochemicals, Vilnius, Lithuania), nitric acid (70%, Eurochemicals, Vilnius, Lithuania), and ammonium hydroxide (30%, Chempur, Karlsruhe, Germany). $Ln(NO_3)_3$ was prepared by dissolving Ln_2O_3 in diluted nitric acid.

All samples were prepared via the hydrothermal synthesis method manipulating only two parameters, i.e., <Gd>:<P> molar ratio and the pH of the reaction mixture. Firstly, two series of undoped $GdPO_4$ samples were produced as follows:

- nine samples were synthesized under a neutral reaction media (pH = 7) using different <Gd>:<P> molar ratios (1:7.5, 1:10, 1:12.5, 1:15, 1:17.5, 1:20, 1:25, 1:30, 1:50);
- nine samples were synthesized under a molar ratio of <Gd>:<P> = 1:10 and different pH of the reaction mixture (2, 3, 4, 5, 6, 8, 9, 10, 11).

The detailed synthesis procedure of $GdPO_4$ samples, doped with Eu^{3+} in alkaline media (pH = 10) at a molar ratio <Gd>:<P> = 1:10, is presented below [13]. Overall, a set of ten $GdPO_4{:}Eu^{3+}$ nanoparticles was prepared where the Eu^{3+} concentration was 0.5, 1, 2.5, 5, 7.5, 10, 20, 50, 75, and 100%.

The synthesis procedure starts with the formation of tartaric acid–Ln^{3+} complex, which was induced by mixing stoichiometric amounts of $Ln(NO_3)_3$ (0.4 M) and tartaric acid (30 mL 0.3 M) aqueous solutions. The obtained mixture was left under magnetic stirring conditions for 30 min at room temperature. Afterward, the pH of the produced solution was adjusted to 10 by adding an NH_4OH solution. Subsequently, 20 mL of freshly prepared aqueous $NH_4H_2PO_4$ solution was poured at once, instantly turning the transparent reaction mixture into the turbid one. The morphology of the $GdPO_4$ nanoparticles depends on the <Gd>:<P> molar ratio; therefore, a different concentration of $NH_4H_2PO_4$ solution was prepared each time since the volume of the solution was kept constant, i.e., 20 mL. Furthermore, the pH of the obtained reaction mixture was again adjusted to 10 using NH_4OH solution and then diluted to 80 mL by adding DI water, followed by adjusting the pH value once again, if required. Consequently, the produced solution was left under magnetic stirring conditions for 30 min at room temperature. Finally, the reaction mixture was poured into a Teflon liner and placed inside the stainless-steel autoclave. The hydrothermal reaction took place at a 160 °C for 24 h. The synthesized particles were centrifuged four times at 10,000 rpm for ten minutes. In between centrifugation cycles, particles were washed under ultrasound conditions using deionized (DI) water. The obtained powders were either dried at 70 °C for 24 h or stored in an aqueous media.

The phase purity of prepared $GdPO_4$ or Eu-doped $GdPO_4$ samples was examined by the X-ray diffraction (XRD) technique. XRD patterns were recorded using a Rigaku MiniFlexII diffractometer operating in Bragg–Brentano geometry in a $5° \leq 2\theta \leq 80°$ range under a Ni-filtered Cu K_α radiation. The scanning step width was 0.02°, and the scanning speed was 5°/min. A zero-diffraction plate made from Si crystal (MTI Corporation, Richmond, CA, USA) was used as a sample holder.

To determine the morphology and size of the synthesized phosphate particles, scanning electron microscope (SEM) images were taken on a field-emission Hitachi SU-70 electron microscope. The electron acceleration voltage was 5 kV. Particle size and particle size distribution (PSD) were evaluated manually (by taking 50 random particles per sample) using ImageJ (v1.8.0) software.

The Eu^{3+}/Gd^{3+} ratio in the $GdPO_4:Eu^{3+}$ samples was determined by inductively coupled plasma–optical emission spectroscopy (ICP—OES), using Perkin-Elmer Optima 7000DV spectrometer. The samples were dissolved in nitric acid (Rotipuran® Supra 69%, Carl Roth) and diluted to the required volume with DI water. The calibration solutions were prepared by the appropriate dilution of the stock standard solutions (single-element ICP standards, 1000 mg/L, Carl Roth).

Excitation and emission spectra were recorded with an Edinburgh Instruments FLS980 spectrometer (double grating Czerny-Turner excitation and emission monochromators, 450 W Xe arc lamp, single-photon counting photomultiplier Hamamatsu R928P). When measuring excitation spectra, λ_{em} was set to 587.5 nm (excitation and emission slits being 0.50 and 3.50 nm, respectively). Analogously, when measuring emission spectra, λ_{ex} was set to 393 nm (excitation and emission slits being 3.50 and 0.50 nm, respectively). Each spectrum was recorded with 0.5 nm step width and 0.2 s dwell (integration) time. Emission spectra were corrected for instrument response using the correction file provided by Edinburgh Instruments. Excitation spectra were corrected by a reference detector.

Color coordinates (in CIE 1931 color space) of the synthesized samples were calculated using F980 Spectrometer Software (v.1.3.1) from Edinburgh Instruments.

PL decay curves were recorded with Edinburgh Instruments FLS980 spectrometer using a μ-flash lamp (μF2) as an excitation source. The pulse repetition rate was 25 Hz; λ_{ex} and λ_{em} were set to 393 and 587.5 nm, respectively.

The temperature-dependent excitation and emission spectra and PL decay curves were also recorded with Edinburgh Instruments FLS980 spectrometer, employing cryostat "MicrostatN" from Oxford Instruments (cooling agent–liquid nitrogen) for the temperature control. All measurements were conducted at 77 K and in the range from 100 to 500 K in 50 K intervals (stabilization time was 120 s, and temperature tolerance was set to ±5 K).

$TQ_{1/2}$ (the temperature at which a luminescent material loses half of its emission intensity) and E_a (activation energy–the amount of energy that must be given to induce thermal quenching) values for the synthesized samples were calculated using the following equations:

$$I(T) = \frac{I_0}{1 + Be^{-\frac{E_a}{k_B T}}} \tag{1}$$

$$TQ_{1/2} = -\frac{E_a}{k_B \ln \frac{1}{B}} \tag{2}$$

where $I(T)$—normalized integrated emission value at a certain temperature (T); I_0—the highest normalized integrated emission value (in this case equal to 1); B—quenching frequency factor; k_B—Boltzmann constant, equal to $8.6173 \cdot 10^{-5}$ eV/K [55].

3. Results and Discussion

XRD patterns of the produced undoped $GdPO_4$ samples match well with the reference pattern at any chosen <Gd>:<P> molar ratio or pH of the reaction mixture (Figure 1). This indicates that phase pure particles with the trigonal crystal structure (space group $P3_121$ (#152)) were obtained.

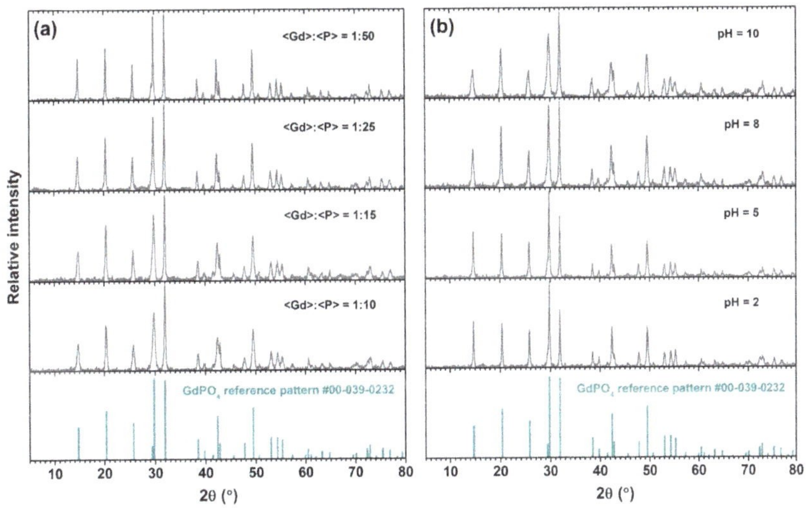

Figure 1. XRD patterns of the synthesized $GdPO_4$ particles under different synthesis conditions: neutral reaction media (pH = 7) with different <Gd>:<P> molar ratio (**a**) and molar ratio <Gd>:<P> = 1:10 with different pH of the reaction mixture (**b**).

Figure 2a presents the SEM images of $GdPO_4$ particles, synthesized at neutral reaction media (pH = 7), changing only the <Gd>:<P> molar ratio. These images show that wider nanorods are obtained with decreasing <Gd>:<P> molar ratio (i.e., increasing <P> concentration). The results obtained from the SEM images are in good agreement with the XRD patterns (see Figure 1a). Clearly, the peaks in the XRD pattern of the smallest particles are broader if compared to the XRD patterns of the larger particles. SEM images depicted in Figure 2b reveal that the length of $GdPO_4$ rods tends to decrease from sub-micro to nano-dimensions, with pH values changing from acidic to alkaline. Thus, the pH of the reaction media has a substantially greater effect on the size of the synthesized phosphates than the effect of the <Gd>:<P> molar ratio.

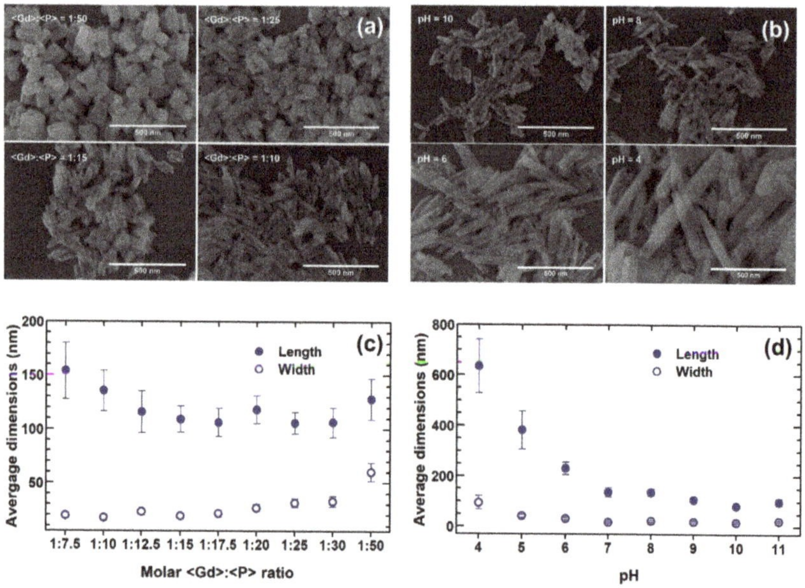

Figure 2. SEM images of $GdPO_4$ nanoparticles prepared under different molar <Gd>:<P> ratios in a neutral reaction medium (**a**). SEM images of $GdPO_4$ nanoparticles prepared at different pH of the reaction mixture at a fixed <Gd>:<P> = 1:10 molar ratio (**b**). Average dimensions of $GdPO_4$ particles as a function of <Gd>:<P> molar ratio (pH = 7) (**c**) and as a function of the reaction mixture's pH (<Gd>:<P> = 1:10) (**d**).

Figure 1 also shows that the ratio of (200) (ca. 30°) and (102) (ca. 32°) peak intensity is sensitive to changes in <Gd>:<P> (please refer to Figure 1a) and synthesis media pH (please refer to Figure 1b). These changes can be explained by analyzing the particle size and shape. For instance, the relative intensity of (200) peak increases with increasing <P> content in the reaction media. The relevant SEM images also show that the particles get wider with increasing <P> concentration. Keeping in mind that the particles grow along the *c*-axis direction [56], the relative (200) peak intensity must increase since there are more facets on the particle surface related to this lattice plane. Furthermore, Figure 2b shows that the relative intensity of the (200) peak increases with decreasing reaction media pH. It was already discussed that the length of the particles increases with decreasing media pH. At the same time, the width of the nanoparticles barely changes. Therefore, more and more facets related to the (200) lattice plane are on the particle surface, resulting in the increase in the (200) peak intensity. One should also keep in mind that the rod-shaped particles are subject to the preferred orientation, which is also in favor of (200) intensity. The double unit cell (along the *c*-axis) of the $GdPO_4$ crystal structure, together with (200) and (102) planes, is shown in Figure S1 for better visualization.

SEM images of the synthesized $GdPO_4$ samples were also used to calculate the average size of the produced particles (please refer to Figures S2 and S3 in Electronic Supplementary Information (ESI)). The obtained results are depicted graphically as a function of <Gd>:<P> molar ratio (please refer to Figure 2c) and as a function of the reaction mixture pH (please refer to Figure 2d). With increasing <Gd>:<P> molar ratio, the average particle length and dispersion slightly decreased. On the contrary, the average particle width and PSD tend to increase with increasing <Gd>:<P> molar ratio (please refer to Table S1). As for increasing the pH value of the reaction media, both the length and the width of the phosphate particles and their PSD tend to decrease (please refer to Table S2). This behavior relies on the complexation ability of different phosphate anion species. It is established that the stronger ability of PO_4^{3-} than $H_2PO_4^-$ and HPO_4^{2-} to coordinate with RE^{3+} leads to the preferential formation of $REPO_4$ nuclei. In alkaline media, the dominant phosphate anion species are HPO_4^{2-} and PO_4^{3-}. Thus, under such conditions, significantly faster formation of $REPO_4$ nuclei occurs compared to nuclei formation under acidic media. This behavior was also observed by Wang et al. [57].

The smallest and the most monodisperse $GdPO_4$ nanorods were obtained under a molar ratio <Gd>:<P> = 1:10 (pH = 10). Their average length is equal to ca. 81 nm, and their average width is ca. 17 nm. Therefore, these conditions were selected for the synthesis of $GdPO_4$ nanoparticles doped with Eu^{3+}. Moreover, such particles are small enough for biomedical applications since studies show that even larger nanoparticles are successfully accumulated in the cells [19,41].

Figure 3a is the XRD patterns of three out of ten synthesized $GdPO_4$:Eu^{3+} samples (the Eu^{3+} concentration in eight samples was 0.5, 1, 2.5, 5, 7.5, 10, 20, 50, 75, and 100%). The given XRD patterns match well with the reference pattern, indicating that produced materials are characterized by trigonal crystal structure with no impurity phases present (the VIII coordinated Gd^{3+} (R = 1.053 Å) are replaced by VIII coordinated Eu^{3+} (R = 1.066 Å), which is only 1.23% larger [58] than Gd^{3+}; therefore, such an ionic radii difference falls within the limits of the solid solution formation range determined by Vegard's law [59]).

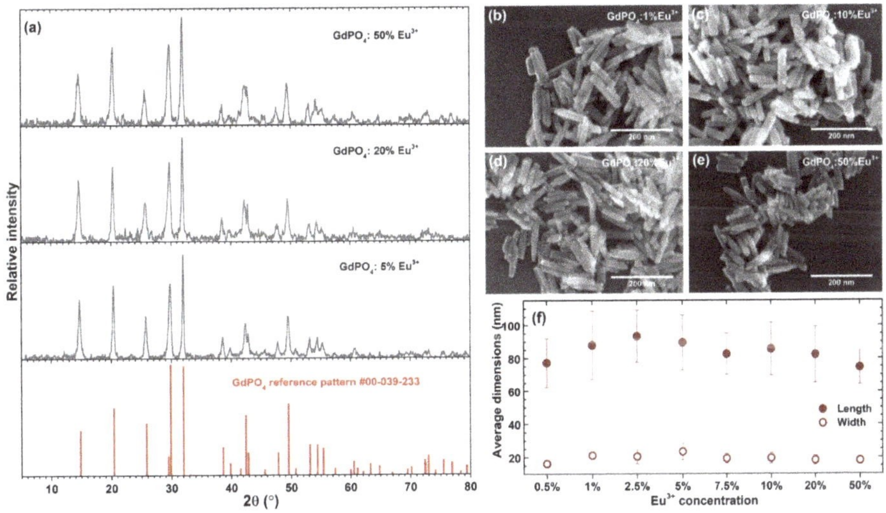

Figure 3. XRD patterns of $GdPO_4$:Eu^{3+} samples as a function of Eu^{3+} concentration (**a**). SEM images of $GdPO_4$:Eu^{3+} samples doped with 1%, 10%, 20%, and 50% Eu^{3+} (**b**–**e**, respectively). The average dimensions of $GdPO_4$:Eu^{3+} particles as a function of Eu^{3+} concentration (**f**). Synthesis conditions: pH = 10; <Gd/Eu>:<P> = 1:10.

SEM images of $GdPO_4$ nanorods containing 1, 10, 20, and 50% Eu^{3+} are shown in Figure 3b–e, respectively. The SEM images of $GdPO_4$ nanorods, doped with other Eu^{3+} concentrations, are provided in Figure S4. The calculated average particle sizes are plotted in Figure 3f and tabulated in Table S3. The average length and the width of orthophosphate nanorods vary between ca. 74 to 93 nm and between ca. 16 to 23 nm, respectively. Nevertheless, particle length and thickness variation fall within the standard deviation limits. Therefore, the incorporation of Eu^{3+} into $GdPO_4$ NPs does not cause significant changes in the size of the obtained nanorods. To confirm that the actual Eu^{3+} concentration in the $GdPO_4$:Eu^{3+} samples is the same as the nominal one, the Eu^{3+} and Gd^{3+} concentrations were determined by ICP—OES. The nominal and measured values of Eu^{3+} and Gd^{3+}

concentrations are given in Table S4 and match well with each other. Therefore, we can conclude that Eu^{3+} easily replaces Gd^{3+} in the $GdPO_4$ structure.

The excitation spectra (λ_{em} = 587.5 nm) of $GdPO_4$:Eu^{3+} samples doped with 1%, 10%, and 50% Eu^{3+} are given in Figure 4a. All spectra contain the typical sets of Eu^{3+} excitation lines originating from the intraconfigurational [Xe]$4f^6 \leftrightarrow$ [Xe]$4f^6$ transitions: ca. 295 nm ($^7F_0 \rightarrow {}^5F_J$), ca. 317 nm ($^7F_0 \rightarrow {}^5H_J$), ca. 360 nm ($^7F_0 \rightarrow {}^5D_4$), ca. 370–390 nm ($^7F_{0,1} \rightarrow {}^5L_{7,8}$;$^5G_J$), ca. 395 nm ($^7F_0 \rightarrow {}^5L_6$) (the strongest transition), ca. 415 nm ($^7F_1 \rightarrow {}^5D_3$), ca. 465 nm ($^7F_0 \rightarrow {}^5D_2$), ca. 525 nm ($^7F_0 \rightarrow {}^5D_1$), and ca. 532 nm ($^7F_1 \rightarrow {}^5D_1$) [36]. The broad excitation band in the range of 250–280 nm is associated with the ligand-to-metal charge transfer (CT) band ($O^{2-} \rightarrow Eu^{3+}$) [60]. Moreover, the optical transitions of Gd^{3+} (ca. 272 nm $^8S \rightarrow {}^6I_J$ and ca. 309 nm $^8S \rightarrow {}^6P_J$) are also observed in the excitation spectra of $GdPO_4$:Eu^{3+} samples when monitoring Eu^{3+} emission. Therefore, it can be concluded that $Gd^{3+} \rightarrow Eu^{3+}$ energy transfer occurs in these phosphors. This can be confirmed by analyzing the intensity of Gd^{3+} lines in the excitation spectra. The intensity of Gd^{3+} excitation lines (ca. 272 nm) is the highest when Eu^{3+} concentration is the lowest, i.e., 1%. Furthermore, the intensity of Gd^{3+} excitation lines gradually decreases with increasing Eu^{3+} concentration. At the same time, the concentration of Gd^{3+} decreases; thus, there is less Gd^{3+} that could transfer the energy to Eu^{3+}, resulting in a decline of Gd^{3+} excitation line intensity. The intensity of Eu^{3+} excitation lines increases with increasing Eu^{3+} concentration and reaches the maximum in 50% Eu^{3+} sample.

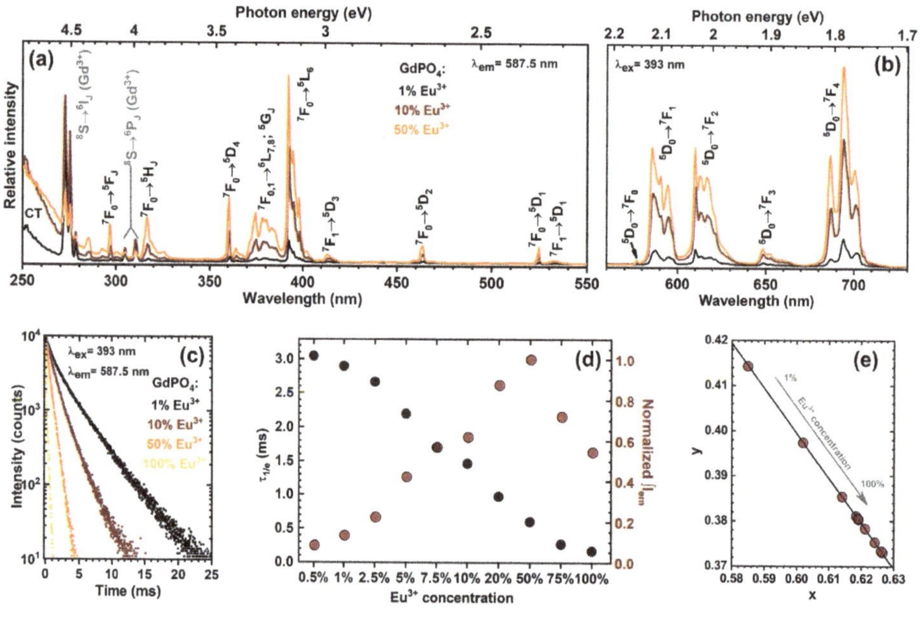

Figure 4. Excitation (λ_{em} = 587.5 nm) (**a**) and emission (λ_{ex} = 393 nm) (**b**) spectra of $GdPO_4$:Eu^{3+} samples. PL decay curves (λ_{ex} = 393 nm, λ_{em} = 587.5 nm) of $GdPO_4$:Eu^{3+} particles (**c**) and calculated $\tau_{1/e}$ values together with normalized integrated emission intensity values as a function of Eu^{3+} concentration (**d**). CIE 1931 color space coordinates of $GdPO_4$:Eu^{3+} phosphors (**e**). Synthesis conditions: pH = 10; <Gd/Eu>:<P> = 1:10.

The emission spectra (λ_{ex} = 393 nm) of $GdPO_4$:Eu^{3+} samples doped with 1%, 10%, and 50% Eu^{3+} are given in Figure 4b (for the emission spectra of all Eu^{3+}-doped samples, please refer to Figure S5). All the spectra contain the typical sets of Eu^{3+} emission lines at ca. 578 nm ($^5D_0 \rightarrow {}^7F_0$), ca. 590 nm ($^5D_0 \rightarrow {}^7F_1$), ca. 615 nm ($^5D_0 \rightarrow {}^7F_2$), ca. 650 nm ($^5D_0 \rightarrow {}^7F_3$), and ca. 695 nm ($^5D_0 \rightarrow {}^7F_4$). Typically, the strongest Eu^{3+} emission transitions are $^5D_0 \rightarrow {}^7F_1$ (magnetic dipole (MD)) and $^5D_0 \rightarrow {}^7F_2$ (electric dipole (ED)). However, in rare-earth orthophosphates, garnets, and some europium complexes (for instance, $Eu(Tp)_3$ (Tp = hydrotris(pyrazol-1-yl)borate), $[Eu(4\text{-picoline-N-oxide})_8](PF_6)_3$, etc.), the strongest intensity is observed for the $^5D_0 \rightarrow {}^7F_4$ transition. The high intensity of $^5D_0 \rightarrow {}^7F_4$ transitions in these materials is attributed to the specific symmetry (like D_{4d}) of the compounds or the optical basicity of these materials [36]. This was also the case in our study. Similar to the excitation spectra, the emission line intensity in emission spectra increased with increasing Eu^{3+} concentration and reached a maximum for the 50% Eu^{3+}-doped sample. This is also the case with the overall emission intensity, which gradually increased, following the same trend (please refer to Figure 4d). Since the excitation spectra of $GdPO_4:Eu^{3+}$ samples (please refer to Figure 4a) also contained the Gd^{3+} lines, we have also measured the Eu^{3+} emission spectra upon Gd^{3+} excitation (λ_{ex} = 273 nm). The recorded spectra are given in Figure S6. The Eu^{3+} emission intensity increases up to 10% Eu^{3+} concentration and then abruptly decreases with a further Eu^{3+} concentration increase. It is worth mentioning that samples doped with 0.5 and 50% Eu^{3+} possess virtually the same emission intensity. Since Gd^{3+} concentration decreases with increasing Eu^{3+} concentration, the decrease in Eu^{3+} emission intensity at higher Eu^{3+} concentrations is caused by lower Gd^{3+} concentration, leading to a less efficient $Gd^{3+} \rightarrow Eu^{3+}$ energy transfer.

The PL decay curves (λ_{ex} = 393 nm, λ_{em} = 587.5 nm) of $GdPO_4:Eu^{3+}$ samples are shown in Figure 4c. The PL lifetime values were calculated using the following equation [61]:

$$\tau_{1/e} = \frac{\int_0^{\infty} I(t)tdt}{\int_0^{\infty} I(t)dt} \quad (3)$$

Here, $I(t)$ stands for PL intensity at time t. The change in average $\tau_{1/e}$ values as a function of Eu^{3+} concentration is plotted in Figure 4d, whereas the exact calculated $\tau_{1/e}$ values are summarized in Table S7. The PL decay curves get steeper with increasing Eu^{3+} concentration, indicating that the PL lifetime values decrease. The average PL lifetime values of Eu^{3+} emission at 587.5 nm decrease from ca. 3043 µs to ca. 173 µs as the concentration of Eu^{3+} in the NPs increase from 0.5 to 100% (please refer to Figure 4d and Table S5).

The color coordinates in CIE 1931 color space were calculated for each sample doped with Eu^{3+}. The obtained color coordinates are located directly on the edge of the CIE 1931 color space diagram (please refer to Figure 4e), indicating that the red emission of the produced NPs would be perceived as a rich and monochromatic light by the human eye. In addition, with increasing Eu^{3+} concentration in the samples, a slight red shift of color coordinates is observed. However, this shift is relatively insignificant (especially in heavier Eu^{3+}-doped samples), and color coordinates can be considered stable regardless of the amount of Eu^{3+} (please refer to Table S5 for the precise calculated color coordinate values).

As was discussed above, the $GdPO_4:50\%Eu^{3+}$ phosphor exhibited the highest emission intensity and, therefore, was selected for temperature-dependent measurements. Figure 5a,b shows that the intensity of both excitation and emission spectra of the $GdPO_4$ $:50\%Eu^{3+}$ sample drops dramatically when the temperature increases from 77 to 500 K. It is also interesting to note that the excitation spectrum recorded at 77 K does not contain any excitation lines originating from the $^7F_1$ level, indicating that thermal population of this level is significantly suppressed at such low temperature. The normalized integrated emission intensity as a function of temperature is presented in the inset of Figure 5b. These data were used to calculate $TQ_{1/2}$ and E_a, and the obtained values are equal to 291 $\pm$ 19 K and 0.049 $\pm$ 0.007 eV, respectively. The $TQ_{1/2}$ value (in the 77 to 500 K range) showed that this phosphor maintained half of its emission intensity even at room temperature.

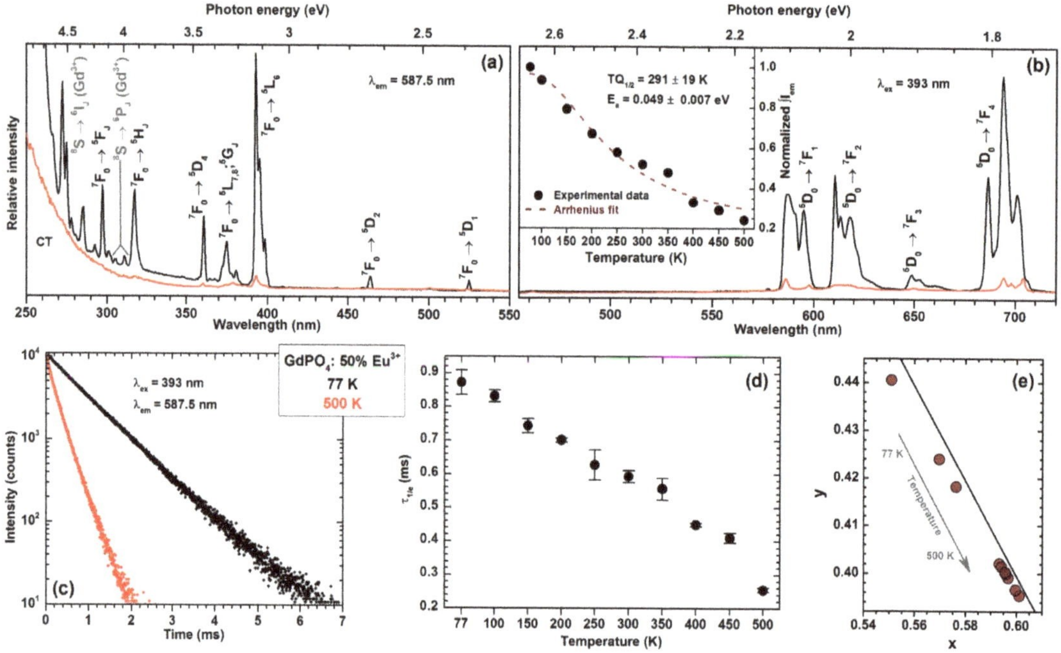

Figure 5. Temperature-dependent excitation (λ_{em} = 587.5 nm) (**a**) and emission (λ_{ex} = 393 nm) (**b**) spectra with normalized integrated emission intensity as a function of temperature (inset); temperature-dependent PL decay curves (λ_{ex} = 393 nm, λ_{em} = 587.5 nm) (**c**), average $\tau_{1/e}$ values (**d**) and CIE 1931 color coordinates (**e**) as a function of the temperature of $GdPO_4$:50%Eu^{3+} phosphor.

The PL decay curves (please refer to Figure 5c) of the $GdPO_4$:50%Eu^{3+} sample got steeper with increasing temperature, indicating the decreasing average PL lifetime values of Eu^{3+}. This indeed was confirmed after calculating the average PL lifetime values, which are plotted in Figure 5d, and their exact values are given in Table S6. It turned out that the average PL lifetime values of $GdPO_4$:50%Eu^{3+} phosphor decreased from ca. 871 µs to ca. 257 µs with the temperature increase from 77 to 500 K, which can be related to the decreasing internal efficiency of the phosphor.

The temperature-dependent emission spectra of $GdPO_4$:50%Eu^{3+} phosphor were also used to calculate the temperature-dependent color coordinates, which are plotted in Figure 5e. The exact calculated values of color coordinates are summarized in Table S6. A slight red shift of calculated color coordinates is observed with the increasing temperature; however, this shift is relatively insignificant, and color coordinates can be considered temperature-stable, especially at higher temperatures.

4. Conclusions

In summary, we have demonstrated that the aqueous hydrothermal synthesis method is highly suitable for preparing $GdPO_4$:Eu^{3+} nanoparticles possessing hydrophilic surfaces. The nanoparticle size and morphology can be controlled by selecting the appropriate pH and <Gd>:<P> molar ratio. The width of $GdPO_4$ nanoparticles decreased from 93 to 23 nm, and the length decreased from 635 to 99 nm when the pH of reaction media was increased from 4 to 11. Furthermore, the width of $GdPO_4$ nanoparticles increased from 19 to 60 nm, and the length decreased from 154 to 128 nm when the <Gd>:<P> molar ratio was changed from 1:7.5 to 1:50. The nanoparticle size and particle size distribution of the Eu^{3+}-doped samples, on the other hand, remained virtually the same, regardless of the Eu^{3+} concentration. The PL measurements showed that the emission intensity of $GdPO_4$:Eu^{3+} samples increased with increasing Eu^{3+} concentration and reached the maximum for the $GdPO_4$:50%Eu^{3+} sample. The temperature-dependent measurements in a 77–500 K range revealed that the $GdPO_4$:50%Eu^{3+} sample possesses relatively high luminescence thermal stability. This sample maintained half of its emission intensity, even at room temperature ($TQ_{1/2}$ = 291 ± 19 K). The determined thermal optical stability of the Eu^{3+}-doped samples is sufficient for various applications, including luminescent security inks, bio-imaging probes, etc. Recent studies also showed the magnetic properties of $GdPO_4$ nanoparticles. Therefore, the combination of Gd^{3+} magnetic properties and Eu^{3+} distinctive luminescence properties extends the possible application field of these nanomaterials even further. Such unique opto-magnetic nanoparticles could be applied in biomedicine as selective bio-imaging probes or MRI contrast materials.

Supplementary Materials: The following supporting information can be downloaded at: https://www.mdpi.com/article/10.3390/cryst13020174/s1, Figure S1: The double unit cell (along *c*-axis) of $GdPO_4$ crystal structure with (102) plane and (200) plane family (b); Figure S2: SEM images of $GdPO_4$ nanoparticles prepared under different molar <Gd>:<P> ratio at a neutral reaction media (pH = 7); Figure S3: SEM images of $GdPO_4$ nanoparticles prepared under different pH of the reaction media at a fixed <Gd>:<P> = 1:10 molar ratio; Figure S4: SEM images of the $GdPO_4$:Eu^{3+} samples as a function of Eu^{3+} concentration. Synthesis conditions: pH = 10; <Gd/Eu>:<P> = 1:10; Figure S5: Emission spectra (λ_{ex} = 393 nm) of $GdPO_4$:Eu^{3+} phosphors (a) and $^5D_0 \rightarrow {}^7F_4$ optical transition zoomed in (b); Figure S6: Emission spectra (λ_{ex} = 273 nm) of $GdPO_4$:Eu^{3+} phosphors (a) and integrated emission intensity as a function of Eu^{3+} concentration (b); Table S1: The average dimensions of produced $GdPO_4$ samples at a neutral reaction media (pH = 7) as a function of <Gd>:<P> molar ratio; Table S2: The average dimensions of produced $GdPO_4$ samples using molar ratio <Gd>:<P> = 1:10 at different pH of the reaction media.; Table S3: The average dimensions of produced $GdPO_4$:Eu^{3+} nanoparticles as a function of Eu^{3+} concentration. Synthesis conditions: pH = 10; <Gd/Eu>:<P> = 1:10; Table S4: Theoretical and actual (detected using ICP−OES) amounts of Gd^{3+} and Eu^{3+} in the prepared samples.; Table S5: PL lifetime values (λ_{ex} = 393 nm, λ_{em} = 587.5 nm) and CIE 1931 color coordinates of $GdPO_4$:Eu^{3+} nanophosphors.; Table S6: The temperature-dependent PL lifetime values (λ_{ex} = 393 nm, λ_{em} = 587.5 nm) and CIE 1931 color coordinates of $GdPO_4$:50%Eu^{3+} nanophosphor.

Author Contributions: Conceptualization, V.K. and A.K.; methodology, E.E. and V.K.; investigation, E.E. and A.Z.; resources, A.K.; data curation, E.E.; writing—original draft preparation, E.E. and V.K.; writing—review and editing, A.Z. and A.K.; visualization, E.E. and V.K.; supervision, V.K. and A.K.; funding acquisition, A.K. All authors have read and agreed to the published version of the manuscript.

Funding: This project has received funding from the Research Council of Lithuania (LMTLT), agreement No [S-MIP-22-68].

Institutional Review Board Statement: Not applicable.

Informed Consent Statement: Not applicable.

Data Availability Statement: The data presented in this study are available on request from the corresponding author.

Conflicts of Interest: The authors declare no conflict of interest.

References

1. Ryoo, R.; Kim, J.; Jo, C.; Han, S.W.; Kim, J.C.; Park, H.; Han, J.; Shin, H.S.; Shin, J.W. Rare-earth-platinum alloy nanoparticles in mesoporous zeolite for catalysis. *Nature* **2020**, *585*, 221–224. [CrossRef] [PubMed]
2. Lu, Y.F.; Li, J.; Ye, T.N.; Kobayashi, Y.; Sasase, M.; Kitano, M.; Hosono, H. Synthesis of Rare-Earth-Based Metallic Electride Nanoparticles and Their Catalytic Applications to Selective Hydrogenation and Ammonia Synthesis. *ACS Catal.* **2018**, *8*, 11054–11058. [CrossRef]
3. Runowski, M.; Wozny, P.; Martin, I.R. Optical pressure sensing in vacuum and high-pressure ranges using lanthanide-based luminescent thermometer-manometer. *J. Mater. Chem. C* **2021**, *9*, 4643–4651. [CrossRef]
4. Sahu, N.K.; Singh, N.S.; Ningthoujam, R.S.; Bahadur, D. Ce^{3+}-Sensitized $GdPO_4$:Tb^{3+} Nanorods: An Investigation on Energy Transfer, Luminescence Switching, and Quantum Yield. *ACS Photonics* **2014**, *1*, 337–346. [CrossRef]

5. Wei, Z.; Liu, Y.W.; Li, B.; Li, J.J.; Lu, S.; Xing, X.W.; Liu, K.; Wang, F.; Zhang, H.J. Rare-earth based materials: An effective toolbox for brain imaging, therapy, monitoring and neuromodulation. *Light-Sci. Appl.* **2022**, *11*, 19. [CrossRef]
6. Wu, Y.L.; Xu, X.Z.; Li, Q.L.; Yang, R.C.; Ding, H.X.; Xiao, Q. Synthesis of bifunctional Gd_2O_3:Eu^{3+} nanocrystals and their applications in biomedical imaging. *J. Rare Earth* **2015**, *33*, 529–534. [CrossRef]
7. Tang, Y.X.; Mei, R.; Yang, S.K.; Tang, H.X.; Yin, W.Z.; Xu, Y.C.; Gao, Y.P. Hollow $GdPO_4$:Eu^{3+} microspheres: Luminescent properties and applications as drug carrier. *Superlattices Microstruct.* **2016**, *92*, 256–263. [CrossRef]
8. Lin, F.; Sun, Z.; Jia, M.C.; Zhang, A.Q.; Fu, Z.L.; Sheng, T.Q. Core-shell mutual enhanced luminescence based on space isolation strategy for anti-counterfeiting applications. *J. Lumin.* **2020**, *218*, 6. [CrossRef]
9. Pushpendra; Suryawanshi, I.; Srinidhi, S.; Singh, S.; Kalia, R.; Kunchala, R.K.; Mudavath, S.L.; Naidu, B.S. Downshifting and upconversion dual mode emission from lanthanide doped $GdPO_4$ nanorods for unclonable anti-counterfeiting. *Mater. Today Commun.* **2021**, *26*, 10. [CrossRef]
10. Wang, G.F.; Peng, Q.; Li, Y.D. Lanthanide-Doped Nanocrystals: Synthesis, Optical-Magnetic Properties, and Applications. *Acc. Chem. Res.* **2011**, *44*, 322–332. [CrossRef]
11. Choi, S.J.; Lee, J.K.; Jeong, J.; Choy, J.H. Toxicity evaluation of inorganic nanoparticles: Considerations and challenges. *Mol. Cell. Toxicol.* **2013**, *9*, 205–210. [CrossRef]
12. Klimkevicius, V.; Janulevicius, M.; Babiceva, A.; Drabavicius, A.; Katelnikovas, A. Effect of Cationic Brush-Type Copolymers on the Colloidal Stability of $GdPO_4$ Particles with Different Morphologies in Biological Aqueous Media. *Langmuir* **2020**, *36*, 7533–7544. [CrossRef] [PubMed]
13. Janulevicius, M.; Klimkevicius, V.; Vanetsev, A.; Plausinaitiene, V.; Sakirzanovas, S.; Katelnikovas, A. Controlled hydrothermal synthesis, morphological design and colloidal stability of $GdPO_4 \cdot nH_2O$ particles. *Mater. Today Commun.* **2020**, *23*, 8. [CrossRef]
14. Cui, X.X.; Fan, Q.; Shi, S.J.; Wen, W.H.; Chen, D.F.; Guo, H.T.; Xu, Y.T.; Gao, F.; Nie, R.Z.; Ford, H.D.; et al. A novel near-infrared nanomaterial with high quantum efficiency and its applications in real time in vivo imaging. *Nanotechnology* **2018**, *29*, 11. [CrossRef]
15. Ezerskyte, E.; Grigorjevaite, J.; Minderyte, A.; Saitzek, S.; Katelnikovas, A. Temperature-Dependent Luminescence of Red-Emitting $Ba_2Y_5B_5O_{17}$: Eu^{3+} Phosphors with Efficiencies Close to Unity for Near-UV LEDs. *Materials* **2020**, *13*, 13. [CrossRef]
16. Priya, R.; Mariappan, R.; Karthikeyan, A.; Palani, E.; Krishnamoorthy, E.; Gowrisankar, G. Review on rare earth metals doped $LaPO_4$ for optoelectronic applications. *Solid State Commun.* **2021**, *339*, 26. [CrossRef]
17. Zheng, B.Z.; Fan, J.Y.; Chen, B.; Qin, X.; Wang, J.; Wang, F.; Deng, R.R.; Liu, X.G. Rare-Earth Doping in Nanostructured Inorganic Materials. *Chem. Rev.* **2022**, *122*, 5519–5603. [CrossRef]
18. Zhou, J.; Liu, Q.; Feng, W.; Sun, Y.; Li, F.Y. Upconversion Luminescent Materials: Advances and Applications. *Chem. Rev.* **2015**, *115*, 395–465. [CrossRef]
19. Klimkevicius, V.; Voronovic, E.; Jarockyte, G.; Skripka, A.; Vetrone, F.; Rotomskis, R.; Katelnikovas, A.; Karabanovas, V. Polymer brush coated upconverting nanoparticles with improved colloidal stability and cellular labeling. *J. Mat. Chem. B* **2022**, *10*, 625–636. [CrossRef]
20. Bagheri, A.; Arandiyan, H.; Boyer, C.; Lim, M. Lanthanide-Doped Upconversion Nanoparticles: Emerging Intelligent Light-Activated Drug Delivery Systems. *Adv. Sci.* **2016**, *3*, 25. [CrossRef]
21. Cichos, J.; Karbowiak, M. Spectroscopic characterization of ligands on the surface of water dispersible $NaGdF_4$:Ln^{3+} nanocrystals. *Appl. Surf. Sci.* **2012**, *258*, 5610–5618. [CrossRef]
22. Alonso-de Castro, S.; Ruggiero, E.; Fernandez, A.L.; Cossio, U.; Baz, Z.; Otaegui, D.; Gomez-Vallejo, V.; Padro, D.; Llop, J.; Salassa, L. Functionalizing $NaGdF_4$:Yb,Er Upconverting Nanoparticles with Bone-Targeting Phosphonate Ligands: Imaging and In Vivo Biodistribution. *Inorganics* **2019**, *7*, 12. [CrossRef]
23. Liu, C.Y.; Gao, Z.Y.; Zeng, J.F.; Hou, Y.; Fang, F.; Li, Y.L.; Qiao, R.R.; Shen, L.; Lei, H.; Yang, W.S.; et al. Magnetic/Upconversion Fluorescent $NaGdF_4$:Yb,Er Nanoparticle-Based Dual-Modal Molecular Probes for Imaging Tiny Tumors in Vivo. *ACS Nano* **2013**, *7*, 7227–7240. [CrossRef]
24. Dong, C.H.; Korinek, A.; Blasiak, B.; Tomanek, B.; van Veggel, F. Cation Exchange: A Facile Method To Make $NaYF_4$:Yb,Tm-$NaGdF_4$ Core-Shell Nanoparticles with a Thin, Tunable, and Uniform Shell. *Chem. Mat.* **2012**, *24*, 1297–1305. [CrossRef]
25. Song, H.J.; Zhou, L.Q.; Li, L.; Hong, F.; Luo, X.R. Hydrothermal synthesis, characterization and luminescent properties of $GdPO_4 \cdot H_2O$:Tb^{3+} nanorods and nanobundles. *Mater. Res. Bull.* **2013**, *48*, 5013–5018. [CrossRef]
26. Janulevicius, M.; Klimkevicius, V.; Mikoliunaite, L.; Vengalis, B.; Vargalis, R.; Sakirzanovas, S.; Plausinaitiene, V.; Zilinskas, A.; Katelnikovas, A. Ultralight Magnetic Nanofibrous $GdPO_4$ Aerogel. *ACS Omega* **2020**, *5*, 14180–14185. [CrossRef] [PubMed]
27. Cao, Y.Y.; Sun, P.; Liang, Y.M.; Wang, R.R.; Zhang, X. Sol-precipitation-hydrothermal synthesis and luminescence of $GdPO_4$:Tb^{3+} submicron cubes. *Chem. Phys. Lett.* **2016**, *651*, 80–83. [CrossRef]
28. Yang, R.; Qin, J.; Li, M.; Liu, Y.H.; Li, F. Redox hydrothermal synthesis of cerium phosphate microspheres with different architectures. *Crystengcomm* **2011**, *13*, 7284–7292. [CrossRef]
29. Feng, S.H.; Xu, R.R. New materials in hydrothermal synthesis. *Acc. Chem. Res.* **2001**, *34*, 239–247. [CrossRef]
30. Buissette, V.; Moreau, M.; Gacoin, T.; Boilot, J.P.; Chane-Ching, J.Y.; Le Mercier, T. Colloidal synthesis of luminescent rhabdophane $LaPO_4$:$Ln^{3+} \cdot xH_2O$ (Ln = Ce, Tb, Eu; $x \approx 0.7$) nanocrystals. *Chem. Mater.* **2004**, *16*, 3767–3773. [CrossRef]
31. Huong, N.T.; Van, N.D.; Tien, D.M.; Tung, D.K.; Binh, N.T.; Anh, T.K.; Minh, L.Q. Structural and luminescent properties of (Eu,Tb)$PO_4 \cdot H_2O$ nanorods/nanowires prepared by microwave technique. *J. Rare Earths* **2011**, *29*, 1170–1173. [CrossRef]

32. Li, J.G.; Sakka, Y. Recent progress in advanced optical materials based on gadolinium aluminate garnet ($Gd_3Al_5O_{12}$). *Sci. Technol. Adv. Mater.* **2015**, *16*, 18. [CrossRef] [PubMed]
33. Matos, M.G.; Calefi, P.S.; Ciuffi, K.J.; Nassar, E.J. Synthesis and luminescent properties of gadolinium aluminates phosphors. *Inorg. Chim. Acta* **2011**, *375*, 63–69. [CrossRef]
34. Maldiney, T.; Doan, B.T.; Alloyeau, D.; Bessodes, M.; Scherman, D.; Richard, C. Gadolinium-Doped Persistent Nanophosphors as Versatile Tool for Multimodal In Vivo Imaging. *Adv. Funct. Mater.* **2015**, *25*, 331–338. [CrossRef]
35. Mahakhode, J.G.; Nande, A.; Dhoble, S.J. A review: X-ray excited luminescence of gadolinium based optoelectronic phosphors. *Luminescence* **2021**, *36*, 1344–1353. [CrossRef]
36. Binnemans, K. Interpretation of europium(III) spectra. *Coordin. Chem. Rev.* **2015**, *295*, 1–45. [CrossRef]
37. Sun, C.; Pratx, G.; Carpenter, C.M.; Liu, H.G.; Cheng, Z.; Gambhir, S.S.; Xing, L. Synthesis and Radioluminescence of PEGylated Eu^{3+}-doped Nanophosphors as Bioimaging Probes. *Adv. Mater.* **2011**, *23*, H195–H199. [CrossRef]
38. Grigorjevaite, J.; Ezerskyte, E.; Minderyte, A.; Stanionyte, S.; Juskenas, R.; Sakirzanovas, S.; Katelnikovas, A. Optical Properties of Red-Emitting $Rb_2Bi(PO_4)(MoO_4):Eu^{3+}$ Powders and Ceramics with High Quantum Efficiency for White LEDs. *Materials* **2019**, *12*, 14. [CrossRef]
39. Díaz García, M.E.; Badía-Laíño, R. Fluorescence | Time-Resolved Fluorescence ☆. In *Encyclopedia of Analytical Science*, 3rd ed.; Worsfold, P., Poole, C., Townshend, A., Miró, M., Eds.; Academic Press: Oxford, UK, 2019; pp. 327–340.
40. Skripka, A.; Cheng, T.; Jones, C.M.S.; Marin, R.; Marques-Hueso, J.; Vetrone, F. Spectral characterization of $LiYbF_4$ upconverting nanoparticles. *Nanoscale* **2020**, *12*, 17545–17554. [CrossRef]
41. Kim, D.K.; Dobson, J. Nanomedicine for targeted drug delivery. *J. Mater. Chem.* **2009**, *19*, 6294–6307. [CrossRef]
42. Zhang, Y.W.; Yan, Z.G.; You, L.P.; Si, R.; Yan, C.H. General synthesis and characterization of monocrystalline lanthanide orthophosphate nanowires. *Eur. J. Inorg. Chem.* **2003**, *2003*, 4099–4104. [CrossRef]
43. Hernandez, A.G.; Boyer, D.; Potdevin, A.; Chadeyron, G.; Murillo, A.G.; Romo, F.D.C.; Mahiou, R. Hydrothermal synthesis of lanthanide-doped $GdPO_4$ nanowires and nanoparticles for optical applications. *Phys. Status Solidi A Appl. Mat.* **2014**, *211*, 498–503. [CrossRef]
44. Yan, B.; Gu, J.F.; Xiao, X.Z. $LnPO_4:RE^{3+}$ (La = La, Gd; RE = Eu, Tb) nanocrystals: Solvo-thermal synthesis, microstructure and photoluminescence. *J. Nanopart. Res.* **2010**, *12*, 2145–2152. [CrossRef]
45. Ren, W.L.; Tian, G.; Zhou, L.J.; Yin, W.Y.; Yan, L.; Jin, S.; Zu, Y.; Li, S.J.; Gu, Z.J.; Zhao, Y.L. Lanthanide ion-doped $GdPO_4$ nanorods with dual-modal bio-optical and magnetic resonance imaging properties. *Nanoscale* **2012**, *4*, 3754–3760. [CrossRef]
46. Budrevicius, D.; Skaudzius, R. Volume dependence of the size of $GdPO_4$:15%Eu particles synthesized by the hydrothermal method. *J. Alloys Compd.* **2022**, *911*, 5. [CrossRef]
47. Yaiphaba, N.; Ningthoujam, R.S.; Singh, N.R.; Vatsa, R.K. Luminescence Properties of Redispersible Tb^{3+}-Doped $GdPO_4$ Nanoparticles Prepared by an Ethylene Glycol Route. *Eur. J. Inorg. Chem.* **2010**, *2010*, 2682–2687. [CrossRef]
48. Huo, Z.Y.; Chen, C.; Chu, D.; Li, H.H.; Li, Y.D. Systematic synthesis of lanthanide phosphate nanocrystals. *Chem. Eur. J.* **2007**, *13*, 7708–7714. [CrossRef]
49. Rodriguez-Liviano, S.; Becerro, A.I.; Alcantara, D.; Grazu, V.; de la Fuente, J.M.; Ocana, M. Synthesis and Properties of Multifunctional Tetragonal Eu:$GdPO_4$ Nanocubes for Optical and Magnetic Resonance Imaging Applications. *Inorg. Chem.* **2013**, *52*, 647–654. [CrossRef]
50. Di, W.H.; Wang, X.J.; Zhao, H.F. Synthesis and characterization of $LnPO_4{\cdot}nH_2O$ (Ln = La, Ce, Gd, Tb, Dy) nanorods and nanowires. *J. Nanosci. Nanotechnol.* **2007**, *7*, 3624–3628. [CrossRef]
51. Huang, C.C.; Lo, Y.W.; Kuo, W.S.; Hwu, J.R.; Su, W.C.; Shieh, D.B.; Yeh, C.S. Facile preparation of self-assembled hydrogel-like $GdPO_4{\cdot}H_2O$ nanorods. *Langmuir* **2008**, *24*, 8309–8313. [CrossRef]
52. Xu, Z.H.; Cao, Y.; Li, C.X.; Ma, P.A.; Zhai, X.F.; Huang, S.S.; Kang, X.J.; Shang, M.M.; Yang, D.M.; Dai, Y.L.; et al. Urchin-like $GdPO_4$ and $GdPO_4:Eu^{3+}$ hollow spheres—hydrothermal synthesis, luminescence and drug-delivery properties. *J. Mater. Chem.* **2011**, *21*, 3686–3694. [CrossRef]
53. Camacho, A.R.; Romo, F.D.C.; Murillo, A.G.; Oliva, J.; Garcia, C.R. Sol-gel synthesis and up-conversion luminescence of $GdPO_4{\cdot}Gd_3PO_7:Yb^{3+}$, Ln^{3+} (Ln = Er, Ho, Tm) phosphor. *Mater. Lett.* **2018**, *226*, 34–37. [CrossRef]
54. Kumar, V.; Rani, P.; Singh, D.; Chawla, S. Efficient multiphoton upconversion and synthesis route dependent emission tunability in $GdPO_4:Ho^{3+},Yb^{3+}$ nanocrystals. *RSC Adv.* **2014**, *4*, 36101–36105. [CrossRef]
55. Baur, F.; Glocker, F.; Jüstel, T. Photoluminescence and energy transfer rates and efficiencies in Eu^{3+} activated $Tb_2Mo_3O_{12}$. *J. Mater. Chem. C* **2015**, *3*, 2054–2064. [CrossRef]
56. Wang, X.J.; Gao, M.Y. A facile route for preparing rhabdophane rare earth phosphate nanorods. *J. Mater. Chem.* **2006**, *16*, 1360–1365. [CrossRef]
57. Wang, Z.H.; Shi, X.F.; Wang, X.J.; Zhu, Q.; Kim, B.N.; Sun, X.D.; Li, J.G. Breaking the strong 1D growth habit to yield quasi-equiaxed $REPO_4$ nanocrystals (RE = La-Dy) via solvothermal reaction and investigation of photoluminescence. *Crystengcomm* **2018**, *20*, 796–806. [CrossRef]
58. Shannon, R.D. Revised effective ionic radii and systematic studies of interatomic distances in halides and chalcogenides. *Acta Crystallogr. A* **1976**, *32*, 751–767. [CrossRef]
59. Ropp, R.C. *Luminescence and the Solid State*; Elsevier Science: Amsterdam, The Netherlands, 2004.

60. Jubera, V.; Chaminade, J.P.; Garcia, A.; Guillen, F.; Fouassier, C. Luminescent properties of Eu^{3+}-activated lithium rare earth borates and oxyborates. *J. Lumin.* **2003**, *101*, 1–10. [CrossRef]
61. Lahoz, F.; Martin, I.R.; Mendez-Ramos, J.; Nunez, P. Dopant distribution in a Tm^{3+}-Yb^{3+} codoped silica based glass ceramic: An infrared-laser induced upconversion study. *J. Chem. Phys.* **2004**, *120*, 6180–6190. [CrossRef]

crystals

MDPI

Article

Synthetic Conditions for Obtaining Different Types of Amine-Holding Silica Particles and Their Sorption Behavior

Inna Melnyk [1,2,*], Veronika Tomina [2] and Nataliya Stolyarchuk [2]

[1] Department of Physical and Physico-Chemical Methods of Mineral Processing, Institute of Geotechnics SAS, 45, Watsonova, 04001 Košice, Slovakia
[2] Department of Chemisorption and Hybrid Materials, Chuiko Institute of Surface Chemistry NASU, 17, Generala Naumova, 03164 Kyiv, Ukraine
* Correspondence: melnyk@saske.sk; Tel.: +421-55-7922612

Abstract: The Stöber version of a sol-gel method of co-condensation of two alkoxysilanes (structuring tetraethoxysilane (TEOS) and functionalising N-containing silane) in an ammonia medium was used for the one-pot synthesis of spherical silica particles with $\equiv Si(CH_2)_3NH_2$, $\equiv Si(CH_2)_3NH(CH_2)_2NH_2$, and $\equiv [Si(CH_2)_3NH]_2$ functional groups with available groups content of 1.3–2.3 mmol/g. The materials were researched by a range of methods, including SEM, TEM, IR spectroscopy, ^{13}C, and ^{29}Si solid-state NMR spectroscopy, acid-base titration, and thermal analysis to identify the peculiarities of the morphology, functional groups content, composition, and thermal resistance of the surface layers in the synthesised samples. The type of N-containing silane was shown to affect the structure and properties of the synthesised spherical particles. The silane with the simplest, 3-aminopropyl, functional group caused the formation of nonporous material composed of large 600–800 nm spherical microparticles. Meanwhile, the complication of functional groups enhanced the emergence of small 15 nm primary particles and the origination of porosity, generated by the slits between particles and particle agglomerates. Thereafter, the sorption properties of the synthesised hybrid materials for nickel(II) and copper(II)ions, and bovine serum albumin (BSA) were also found to be dependent on the structure of the materials and the type of incorporated functional group. The maximal static sorption capacity values towards the targeted adsorbates were shown by the samples with 3-aminopropyl groups (1.27 mmol Ni/g), diamine groups (1.09 mmol Cu/g), and secondary amine groups (204.6 mg BSA/g). The conducted research opens up the prospects of directed one-pot synthesis of amino-functionalised hybrid organosilica materials for different applications.

Keywords: sol-gel technique; one-pot synthesis; 3-aminopropyl groups; ethylenediamine groups; secondary amine groups; Cu(II) ions sorption; Ni(II) ions sorption; bovine serum albumin sorption

Citation: Melnyk, I.; Tomina, V.; Stolyarchuk, N. Synthetic Conditions for Obtaining Different Types of Amine-Holding Silica Particles and Their Sorption Behavior. *Crystals* **2023**, *13*, 190. https://doi.org/10.3390/cryst13020190

Academic Editors: Aleksej Zarkov, Aivaras Kareiva and Loreta Tamasauskaite-Tamasiunaite

Received: 30 December 2022
Revised: 17 January 2023
Accepted: 18 January 2023
Published: 21 January 2023

1. Introduction

The influence of anthropogenic factors on environmental pollution has aggravated recently. Heavy metals that are widely used in various industries form a special group of pollutants [1]. Heavy metal ions are usually distributed in the environment by means of water [2]. Therefore, their removal from water is an urgent problem that constantly causes concern [3]. Respecting sorption technologies, the search for approaches to synthesise novel materials with a well-developed porous structure (ensuring good kinetic characteristics) and a high content of available functional groups is worthy of attention.

Silica-based materials can be used for catalysis and environmental applications [4]. However, the efficiency of pure silica sorbents is relatively low [5]. To improve the sorption properties of these materials, oxygen-, nitrogen-, or sulfur-containing functional groups are incorporated into them [6]. Recently, silicas functionalised with amino groups have been of considerable interest, because their surface layer allows binding ions of such metals as copper and nickel into stable complex compounds [7,8]. Nitrogen-containing polysiloxane sorbents have been well-studied, described in detail in [9,10], and even produced as commercial materials [11,12].

It should be mentioned that sorption materials formed by spherical particles are more effective for sorption processes from a technological point of view. Such materials are characterised by greater availability of complex-forming functional groups for interaction with metal ions, which improves the process of extraction of metal ions from their aqueous solutions. Sorption materials in the form of spherical particles can be prepared using the Stöber version of sol-gel synthesis [13]. Spherical silica particles with amino groups have recently been very popular as carriers for various applications, such as drug delivery [14], chemotherapy [15], biomedicine [16], antibacterial therapy [17], liquid chromatography [18], catalysis [19]; as fillers for the flame retardants [20]; as sorbents of organic molecules [21,22], gasses [23,24], and metal ions [25–27]. Therefore, their economically feasible production is an urgent task. In many cases in the above-mentioned references, surface modification was used to attach amino groups to the silica surface, while a template method was applied to create mesoporous materials. Functionalising alkoxysilanes of different compositions can be used to affix different types of nitrogen-containing groups onto the surface of the materials. Previously, we described a one-step synthesis of aminosilica particles [28] and the possibility of influencing their size and surface chemistry [29]. Moreover, for the certain tasks, silanes with diamine groups [30,31] and bridged silanes with amino groups [32,33] are increasingly used, so the research data will be of practical interest to the scientists in this area. The main idea of this work is to use the Stöber method for the one-component system (TEOS) and apply it to the two-component system, where the second component is a silane with amino groups of different structure. In this study, it is proposed to obtain in one step silica spheres with 3-aminopropyl groups, diamine groups (which increases the number of heteroatoms on the surface), as well as to synthesise materials with bridging silanes containing a secondary amino group and affecting the porosity of the obtained materials. It was shown that the suggested method of synthesis promotes the fixation of a large number of amino-containing groups and they are available for interactions with metal cations and protein molecules on the example of bovine serum albumin.

2. Materials and Methods

2.1. Reagents

The following substances were used to prepare the samples: tetraethoxysilane, $Si(OC_2H_5)_4$ (TEOS, 98%, Aldrich, St.Louis, MO, USA); 3-aminopropyltriethoxysilane, $C_9H_{23}NO_3Si$ (APTES, 98%, Fluka); N-[3-(trimethoxysilyl)propyl]ethylenediamine, $C_8H_{22}N_2O_3Si$ (TMPED, 97%, Aldrich, USA); bis [3-(trimethoxysilyl)propyl]amine, $[(CH_3O)_3Si(CH_2)_3]_2NH$ (BTMPA, 95%, ABCR); ethanol (96%, Ukrspirt); ammonia solution, 25% (NH_4OH, p.a., Macrochem, Ukraine).

Reagents used for acid-base titration and sorption: $Cu(NO_3)_2 \cdot 3H_2O$ (Merck); $Ni(NO_3)_2 \cdot 3H_2O$ (Macrochem, Ukraine); bovine serum albumin (BSA, for biochemistry, protease free, Acros organics); buffer aqueous solution (phosphate buffered saline tablets, Fisher BioReagents, pH = 7.4); solutions of 0.1 M HNO_3, 0.1 M NaOH, and 0.1 M HCl prepared from fixanal concentrates; 0.1 M $NaNO_3$ solution prepared from $NaNO_3$ salt (chem.pure, Macrochem, Ukraine), methyl orange (1% solution prepared from methyl orange powder, chem.pure, Macrochem, Ukraine).

2.2. Synthesis of Silica-Based Particles with Amino Groups in the Surface Layer

For the synthesis of TEOS/APTES, TA (and TEOS/TMPED, TEDA) samples, 100 cm^3 (70 cm^3) of ethanol, 21.7 cm^3 of NH_4OH and 14 cm^3 of distilled water were mixed in a glass at a constant stirring with a magnetic stirrer. In a few minutes, 2 cm^3 (0.0083 mol) of APTES (1.9 cm^3 (0.0085 mol) of TMPED) were added, followed by 6 cm^3 (0.0263 mol) of TEOS. The precipitation started in 2 min (after that for TEDA, another 30 cm^3 of ethanol were added).

For the synthesis of TEOS/BTMPA, TBA sample, 30 cm^3 of ethanol, 21.7 cm^3 of NH_4OH and 14 cm^3 of distilled water were mixed in a glass at a constant stirring with a magnetic stirrer. Then, 2.8 cm^3 (0.0081 mol) of BTMPA in 40 cm^3 of ethanol were added to the reaction, followed by 6 cm^3 (0.0263 mol) of TEOS. In 2 min, the precipitation started, and when in several more minutes the precipitate turned curd-like, another 30 cm^3 of ethanol was added to the reaction.

In general, the molar ratio of TEOS/N-containing silane was ~3/1 (Table 1). The reactions continued for 1 h. The particles were separated from the solutions by centrifugation (10 min, 5000 rpm), and washed 3 times with ethanol. The samples were dried at 120 °C for 3 days.

Table 1. The content of functional groups in the studied samples and the value of the specific surface.

Sample Composition and Its Label	Molar Ratio TEOS/Amino-Silane	C_{groups} Theor., mmol/g	C_{groups} Titrat., mmol/g	C_{groups} Therm.an., mmol/g	S_{sp}, m^2/g
$SiO_2/{\equiv}Si(CH_2)_3NH_2$ (TA)	3/1	3.3	1.3	1.9	17
$SiO_2/{\equiv}Si(CH_2)_3NH(CH_2)_2NH_2$ (TEDA)	3/1	3.0	2.3	2.9	174
$SiO_2/[{\equiv}Si(CH_2)_3]_2NH$ (TBA)	3/1	2.6	2.0	2.1	179

2.3. Analysis Methods

The content of amine functional groups on the surface of the synthesised silica particles was determined by the method of acid-base back titration [34]. It is based on the determination of the number of protons absorbed by the amino group after holding the sorbent batch (0.1 g) in an excess of 0.1 M HCl solution (with periodic stirring) until equilibrium is established (72 h). The excess of hydrochloric acid was titrated with 0.1 M NaOH and methyl orange as an indicator.

Scanning electron microscopy (SEM) is widely used to study the morphology of sorbents and determine their particle sizes. The sample was mounted on the surface of the table with a pre-applied adhesive coating and sputtered with gold by the method of cathodic sputtering. The SEM images were collected with a scanning electron microscope JSM 6060 LA (Jeol, Tokyo, Japan) in the secondary electron mode at an accelerating voltage of 30 kV. The materials (previously deposited on a transparent carbon film on a copper substrate) were also analysed by transmission electron microscopy (TEM) using TEM/HRTEM (JEOL JEM 2010-Fx) at 200 kV and energy-dispersive X-ray spectroscopy (EDXS).

Nitrogen adsorption-desorption isotherms for all samples were collected using a Kelvin-1042 sorptometer. The samples were previously degassed in a helium stream at 110 °C for 1 h. The specific surface area (S_{sp}) was calculated according to BET [35] in the range of relative pressures 0.05–0.35.

Infrared (IR) reflectance spectra of the samples ground with KBr (FT-IR grade, Sigma-Aldrich) at 1/20 weight ratio were recorded on a Thermo Nicolet Nexus FTIR spectrometer using the "SMART Collector" diffuse reflectance attachment in the region of 4000–400 cm^{-1} at a resolution of 8 cm^{-1}.

The thermal analysis (differential thermal analysis (DTA), differential thermal gravimetry (DTG), and thermogravimetry (TG)) was performed on the MOM Q-1500 D (Paulik–Paulik–Erdey) derivatograph operating in the range of 20–1000 °C, with a heating rate of 10 °C·min^{-1}.

The ^{13}C and ^{29}Si MAS NMR experiments were conducted on a Bruker Avance II 400 spectrometer with 4 mm rotors (ZrO_2) spun at 10 kHz. ^{13}C NMR spectra were measured using $^1H{\rightarrow}^{13}C$ CP/MAS, ^1H decoupling during acquisition, a contact time of 3 ms, and 5 s of recycle delay. The number of scans was between 768 and 2300. ^{29}Si NMR spectra were recorded with 3 μs excitation pulses, a contact time of 2 ms, and 5 s of recycle delay. The number of scans was between 640 and 1024. ^{13}C and ^{29}Si chemical shifts are referenced towards 4,4-dimethyl-4-silapentane-1-sulfonic acid (DSS).

2.4. Adsorption Studies

The sorption of heavy metals nickel(II) and copper(II) was studied under static conditions. A batch of sorbent (0.05 g) was poured with 25 cm^3 of working solutions of metal ions. The content of copper and nickel in the aqueous phase was determined by the atomic absorption method on a C-115-M1 spectrophotometer along the resonance line of 324.7 nm (for copper) and 232.0 nm in a depleted (oxidizing) flame of an acetylene/air mixture. The source of resonant radiation was the LS-2 spectral lamp. The detection limit for metals was 0.01 $\mu g/cm^3$.

To study the physical adsorption of bovine serum albumin (BSA) from aqueous solution on functionalised nanoparticles, there was used 1% BSA stock solution in phosphate saline buffer and its diluted solutions 0.1–1.0%. UV–Vis spectrophotometer Helios Gamma (Thermo Electron Corporation, Great Britain) was used to register absorption spectra of BSA in solution. Experimental conditions: m = 0.02 g, V (working volume of albumin solution) = 10 cm^3, the contact time of phases was 1.0 h, t = 36°C. The equilibrium concentration of albumin was determined spectrophotometrically at λ = 278 nm (the thickness of the cell = 1 cm) after removing the sorbent.

3. Results and Discussion

In the current research we synthesised three different types of particles via co-condensation of TEOS (as the main structuring agent) and divergent amino-containing silanes (as functionalising agents), as schematically depicted in Figure 1 and summarized in Table 1. The first sample (**TA**) is carrying the usual 3-aminopropyl groups incorporated with APTES. The second sample (**TEDA**) bears (propyl)ethylenediamine functional groups, i.e., it has lengthened aminoorganic fragments with larger number of nitrogen donor centres. Finally, the third sample (**TBA**) carries secondary amino groups introduced by means of a bissilane having an amino group in the structure of the bridge. The syntheses of sorption materials were carried out in the same conditions of a water-ethanol environment with a molar ratio of reacting components TEOS/functionalising alkoxysilane of 3/1. Ammonium hydroxide was used as a catalyst for the hydrolytic polycondensation reaction [13].

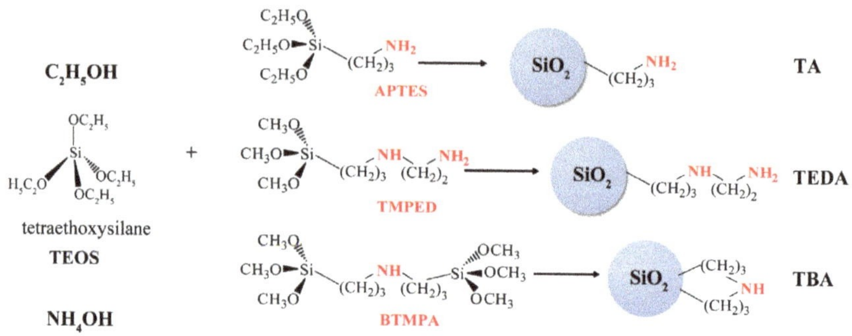

Figure 1. Synthesis of amino-containing spherical silica particles.

3.1. Morphology

The morphology and sizes of the synthesised particles were analysed with SEM microphotographs. In keeping with the SEM and TEM images of the **TA** sample presented in Figure 2, the material is formed by clearly visible spherical particles from 600 to 800 nm in size. Meanwhile, samples **TEDA** and **TBA** (Figure 2) are formed by 220–275 nm agglomerates and aggregates, whereas their primary particles are much smaller in size ~15 nm. That is, by the developed technique of synthesis, the emerging silica spheres with 3-aminopropyl groups are spherical microparticles with a size of 500–800 nm. In the same conditions of synthesis, the introduction of a larger functional group, as well as a group with a higher basicity, leads to the formation of much smaller-sized primary particles, which tend to agglomerate. Such a difference is explained by the rates of hydrolysis and condensation. The kinetics of hydrolysis of methoxy derivatives (functionalising methoxy silanes TMPED and BTMPA used in the synthesis of **TEDA** and **TBA** samples, respectively) is very fast, faster than that of ethoxy derivatives (functionalising ethoxy silanes APTES used in the synthesis of **TA** sample). However, the rate of condensation decreases with the increase in the size of the organic group near the silicon atom [36]. Thus, slow hydrolysis and fast condensation of APTES in the case of **TA** sample result in the formation of large microspheres. Meanwhile, fast hydrolysis and slow condensation of TMPED and BTMPA, as well as electrosteric barriers prevent the primary particles from uniting into spherical secondary particles and facilitating dendromer formation [37], lead to the emergence of 15 nm particles of **TEDA** and **TBA** samples.

Figure 2. SEM (the **left** column) and TEM (the **right** column) images of the synthesised samples.

3.2. The Composition and Structure of the Obtained Samples

Table 1 shows the values of the specific surface area for silica-based samples, which were calculated from the isotherms of low-temperature nitrogen adsorption-desorption (Figure 3). These data correlate with TEM data and are related to the particle sizes. Observing the results of low-temperature nitrogen adsorption-desorption analysis, 3-aminopropyl-bearing **TA** is a non-porous sample with the value of the specific surface $S_{sp.}$ = 17 m^2/g (Table 1), which is consistent with its large particle sizes (about 600–800 nm). Meanwhile, **TEDA** material with ethylenediamine groups and **TBA** material with secondary amine groups, formed by much smaller primary particles, are characterised by specific surfaces of 174 and 179 m^2/g. Calculations based on the SCV/SCR method [38] have shown that most pores are formed by slits between particles and particle agglomerates, and belong to mesopores for the **TEDA** sample and meso- and macropores for the **TBA** sample (Figure S1, Table S1) (see Supplementary Materials).

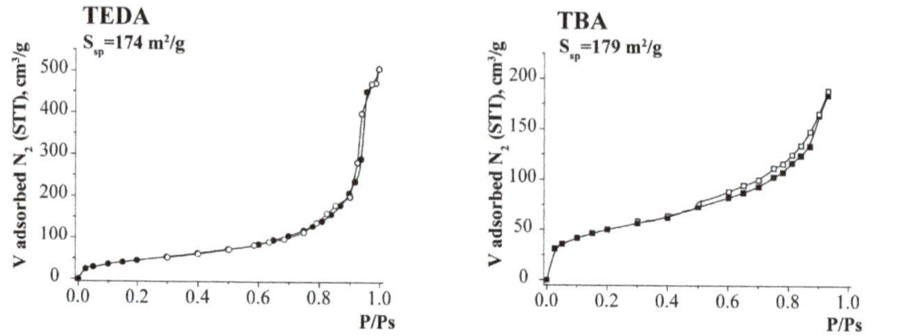

Figure 3. Low-temperature nitrogen adsorption-desorption isotherms for **TEDA** and **TBA** samples.

The IR spectra of the samples are illustrated in Figure 4. All of them contain an intense absorption band in the region of 1000–1200 cm^{-1} with a shoulder in the higher frequencies, which is characteristic of ν(Si-O-Si) stretches and indicates the formation of a network of polysiloxane bonds. A broad absorption band at 3300 cm^{-1} and a band of medium intensity at 1630–1640 cm^{-1} are caused by stretching and bending vibrations of water molecules adsorbed on the surface of the samples. However, there are absorption bands of amino groups in the same region [39] overlapping with water vibrations: two low-intensity absorption bands for **TA** and **TEDA** samples, and one weak band for **TBA** in the region of 3288–3366 cm^{-1} are related to $\nu_{s,as}$(NH) of the amino groups involved in the hydrogen bond. An additional band at ~1540 cm^{-1} refers to deformations $\delta(NH_2)$ of NH_2 groups [31], while a broad strong absorption band around ~800 cm^{-1} belongs to wagging and twisting vibrations of NH_2 (in **TA** and **TEDA**) and NH wagging (in **TBA**).

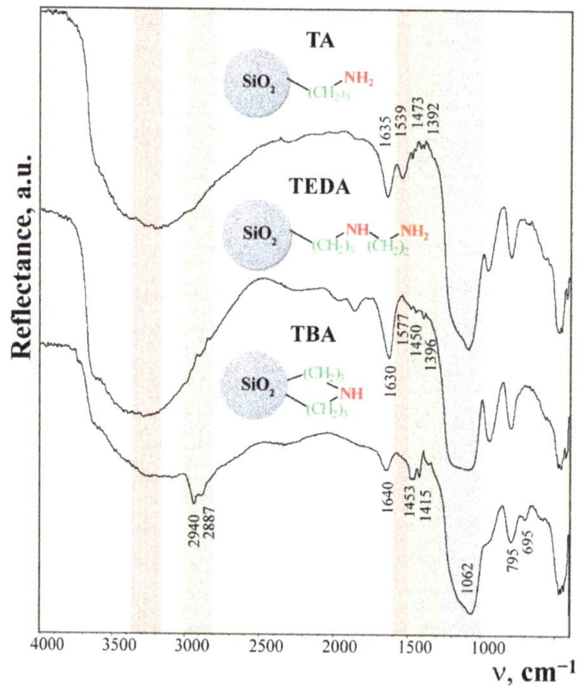

Figure 4. IR spectra of the synthesised samples.

Absorption bands in the region of 2850–2950 cm^{-1} belong to symmetric and asymmetric CH stretches of the aminopropyl radical, whereas a group of bands in the 1350–1480 cm^{-1} region correspond to bending vibrations of CH_2.

Summarizing the analysis of the IR spectra, the particles that were synthesised consist of a well-developed polysiloxane network with affixed targeted amine functional groups.

Furthermore, EDXS data (Figure S2) also confirm the existence of nitrogen-containing functional groups in the surface layers of the silica particles (for the **TA** sample, due to the small number of functional groups, signal N-K, apparently overlaps with oxygen signal).

The quantitative content of the organic component in the prepared samples was additionally assessed on the basis of thermal analysis data (Figure 5). Analysing thermoanalytical curves, all the samples lose weight when heated to 100 °C (**TA**—7.9%; **TEDA**—5.5%; **TBA**—12.2%), which can be associated with the removal of the remains of sorbed water, the evidence of which in the samples was also recorded by IR spectra. The **TEDA** and **TBA** samples have similar thermograms. For the **TEDA** sample, the exoeffect at 230 °C is more pronounced than for the **TBA** sample at 275 °C, and it is completely absent in the **TA** sample. The reason for such a difference may lie both in the porosity and in the more complex structure of the functional groups in **TEDA** and **TBA**, due to which the organic component burns out unevenly in several stages. The lack of endo effects above 200 °C in the DTG curve and the presence of exoeffects in the DTA curve of **TA** sample indicate processes of oxidation of organic groups, accompanied by the weight loss at a constant speed, due to the mutually compensating processes of formation and collapse of decomposition compounds [40]. Taking into account the weight loss for the combustion of the organic groups, the number of functional groups in the samples were calculated, and they agree with the titration data (Table 1). Following Table 1, amino groups content in the spherical silica microparticles **TA** is about 2 times less than in samples **TEDA** and **TBA**. Again, such a relatively high content of functional groups (**TEDA** and **TBA**) at the taken ratio of reacting components is more typical for xerogels with amino groups [41].

The ^{13}C NMR spectra (Figure 6, Table 2) of the synthesised samples confirm the presence of functional groups introduced during the synthesis, as well as the presence of residual ethoxy and methoxy groups [42,43]. These results correlate with the data of ^{29}Si NMR spectroscopy (Figure 6, Table 3). The presence of methoxy groups in the prepared samples demonstrates an incompleteness of hydrolysis of nitrogen-containing alkoxysilanes (TMPED or BTMPA) and, accordingly, explains the appearance of signals in ^{29}Si NMR spectra at ~57–58 ppm related to the T^2 structural units [32]. In addition, in the ^{29}Si NMR spectrum of the **TBA** sample (Figure 6), a signal at −50.5 ppm corresponding to the T^1 structural unit is observed (Table 3). The appearance of such a structural unit may be related to the nature of functionalising alkoxysilane (bissilane BTMPA), when the hydrolysis and polycondensation processes are hindered by the size of the bridging group. Such hindered hydrolysis and polycondensation of BTMPA may be one of the reasons for lower content of functional groups in **TBA** compared to **TEDA** sample (Table 1). However, as secondary amino groups of **TBA** are located within polysiloxane network, some of them may be both unavailable for titration and their burnout may be complicated. The presence in the ^{29}Si NMR spectra of **TA** of only Q^4 and Q^3 structural units indicates the completeness of hydrolysis in the TEOS/APTES system [44].

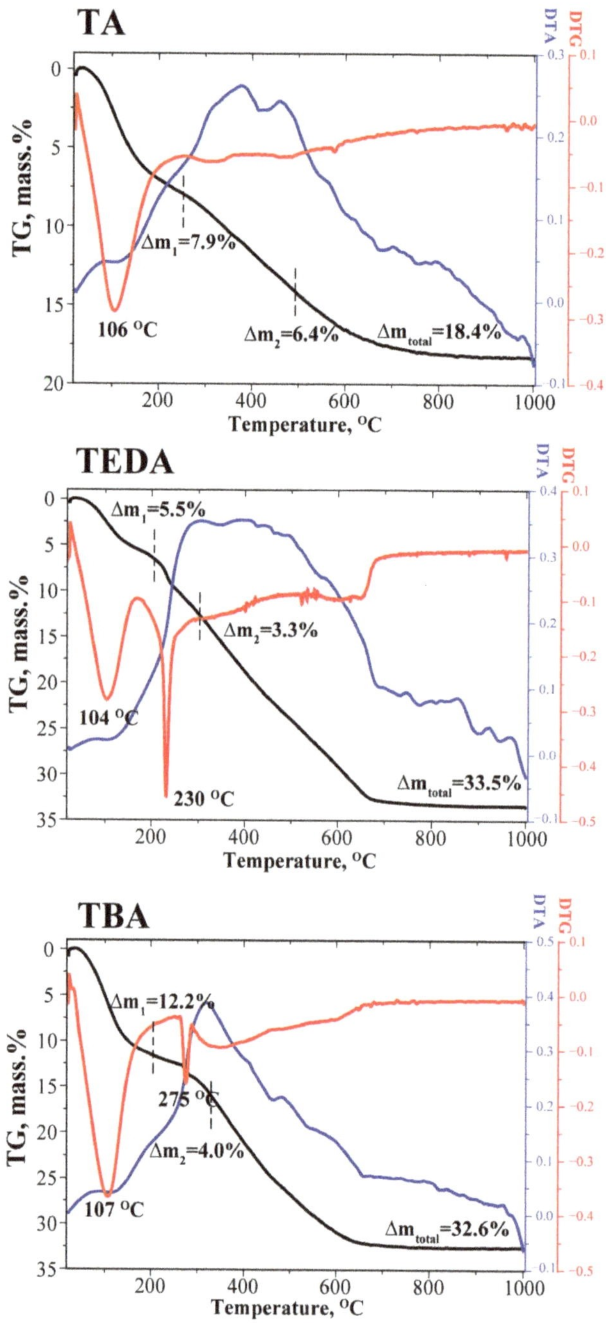

Figure 5. Thermoanalysis curves for the amino-containing samples.

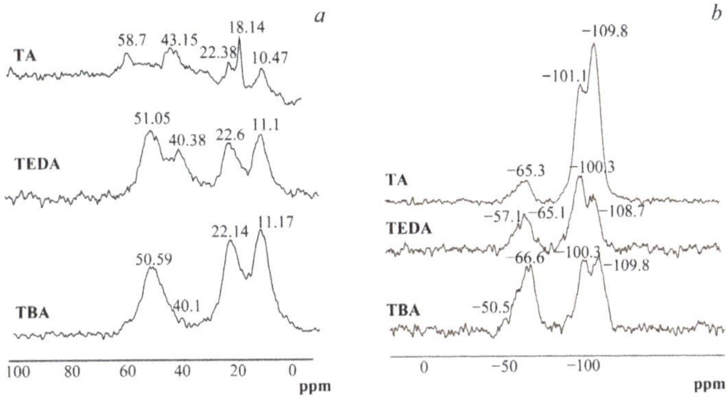

Figure 6. ^{13}C NMR spectra (**a**) and ^{29}Si NMR spectra (**b**) of synthesised samples.

Table 2. ^{13}C CP/MAS NMR spectroscopy data for samples with nitrogen-containing groups.

Peak Assignments	Chemical Shifts, ppm		
	TA	TEDA	TBA
$\equiv Si\text{-}\underline{C}H_2\text{-}CH_2\text{-}CH_2\text{-}N$	10.47	11.1	11.17
$\equiv Si\text{-}O\text{-}\underline{C}H_2\text{-}CH_3$	18.14	-	-
$\equiv Si\text{-}CH_2\text{-}\underline{C}H_2\text{-}CH_2\text{-}N$	22.38	22.6	22.14
$\equiv Si\text{-}CH_2\text{-}CH_2\text{-}\underline{C}H_2\text{-}N$	43.15	40.38	40.1
$\equiv Si\text{-}O\text{-}\underline{C}H_3$	-	51.05	50.59
$\equiv Si\text{-}O\text{-}CH_2\text{-}\underline{C}H_3$	58.7	-	-

Table 3. ^{29}Si CP/MAS NMR spectroscopy data for samples with nitrogen-containing groups.

Structural Units	Chemical Shifts, Ppm		
	TA	TEDA	TBA
Q^4	−109.8	−108.7	−109.8
Q^3	−101.1	−100.3	−100.3
Q^2	-	-	-
T^3	−65.3	−65.1	−66.6
T^2	-	−57.1	−58.5
T^1	-	-	−50.5

3.3. Sorption of Copper(II) and Nickel(II) Ions from Aqueous Solutions

The sorption studies of Cu(II) ions were carried out at pH ~5.5, and Ni(II) at ~6.5 [14], because the higher the pH, the better the sorption of these metals, but copper(II) and nickel(II) hydroxides begin to form at pH values ~5.5 and ~7, respectively [45]. As demonstrated in Figure 7, these materials are not an exception, and the best sorption of nickel(II) by all samples was observed at pH = 6. The addition of acids promotes the protonation of the amino groups reducing their sorption capacity. Meanwhile, the addition of alkali can cause the formation of basic metal salts. Therefore, aqueous solutions of salts with their natural pH were used in the sorption experiments.

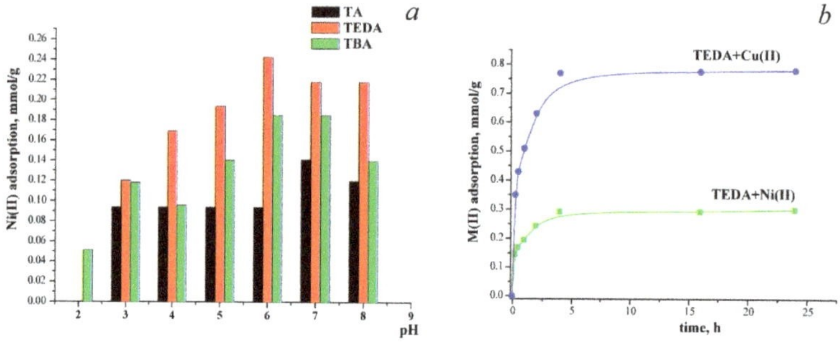

Figure 7. Dependence of nickel(II) ions adsorption on pH for all samples (**a**) and kinetic curves of Ni(II) (green line and symbols) and Cu(II) (blue line and symbols) sorption by **TEDA** sorbent (**b**).

The study of the kinetics of Ni^{2+} and Cu^{2+} sorption by the obtained sorbents was carried out in static mode. Graphically, the kinetics of the Ni^{2+} and Cu^{2+} sorption process by the **TEDA** sample (as an example) is presented in Figure 7. The kinetic curves of Ni(II) and Cu(II) sorption align well with the pseudo-second-order equation (Figure S3, Table S2): the cations sorption process occurs on the same type of active centres on the surface of functional silicas, but with the formation of different complexes. Although the initial uptake of Cu^{2+} by **TEDA** is faster than Ni^{2+}, the rate constant of Ni^{2+} sorption is higher. However, it appeared that 4 h of contact of **TEDA** with either of the mentioned ions was enough to establish a state of dynamic equilibrium. However, relying on our previous research [22,29], the establishment of adsorption equilibrium between copper cations and the samples with 3-aminopropyl groups lasted longer, so 24 h was chosen as the optimal time for further investigations of isothermal sorption.

The experimental isotherms of copper(II) and nickel(II) ions sorption on aminosilica particles, as well as their plots in the coordinates of the linearised Langmuir and Freundlich isotherm model equations, are illustrated in Figure 8, and Table 4 presents the parameters of Ni(II) and Cu(II) sorption. According to these calculations, the adsorption occurs on a heterogeneous surface for a **TA** sample with 3-aminopropyl groups, which are the simplest in structure among the others. This may be due to the fact that amino groups form hydrogen bonds with other amino and silanol groups [22], and adsorption occurs on adsorption centres of different energy. For the samples with diamine groups (**TEDA**) and secondary amines (**TBA**), the isotherms are well fitted by the Langmuir model, which indicates chemisorption [46].

Moreover, the CLINP 2.1 program [47] was used to calculate the composition of the complexes formed on the surface of such sorbents. The composition of copper(II)/N-group complexes is 1/1 for the 3-aminopropyl group and 1/1 and 1/2 for the ethylenediamine and the secondary amine groups, respectively (Figure 8). In the case of nickel(II) complexes, their composition depends on the number of functional groups in the surface layer. If the group content is approximately 1 mmol/g, then complexes of composition 1/1 (**TA** sample) are formed, and if the group content increases, then 1/1 and 1/2 (for **TBA** sample) and 1/2 (for **TEDA** sample) complexes are formed. Thus, the number of nitrogen-containing groups and their structure determine the formation of various complexes on the surface, which affects the total sorption capacity of the samples.

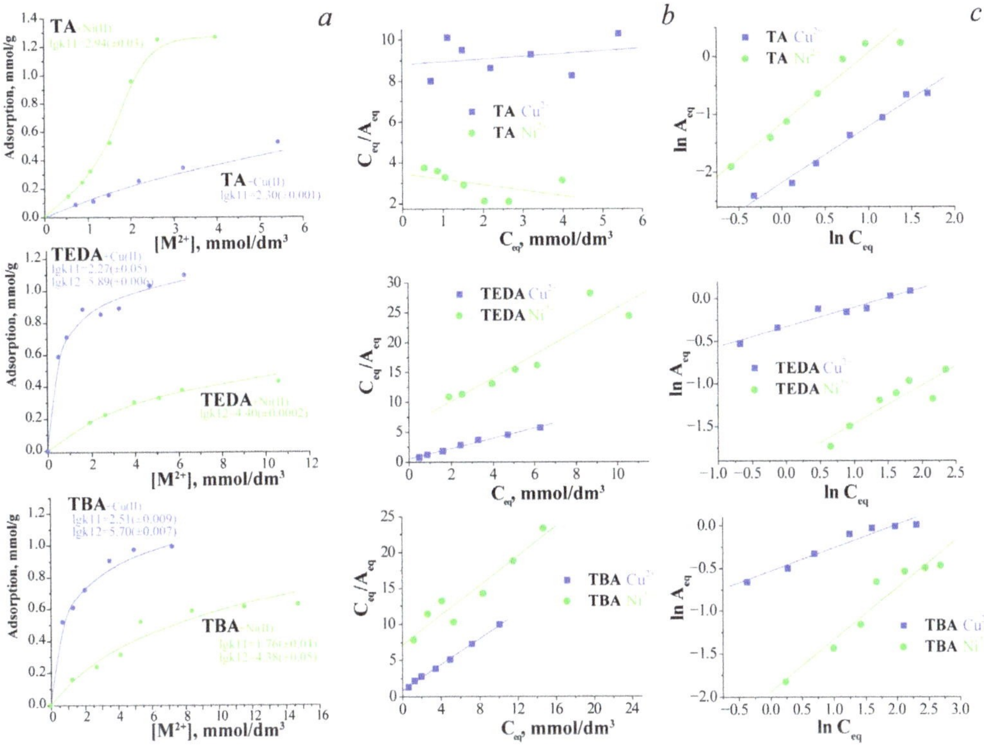

Figure 8. (**a**) Copper(II) and Nickel(II) adsorption isotherms (Adsorption vs. equilibrium concentration of metal ions [M^{2+}]) for amine-containing silica microspheres (points—experimental data, lines—calculated data); adsorption isotherms in the coordinates of the Langmuir (**b**) and Freundlich (**c**) equations.

Table 4. Parameters of Ni(II) and Cu(II) ions sorption in the Langmuir and Freundlich equations.

Sample	Cation	SSC, mmol/g	Langmuir Equation $C_{eq}/A_{eq} = 1/(K_L \cdot A_{max}) + (1/A_{max}) \cdot C_{eq}$			Freundlich Equation $\ln A_{eq} = \ln K_F + (1/n) \cdot \ln C_{eq}$		
			A_{max}, mmol/g	K_L, dm^3/mmol	R^2	K_F, mmol/g	n	R^2
TA	Ni(II)	1.27	−3.69	−0.077	0.2429	0.311	0.82	**0.9541**
	Cu(II)	0.53	7.54	0.015	0.0669	0.112	1.04	**0.9824**
TEDA	Ni(II)	0.43	0.51	0.314	**0.8599**	0.147	2.23	0.8206
	Cu(II)	1.09	1.17	1.512	**0.9888**	0.719	4.40	0.9397
TBA	Ni(II)	0.63	0.96	0.147	**0.9142**	0.143	1.65	**0.9274**
	Cu(II)	1.01	1.12	1.084	**0.9980**	0.589	3.68	0.9458

A_{eq}—adsorption capacity at equilibrium, mmol/g; K_L—the Langmuir constant, which characterises the adsorption energy, dm^3/mmol; C_{eq}—equilibrium concentration of metal ions in the solution, mmol/dm^3; A_{max}—maximal adsorption capacity for complete monolayer covering of the surface, mmol/g; K_F—a Freundlich constant, mmol/g; n—an empirical parameter related to the intensity of adsorption.

3.4. Albumin Adsorption Research

The adsorption of BSA on samples with different types of amino groups was also probed (Figure 9, Table 5). In terms of sorption capacity to BSA molecules, the tested samples can be ranged as the following: **TBA** (204.6 mg/g) > **TEDA** (181.5 mg/g) > **TA** (78.3 mg/g). It was shown that such large protein molecules as BSA have a higher affinity to the materials with higher specific surface area and content of available amino groups for interactions, **TEDA** and **TBA**.

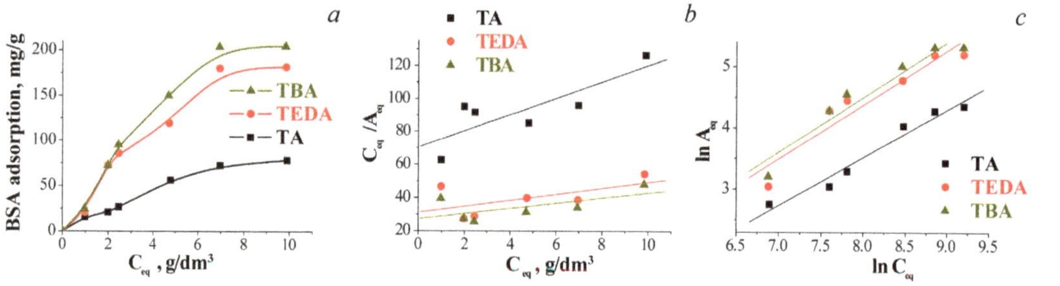

Figure 9. BSA adsorption isotherms for amine-containing silica microspheres (**a**); adsorption isotherms in the coordinates of the Langmuir (**b**) and Freundlich (**c**) equations.

Table 5. Parameters of BSA sorption in the Langmuir and Freundlich equations.

Sample	SSC, mg/g	Langmuir Equation $C_{eq}/A_{eq} = 1/(K_L \cdot A_{max}) + (1/A_{max}) \cdot C_{eq}$			Freundlich Equation $\ln A_{eq} = \ln K_F + (1/n) \cdot \ln C_{eq}$		
		A_{max}, mg/g	K_L, dm³/mg	R^2	K_F, mg/g	n	R^2
TA	78.3	200.8	0.0001	0.682	0.064	1.28	**0.965**
TEDA	181.5	549.5	0.0001	0.356	0.066	1.13	**0.904**
TBA	204.6	641.0	0.0001	0.400	0.067	1.11	**0.927**

A_{eq}—adsorption capacity at equilibrium, mmol/g; K_L—the Langmuir constant, which characterises the adsorption energy, dm³/mmol; C_{eq}—equilibrium concentration of metal ions in the solution, mmol/dm³; A_{max}—maximal adsorption capacity for complete monolayer covering of the surface, mmol/g; K_F—a Freundlich constant, mmol/g; n—an empirical parameter related to the intensity of adsorption.

The higher values of correlation coefficients of isotherms of BSA adsorption by all the studied samples with Freindlich isotherm model indicate adsorption on heterogeneous adsorption centres (Table 5). Thus, a **TBA** material, which apart from amino groups also contains the most significant among the samples hydrocarbon component, is characterised by the highest SSA value towards BSA. Moreover, the molecules of BSA can also be trapped between the aggregates of small particles of **TEDA** and **TBA** in the course of sorption.

4. Conclusions

The sol-gel method and its variant, Stöber technique, were applied for the synthesis of spherical silica particles containing 3-aminopropyl, ethylenediamine and secondary amine groups. It was demonstrated that the size of the obtained particles is affected by the size of the functional group (steric factor), the type of applied (methoxy or ethoxy) functionalising derivative, and, consequently, the hydrolysis rate, leading to the formation of spherical particles with 3-aminopropyl groups with a size up to 800 nm, and dendric silica particles with ethylenediamine and secondary amine groups making aggregates up to 275 nm. The presence of functional groups was qualitatively confirmed by IR spectroscopy, ^{13}C NMR spectroscopy, thermal analysis, and titration, and the quantitative content of functional groups available for interaction was 1.3–2.3 mmol/g. The calculated composition of surface complexes of metal ions indicates the complexation of one Cu(II) and Ni(II) ion with one or two surface amine-containing groups, depending on the concentration and geometry of the surface functional groups. Therefore, current research explains how under similar conditions of synthesis, functional silanes of a similar nature, but of a different structure, affect the morphology and the properties of the obtained materials. Using nitrogen-containing silanes of various configurations, it is possible to form various complexes of copper(II) and nickel(II) on the surface, and to regulate the sorption capacity of the materials. Besides, the amino-containing spherical silicas are suitable for the adsorption of biomolecules, whereby the sorption activity of the samples depends on the number of available functional groups, the value of the specific surface and the presence of heterogeneous sorption centres on the surface.

Supplementary Materials: The following supporting information can be downloaded at: https://www.mdpi.com/article/10.3390/cryst13020190/s1, Figure S1: Pore-size distribution curves for the porous samples: TEDA and TBA plotted based on the SCV/SCR method; Figure S2: The EDXS spectra of the synthesised samples title; Figure S3: Kinetic curves of Ni(II) and Cu(II) sorption by **TEDA** silica in Lagergren equation coordinates for pseudo-first (left) and pseudo-second (right) order processes; Table S1: Structural and adsorption characteristics of the synthesised porous samples; Table S2: Kinetic parameters of Cu(II) and Ni(II) ions adsorption by the **TEDA** sample.

Author Contributions: Conceptualization, I.M.; Methodology, I.M., V.T. and N.S.; Software, I.M.; Formal analysis, V.T., N.S. and I.M.; Investigation, V.T., N.S. and I.M.; Resources, I.M.; Data curation, V.T., N.S. and I.M.; Writing-original draft preparation, N.S., V.T. and I.M.; Writing-review and editing, I.M.; Supervision, I.M.; Project administration, I.M. All authors have read and agreed to the published version of the manuscript.

Funding: The research was carried out within the Slovak Research and Development Agency (APVV-19-0302) and Ministry of Education, Science, Research and Sport of the Slovak Republic (VEGA 2/0116/23) projects.

Data Availability Statement: The data presented in this study are available on request from the corresponding author.

Conflicts of Interest: The authors declare no conflict of interest.

References

1. Briffa, J.; Sinagra, E.; Blundell, R. Heavy metal pollution in the environment and their toxicological effects on humans. *Heliyon* **2020**, *6*, e04691. [CrossRef] [PubMed]
2. Manahan, S.E. *Environmental Chemistry*, 7th ed.; Lewis Publishers: Boca Raton, FL, USA, 2000; p. 898.
3. Sharma, S.; Bhattacharya, A. Drinking water contamination and treatment techniques. *Appl. Water Sci.* **2017**, *7*, 1043–1067. [CrossRef]
4. Singh, B.; Na, J.; Konarova, M.; Wakihara, T.; Yamauchi, Y.; Salomon, C.; Gawande, M.B. Functional mesoporous silica nanomaterials for catalysis and environmental applications. *Bull. Chem. Soc. Jpn.* **2020**, *93*, 1412–1608. [CrossRef]
5. Gomes, A.L.M.; Andrade, P.H.M.; Palhares, H.G.; Dumont, M.R.; Soares, D.C.F.; Volkringer, C.; Houmard, M.; Nunes, E.H.M. Facile sol-gel synthesis of silica sorbents for the removal of organic pollutants from aqueous media. *J. Mater. Res. Technol.* **2021**, *15*, 4580–4594. [CrossRef]
6. Pietras-Ożga, D.; Piątkowska-Sawczuk, K.; Duro, G.; Pawlak, B.; Stolyarchuk, N.; Tomina, V.; Melnyk, I.; Giannakoudakis, D.A.; Barczak, M. Chapter 9-Sol-gel–derived silica xerogels: Synthesis, properties, and their applicability for removal of hazardous pollutants. In *Advanced Materials for Sustainable Environmental Remediation*; Giannakoudakis, D., Meili, L., Anastopoulos, I., Eds.; Elsevier: Amsterdam, The Netherlands, 2022; pp. 261–277; ISBN 9780323904858. [CrossRef]
7. Sarker, M.Z.; Rahman, M.M.; Minami, H.; Suzuki, T.; Ahmad, H. Amine functional silica–supported bimetallic Cu-Ni nanocatalyst and investigation of some typical reductions of aromatic nitro-substituents. *Colloid Polym. Sci.* **2022**, *300*, 279–296. [CrossRef]
8. Tomina, V.V.; Furtat, I.M.; Stolyarchuk, N.V.; Zub, Y.L.; Kanuchova, M.; Vaclavikova, M.; Melnyk, I.V. Surface and structure design of aminosilica nanoparticles for multifunctional applications: Adsorption and antimicrobial studies. In *Micro and Nano Technologies, Biocompatible Hybrid Oxide Nanoparticles for Human Health*; Melnyk, I.V., Vaclavikova, M., Seisenbaeva, G.A., Kessler, V.G., Eds.; Elsevier: Amsterdam, The Netherlands, 2019; pp. 15–31. ISBN 9780128158753. [CrossRef]
9. Miller, P.J.; Shantz, D.F. Covalently functionalized uniform amino-silica nanoparticles. Synthesis and validation of amine group accessibility and stability. *Nanoscale Adv.* **2020**, *2*, 860. [CrossRef]
10. Cuoq, F.; Masion, A.; Labille, J.; Rose, J.; Ziarelli, F.; Prelot, B.; Bottero, J.-Y. Preparation of amino-functionalized silica in aqueous conditions. *Appl. Surf. Sci.* **2013**, *266*, 155–160. [CrossRef]
11. Amino-Functionalized Silica Gel Spherical. Available online: https://www.sigmaaldrich.com/UA/en/product/supelco/79297?gclid=CjwKCAiAnZCdBhBmEiwA8nDQxY8ctBH_7HW0TXpTAb5pzwWL7WszPYaDBmX3zYJPAQ0s61SM9cd8hoC_K0QAvD_BwE&gclsrc=aw.ds (accessed on 30 December 2022).
12. Amine Functionalized Silica Nanoparticles. Available online: https://www.geandr.com/products/amine-functionalized-silica-nanoparticles (accessed on 30 December 2022).

13. Stöber, W.; Fink, A.; Bohn, E. Controlled growth of monodisperse silica spheres in the micron size range. *J. Colloid Interface Sci.* **1968**, *26*, 62–69. [CrossRef]
14. He, Y.; Luo, L.; Liang, S.; Long, M.; Xu, H. Amino-functionalized mesoporous silica nanoparticles as efficient carriers for anticancer drug delivery. *J. Biomater. Appl.* **2017**, *32*, 524–532. [CrossRef]
15. Martínez-Carmona, M.; Ho, Q.P.; Morand, J.; García, A.; Ortega, E.; Erthal, L.C.S.; Ruiz-Hernandez, E.; Santana, M.D.; Ruiz, J.; Vallet-Regí, M.; et al. Amino-functionalized mesoporous silica nanoparticle-encapsulated octahedral organoruthenium complex as an efficient platform for combatting cancer. *Inorg. Chem.* **2020**, *59*, 10275–10284. [CrossRef]
16. Kardys, A.Y.; Bharali, D.J.; Mousa, S.A. Amino-functionalized silica nanoparticles: In vitro evaluation for targeted delivery and therapy of pancreatic cancer. *J. Nanotechnol.* **2013**, *2013*, 768724. [CrossRef]
17. Tomina, V.; Furtat, I.; Lebed, A.; Kotsyuda, S.; Kolev, H.; Kanuchova, M.; Marcin Behunova, D.; Vaclavikova, V.; Melnyk, I. Diverse pathway to obtain antibacterial and antifungal agents based on silica particles functionalized by amino and phenyl groups with Cu(II) ions complexes. *ACS Omega* **2020**, *5*, 15290–15300. [CrossRef] [PubMed]
18. Liu, H.; Guo, Y.; Wang, X.; Liang, X.; Liu, X.; Jiang, S. A novel fullerene oxide functionalized silica composite as stationary phase for high performance liquid chromatography. *RSC Adv.* **2014**, *4*, 17541–17548. [CrossRef]
19. Gao, J.; Wu, S.; Tan, F.; Tian, H.; Liu, J.; Lu, G.Q.M. Nanoengineering of amino-functionalized mesoporous silica nanospheres as nanoreactors. *Progress in Natural Science: Mater. Int.* **2018**, *28*, 242–245. [CrossRef]
20. Cheedarala, R.V.; Chalapathi, K.V.; Song, J.I. Delayed flammability for natural fabrics by deposition of silica core-amine shell microspheres through dip-coating process. *Chem. Eng. J. Adv.* **2021**, *8*, 100164. [CrossRef]
21. Attia, M.F.; Anton, N.; Bouchaala, R.; Didier, P.; Arntz, Y.; Messaddeq, N.; Klymchenko, A.S.; Mely, Y.; Vandamme, T.F. Functionalization of nano-emulsions with an amino-silica shell at the oil–water interface. *RSC Adv.* **2015**, *5*, 74353–74361. [CrossRef]
22. Melnyk, I.V.; Tomina, V.V.; Stolyarchuk, N.V.; Seisenbaeva, G.A.; Kessler, V.G. Organic dyes (acid red, fluorescein, methylene blue) and copper(II) adsorption on amino silica spherical particles with tailored surface hydrophobicity and porosity. *J. Mol. Liq.* **2021**, *336*, 116301. [CrossRef]
23. Peng, S.; Deng, Y.; Li, W.; Chen, J.; Liuac, H.; Chen, Y. Aminated mesoporous silica nanoparticles for the removal of low-concentration malodorous aldehyde gases. *Environ. Sci. Nano* **2018**, *5*, 2663–2671. [CrossRef]
24. Hamdy, L.B.; Goel, C.; Rudd, J.A.; Barron, A.R.; Andreoli, E. The application of amine-based materials for carbon capture and utilisation: An overarching view. *Mater. Adv.* **2021**, *2*, 5843–5880. [CrossRef]
25. Howard, A.G.; Khdary, N.H. Nanoscavenger based dispersion preconcentration; sub-micron particulate extractants for analyte collection and enrichment. *Analyst* **2005**, *130*, 1432–1438. [CrossRef]
26. Li, J.; Miao, X.; Hao, Y.; Zhao, J.; Sun, X.; Wang, L. Synthesis, amino-functionalization of mesoporous silica and its adsorption of Cr(VI). *J. Colloid Interface Sci.* **2008**, *318*, 309–314. [CrossRef] [PubMed]
27. Dobrowolski, R.; Oszust-Cieniuch, M.; Dobrzyńska, J.; Barczak, M. Amino-functionalized SBA-15 mesoporous silicas as sorbents of platinum (IV) ions. *Colloids Surf. A: Physicochem. Eng. Asp.* **2013**, *435*, 63–70. [CrossRef]
28. Melnyk, I.V. Aminosilica nano-and submicrospheres: Study of influence factors on morphology, structure and properties. *Chem. J. Mold.* **2014**, *9*, 123–127. [CrossRef] [PubMed]
29. Tomina, V.; Melnyk, I.; Zub, Y.; Kareiva, A.; Vaclavikova, M.; Seisenbaeva, M.; Kessler, V. Tailoring bifunctional hybrid organic-inorganic nanoadsorbents by the choice of functional layer composition probed by adsorption of Cu^{2+} ions. *Beilstein J. Nanotechnol.* **2017**, *8*, 334–347. [CrossRef] [PubMed]
30. Akdemir, E.; Imamoglu, M. Solid-phase extraction of trace Ni(II) ions on ethylenediamine-silica material synthesized by sol–gel method. *Desalination Water Treat.* **2013**, *52*, 4889–4894. [CrossRef]
31. Khdary, N.H.; Ghanem, M.A. Metal–organic–silica nanocomposites: Copper, silver nanoparticles–ethylenediamine–silica gel and their CO_2 adsorption behaviour. *J. Mater. Chem.* **2012**, *22*, 12032–12038. [CrossRef]
32. Randall, J.P.; Meador, M.A.B.; Jana, S.C. Polymer reinforced silica aerogels: Effects of dimethyldiethoxysilane and bis(trimethoxysilylpropyl)amine as silane precursors. *J. Mater. Chem. A* **2013**, *1*, 6642–6652. [CrossRef]
33. Haghighat, M.; Shirini, F.; Golshekan, M. Periodic mesoporous organosilica containing bridges with N-sulfonic acid groups: A new catalyst for the efficient formylation of amines and alcohols. *Silicon* **2020**, *12*, 2087–2098. [CrossRef]
34. Khatib, I.S.; Parish, R.V. Insoluble ligands and their application. I. A comparison of silica-immobilized ligands and functionalized polysiloxanes. *J. Organomet. Chem.* **1989**, *369*, 9–16. [CrossRef]
35. Brunauer, J.S.; Emmet, P.H.; Teller, E. Adsorption of gases in multimolecular layers. *J. Am. Chem. Soc.* **1938**, *60*, 309–319. [CrossRef]
36. Brinker, C.J. Hydrolysis and condensation of silicates: Effects on structure. *J. Non-Cryst. Solids* **1988**, *100*, 31–50. [CrossRef]
37. Brinker, C.J.; Scherer, G.W. *The Physics and Chemistry of Sol-Gel Processing*; Academic Press: Cambridge, MA, USA, 1990; p. 908. ISBN 978-0-08-057103-4. [CrossRef]
38. Gun'ko, V.M. Composite materials: Textural characteristics. *Appl. Surf. Sci.* **2014**, *307*, 444–454. [CrossRef]
39. Colthup, N.B.; Daly, L.H.; Wiberley, S.E. Amines, C=N, And N=O Compounds. In *Introduction to Infrared and Raman Spectroscopy*; Academic Press: Cambridge, MA, USA, 1990; pp. 339–354. [CrossRef]
40. Rymuszka, D.; Terpiłowski, K.; Sternik, D.; Tomczyńska-Mleko, M.; Goncharuk, O. Wettability and thermal analysis of hydrophobic poly(methyl methacrylate)/silica nanocomposites. *Adsorpt. Sci. Technol.* **2017**, *35*, 560–571. [CrossRef]

41. Zub, Y.L.; Melnyk, I.V.; Stolyarchuk, N.V.; Dobryanska, H.I.; Barczak, M.; Dabrowski, A. Comparative characteristics of texture and properties of hybrid organic–inorganic adsorbents functionalized by amine and thiol groups. *Prog. Solid State Chem.* **2005**, *33*, 179–186. [CrossRef]
42. Ostwal, M.; Singh, R.P.; Dec, S.F.; Lusk, M.T.; Way, J.D. 3-Aminopropyltriethoxysilane functionalized inorganic membranes for high temperature CO_2/N_2 separation. *J. Membr. Sci.* **2011**, *369*, 139–147. [CrossRef]
43. Wahab, M.A.; Kim, I.; Ha, C.-K. Bridged amine-functionalized mesoporous organosilica materials from 1,2-bis(triethoxysilyl)ethane and bis[(3-trimethoxysilyl)propyl]amine. *J. Solid State Chem.* **2004**, *177*, 3439–3447. [CrossRef]
44. Park, K.-W.; Jeong, S.-Y.; Kwon, O.-Y. Interlamellar silylation of H-kenyaite with 3-aminopropyltriethoxysilane. *Appl. Clay Sci.* **2004**, *27*, 21–27. [CrossRef]
45. Stolyarchuk, N.; Tomina, V.; Biswajit, M.; Tripathi, B.P.; Václavíková, M.; Dudarko, O.; Melnyk, I. Direct synthesis of efficient silica-based adsorbents carrying EDTA groups for the separation of Cu(II) and Ni(II) ion. *Colloids Surf. A: Physicochem. Eng. Asp.* **2022**, *650*, 129538. [CrossRef]
46. Tomina, V.; Stolyarchuk, N.; Semeshko, O.; Barczak, M.; Melnyk, I. Diamine groups on the surface of silica particles as complex-forming linkers for metal cations. *Molecules* **2023**, *28*, 430. [CrossRef]
47. CLINP 2.1, a Software Program for Computation of Stability Constants and Physicochemical Parameters of Complex Compounds in Solutions, Extractional and Sorptional Systems on the Base of Composition-Property Dependencies. Available online: http://chemo.univer.kharkov.ua/kholin/clinp.html (accessed on 30 December 2022).

crystals

MDPI

Article

Solvothermal Synthesis of Calcium Hydroxyapatite via Hydrolysis of Alpha-Tricalcium Phosphate in the Presence of Different Organic Additives

Rasa Karalkeviciene [1], Eva Raudonyte-Svirbutaviciene [1], Aleksej Zarkov [1,*], Jen-Chang Yang [2], Anatoli I. Popov [3] and Aivaras Kareiva [1]

1 Institute of Chemistry, Faculty of Chemistry and Geosciences, Vilnius University, Naugarduko 24, LT-03225 Vilnius, Lithuania
2 Graduate Institute of Nanomedicine and Medical Engineering, College of Biomedical Engineering, Taipei Medical University, 250 Wu-Hsing St., Taipei 11052, Taiwan
3 Institute of Solid State Physics, University of Latvia, Kengaraga 8, LV-1063 Riga, Latvia
* Correspondence: aleksej.zarkov@chf.vu.lt

Abstract: In this study, the effects of sodium lauryl sulfate and various amino acids (DL-aspartic acid, dodecanedioic acid, and suberic acid) on the formation of calcium-deficient hydroxyapatite via hydrolysis of α-tricalcium phosphate (α-TCP) were investigated; moreover, a combined effect of these additives and ethylene glycol as a synthesis medium was also estimated. The hydrolysis reaction was performed in solutions containing different concentrations of additives in aqueous and mixed aqueous–organic media under solvothermal conditions. It was demonstrated that the nature and the concentration of organic additives influence the phase purity and morphology of the final product. Higher concentrations of sodium lauryl sulfate and dodecanedioic acid induced the formation of impurities in addition to hydroxyapatite, while aspartic and suberic acid did not affect the phase purity. The morphology of the samples varied from plate- to rod-like depending on the concentrations of specific organic additive.

Keywords: calcium hydroxyapatite; α-tricalcium phosphate; amino acids; organic additives; solvothermal synthesis

Citation: Karalkeviciene, R.; Raudonyte-Svirbutaviciene, E.; Zarkov, A.; Yang, J.-C.; Popov, A.I.; Kareiva, A. Solvothermal Synthesis of Calcium Hydroxyapatite via Hydrolysis of Alpha-Tricalcium Phosphate in the Presence of Different Organic Additives. *Crystals* **2023**, *13*, 265. https://doi.org/10.3390/cryst13020265

Academic Editor: Ruikang Tang

Received: 5 January 2023
Revised: 30 January 2023
Accepted: 1 February 2023
Published: 3 February 2023

1. Introduction

Calcium phosphates (CPs) are the group of materials which are widely used in different areas, including medicine and bone regeneration [1], catalysis [2], sorption of organic pollutants [3], and heavy metals [4], as host matrices for the development of optical materials [5] and many others. Probably, the most popular member of the CPs family is calcium hydroxyapatite ($Ca_{10}(PO_4)_6(OH)_2$, HAp), which is the main inorganic component of human hard tissues.

HAp crystallizes in a hexagonal crystal structure with a space group $P6_3/m$. The crystal structure of HAp is notable due to the fact that it has two types of crystal planes, namely, *a*-face and *c*-face. The *a*-face is positively charged due to calcium ions, while the *c*-face is negatively charged due to oxygen atoms belonging to phosphate ions [6]. Anisotropic growth of HAp crystals in one or another crystallographic direction is an important factor from the point of view of specific interaction of oriented crystals with other substances, which may lead to enhanced performance and specific application of this material. For example, the adsorption of proteins on HAp crystals depends on their morphology. Different types of proteins, such as basic and acidic proteins, can adsorb selectively on the crystal planes of HAp [7,8]. Another effect of crystallographically oriented HAp particles was shown by Goto et al. [3] who demonstrated that tuning the morphology of HAp crystals may lead to enhanced or selective adsorption of cationic or anionic dyes. In this light, the controllable synthesis of plate-like or rod-like crystals with different aspect ratios and surface charges is an important task for many industries and purification technologies.

Synthetic HAp can be obtained in a variety of ways, including solid-state reaction [9], sol-gel method [10], as well as wet precipitation [11], hydrothermal [12] and solvothermal syntheses [13], and others. Another group of synthetic approaches employed for the synthesis of HAp considers the phase conversion from other less stable CPs. For instance, Ca-deficient HAp can be obtained through the transformation from alpha-tricalcium phosphate ($Ca_3(PO_4)_2$, α-TCP) or amorphous CP in aqueous medium [14]. Synthetic pathways employing high-temperature treatment, such as solid-state reaction or sol-gel, usually result in the formation of large agglomerated particles and do not lead to the anisotropic growth of the crystals. Crystallographically oriented products are usually obtained by low-temperature wet chemical methods.

Functional groups of organic substances, such as carboxyl or amino, can specifically interact with the planes of HAp particles during the crystal growth process and affect particle growth in one or another direction. Different additives were previously employed in order to investigate their effect on the crystal growth, morphology, and properties of HAp. For instance, Suchanek et al. [15] successfully synthesized HAp nanofibers in the presence of monoethanolamine under hydrothermal conditions. In et al. [16] investigated the role of sebacic acid on the formation of HAp and demonstrated that sebacic acid can accelerate nucleation and act as an inhibitor in the specific direction of HAp crystal growth. Jiang et al. [17] showed that poly(acrylic acid), depending on its concentration, can promote or inhibit HAp crystallization; moreover, the composition of mixed aqueous–organic reaction medium also has an effect on the crystallization and morphology of HAp [18].

In the present work, a comprehensive study has been performed to investigate the influence of various organic additives on the phase purity and morphology of HAp synthesized via a hydrolysis of α-TCP under hydrothermal/solvothermal conditions. The effects of different concentrations of sodium lauryl sulfate (SLS, $CH_3(CH_2)_{11}OSO_3Na$) and three amino acids—dodecanedioic acid ($C_{12}H_{22}O_4$), DL-aspartic acid ($C_4H_7NO_4$), and suberic acid ($C_8H_{14}O_4$)—were studied and compared. To the best of our knowledge, the effect of these additives on the hydrolysis of low-temperature synthesized metastable α-TCP and simultaneous formation of HAp had never been investigated. These compounds have different functional groups and different carbon chain length. Two of the selected molecules contain only carboxyl groups (dodecanedioic and suberic acids); however, they have different chains containing 12 and 8 carbon atoms, respectively. Aspartic acid has a very short chain having only 4 carbon atoms. At the same time, in addition to the carboxyl group, there is a $-NH_2$ group in this molecule. Finally, sodium lauryl sulfate has a long chain (C12) and an SO_4^- functional group, which is absent in other selected molecules; moreover, a complex effect of amino acids and ethylene glycol was studied as well.

2. Materials and Methods

2.1. Synthesis

Starting α-TCP powders were synthesized following the procedure described elsewhere [19]. Briefly, a certain amount (3.42 g) of calcium nitrate tetrahydrate ($Ca(NO_3)_2 \cdot 4H_2O$, 99%, Roth, Karlsruhe, Germany) was dissolved in 20 mL of deionized water and a portion of diammonium hydrogen phosphate ($(NH_4)_2HPO_4$, 98%, Roth, Karlsruhe, Germany) (1.27 g) was dissolved in a separate beaker in 15 mL of deionized water. After a full dissolution, 5 mL of concentrated ammonium hydroxide (NH_4OH, 25%, Roth, Karlsruhe, Germany) was added to the $(NH_4)_2HPO_4$ solution and stirred; subsequently, the Ca-containing solution was added, rapidly resulting in the formation of white precipitates, which were aged in the solution for 10 min. Next, the precipitates were filtered and washed with 100 mL of deionized water and 150 mL of isopropyl alcohol. The synthesis product was dried overnight in an oven at 50 °C and annealed at 700 °C for 5 h with a heating rate of 5 °C/min.

The α-TCP was further used as a precursor for the synthesis of hydroxyapatite (HAp) via a hydrolysis reaction under hydrothermal conditions. The reactions were performed in the presence of different concentrations of sodium lauryl sulfate (SLS, $NaC_{12}H_{25}SO_4$, ≥99%, Roth, Karlsruhe, Germany) and amino acids—dodecanedioic acid (DDDA, $C_{12}H_{22}O_4$, 99%, Sigma Aldrich, Darmstadt, Germany), DL-aspartic acid (Asp, $C_4H_7NO_4$, 99%, Sigma Aldrich, Darmstadt, Germany), and suberic acid (Sa, $C_8H_{14}O_4$, 99%, Sigma Aldrich, Darmstadt, Germany). In a typical procedure, 0.3 g of α-TCP powder and appropriate amount of SLS or amino acids were placed into a 90 mL polytetrafluoroethylene-lined stainless-steel pressure vessel, diluted with 20 mL of water, and treated at 200 °C for 5 h.

Solvothermal reactions were performed in the presence of different concentrations of DL-aspartic acid and suberic acid in water and ethylene glycol (W:EG) mixture (*v*/*v* ratio of 40:60). The synthetic procedure was analogical to that with aqueous solutions.

After the hydrothermal treatment the resulting powders were filtered, washed with ethyl alcohol, and dried at 50 °C overnight. The sample notations and concentrations of additives in the reaction solution are given in Table 1.

Table 1. Sample codes, used organic additive, concentration of additives, and reaction media.

Notation	Organic Additive	Concentration of Additive	Water to Ethylene Glycol Ratio (*v*/*v*)
SLS:005	Sodium lauryl sulfate	0.005 mol/L	100:0
SLS:025	Sodium lauryl sulfate	0.025 mol/L	100:0
SLS:05	Sodium lauryl sulfate	0.05 mol/L	100:0
SLS:075	Sodium lauryl sulfate	0.075 mol/L	100:0
SLS:1	Sodium lauryl sulfate	0.1 mol/L	100:0
DDDA:005	Dodecanedioic acid	0.005 mol/L	100:0
DDDA:025	Dodecanedioic acid	0.025 mol/L	100:0
DDDA:05	Dodecanedioic acid	0.05 mol/L	100:0
DDDA:075	Dodecanedioic acid	0.075 mol/L	100:0
DDDA:1	Dodecanedioic acid	0.1 mol/L	100:0
Asp:005	DL-Aspartic acid	0.005 mol/L	100:0
Asp:025	DL-Aspartic acid	0.025 mol/L	100:0
Asp:05	DL-Aspartic acid	0.05 mol/L	100:0
Asp:075	DL-Aspartic acid	0.075 mol/L	100:0
Asp:1	DL-Aspartic acid	0.1 mol/L	100:0
Sa:005	Suberic acid	0.005 mol/L	100:0
Sa:025	Suberic acid	0.025 mol/L	100:0
Sa:05	Suberic acid	0.05 mol/L	100:0
Sa:075	Suberic acid	0.075 mol/L	100:0
Sa:1	Suberic acid	0.1 mol/L	100:0
Asp:005:EG	DL-Aspartic acid	0.005 mol/L	40:60
Asp:025:EG	DL-Aspartic acid	0.025 mol/L	40:60
Asp:05:EG	DL-Aspartic acid	0.05 mol/L	40:60
Asp:075:EG	DL-Aspartic acid	0.075 mol/L	40:60
Asp:1:EG	DL-Aspartic acid	0.1 mol/L	40:60
Sa:005:EG	Suberic acid	0.005 mol/L	40:60

Table 1. *Cont.*

Notation	Organic Additive	Concentration of Additive	Water to Ethylene Glycol Ratio (*v*/*v*)
Sa:025:EG	Suberic acid	0.025 mol/L	40:60
Sa:05:EG	Suberic acid	0.05 mol/L	40:60
Sa:075:EG	Suberic acid	0.075 mol/L	40:60
Sa:1:EG	Suberic acid	0.1 mol/L	40:60

2.2. Characterization

Powder X-ray diffraction data were collected on a Rigaku miniFlex II diffractometer (Rigaku, The Woodlands, TX, USA) working in Bragg–Brentano ($\theta/2\theta$) geometry, using Ni-filtered Cu $K\alpha_1$ radiation. The data were collected within a 2θ angle range from 10 to 60° at a step width of 0.01° and speed of 5°/min. Infrared (FTIR) spectra were taken in the range of 4000–400 cm^{-1} with Bruker Alpha ATR spectrometer (Bruker, Billerica, MA, USA). To study the morphological features of the samples, a field-emission scanning electron microscope (FE-SEM), the Hitachi SU-70 (FE-SEM, Hitachi, Tokyo, Japan), was employed.

3. Results

The XRD pattern and FTIR spectrum of starting α-TCP powder used for the synthesis of CDHA are shown in Figures S1 and S2 (see Supplementary Materials), respectively. The XRD pattern did not reveal the presence of neighboring crystalline CP, such as β-TCP or others, and all diffraction peaks were ascribed to α-TCP. The FTIR spectrum confirms the results of XRD analysis. The shape of the spectrum is in a good agreement with those reported in literature [20]; moreover, the absence of absorption band at around 729 cm^{-1} indicates the absence of calcium pyrophosphate, which is a commonly observed impurity in TCP powders synthesized by wet precipitation [21].

In aqueous medium α-TCP reacts with water and converts to calcium-deficient hydroxyapatite ($Ca_9(HPO_4)(PO_4)_5(OH)$, CDHA), as described by the following equation [13]:

$$3\alpha\text{-}Ca_3(PO_4)_2 + H_2O \rightarrow Ca_9(HPO_4)(PO_4)_5(OH) \quad (1)$$

The characteristics of the CDHA sample prepared in aqueous solution under hydrothermal conditions without any additives are given in Figure 1. As seen from the XRD pattern, under selected synthesis conditions α-TCP was fully converted to CDHA, and all diffraction peaks matched the standard XRD data of $Ca_{10}(PO_4)_6(OH)_2$ (ICDD #00-76-0694). There were no peaks associated with the starting material. The SEM image revealed that the morphology of the sample was dominated by plate-like particles of micrometric dimensions. HAp of slightly different morphology was obtained from commercial α-TCP by Goto et al. [13]. The difference might be due to the different starting α-TCP, since the authors used commercial α-TCP of micrometric size, whereas our α-TCP consisted of smaller particles [20].

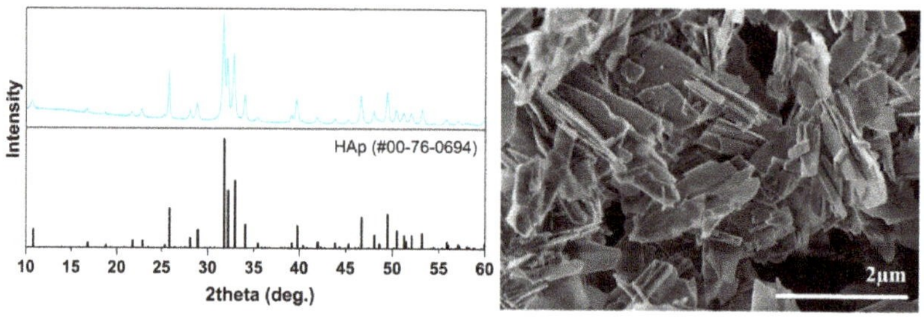

Figure 1. XRD pattern (**left**) and SEM image (**right**) of HAp prepared from α-TCP without additives.

Different results were observed when varying concentrations of SLS and amino acids were applied as additives. The XRD patterns of the final products are given in Figure 2. The lowest concentration of SLS (0.005 mol/L) did not affect the formation of CDHA in terms of phase purity; however, increasing the SLS amount in the reaction solution resulted in the formation of neighboring phase (Figure 2a). In addition to the typical CDHA diffraction pattern, a sharp peak at ca. 25.4° could be seen, and the intensity of this peak gradually increased with an increase in the SLS concentration. This peak was attributed to calcium sulfate ($CaSO_4$, ICDD #00-072-0503). The observed results suggest that under selected synthesis conditions the released Ca^{2+} ions were more easily precipitated by $SO_4{}^{2-}$ counterions rather than phosphate species. A similar trend was observed for DDDA: the lowest concentration of DDDA (0.005 mol/L) did not affect the formation of CDHA, but higher DDDA content resulted in the formation of secondary crystal phase (Figure 2c). The diffraction peak centered at ca. 13.6° emerged in the XRD pattern of the sample prepared using 0.025 mol/L DDDA. The intensity of this peak increased significantly with an increase in DDDA concentration. This peak corresponded to calcium hydrogen phosphate hydrate (CaH_3O_5P, ICDD #00-046-0494). Hydrothermal reactions in the presence of DL-aspartic and suberic acids resulted in the formation of single-phase CDHA, regardless of the concentration of additives (Figure 2b,d). No formation of impurities or traces of TCP were observed in these cases. It should be noted that in the case of suberic acid (Figure 2d) the intensity of three major peaks varied depending on the concentration of the additive, which could suggest the change in the powders' morphology. To summarize, two of four selected additives led to the formation of secondary phases, while two others did not affect the phase purity.

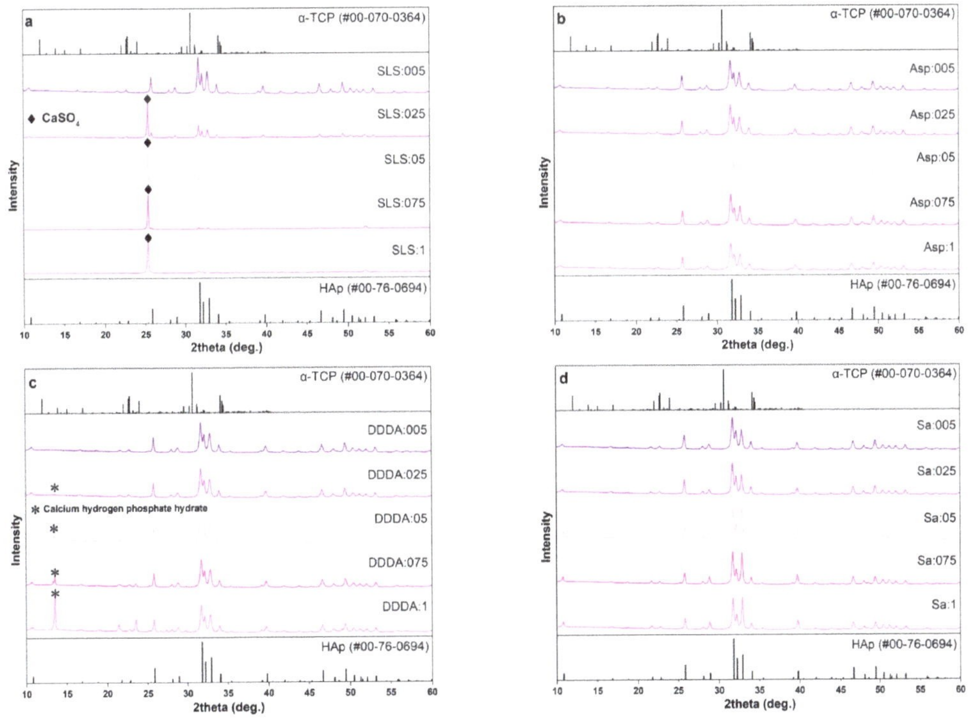

Figure 2. XRD patterns of the samples prepared using various concentrations of SLS (**a**), DL-aspartic acid (**b**), dodecanedioic acid (**c**), and suberic acid (**d**) after a hydrothermal treatment at 200 °C for 5 h.

In this study, we also aimed to check the combined effect of amino acids and organic solvents on the phase purity and morphology of the sample. Since single-phase CDHA was obtained using DL-aspartic and suberic acids, these two additives were used for further experiments. Our previous study revealed that among various organic solvents, ethylene glycol had the greatest influence on the sample morphology [18]. The highest effect on morphology was observed with water to ethylene glycol ratio of 40:60; hence, solvothermal synthesis with DL-aspartic and suberic acids was performed with this mixture. The XRD patterns of the reaction products are demonstrated in Figure 3.

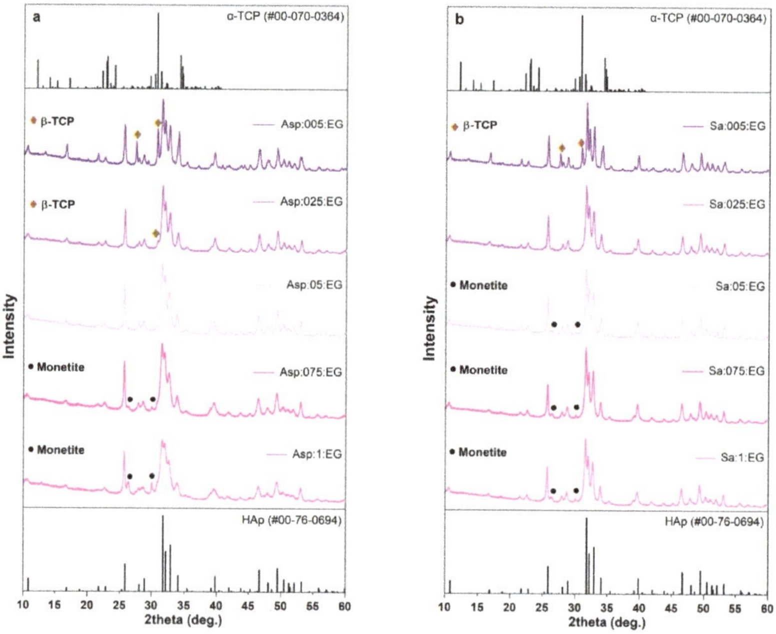

Figure 3. XRD patterns of the samples prepared using various concentrations of DL-aspartic (**a**) and suberic (**b**) acids in water–ethylene glycol mixture (40:60, v/v) after a solvothermal treatment at 200 °C for 5 h.

It is seen that phase composition strongly depends on the concentration of additive. With the lowest concentration of DL-aspartic acid (0.005 mol/L), a significant amount of neighboring β-TCP phase was obtained along with a major CDHA phase; however, the amount of β-TCP was reduced with an increase in the concentration of additive (Figure 3a). The XRD pattern of the sample synthesized with 0.05 mol/L of aspartic acid did not reveal the presence of any secondary crystal phase. At the same time, the diffraction peaks attributed to monetite ($CaHPO_4$) emerged in the XRD patterns with the highest concentrations of acid (0.075 and 0.1 mol/L). The presence of suberic acid demonstrated a very similar effect (Figure 3b) with a minor difference in phase composition. In this case, a negligible amount of monetite was detected already in the sample with the concentration of acid of 0.05 mol/L. In our previous study [18], the formation of monetite in the presence of organic solvents was also observed at higher temperatures (200 °C) of the solvothermal treatment; however, in this case, increasing concentrations of amino acids promoted the formation of monetite revealing the complex nature of phase transitions in CPs.

Figures 4 and 5 present the FTIR spectra of the samples prepared by hydrothermal and solvothermal treatments.

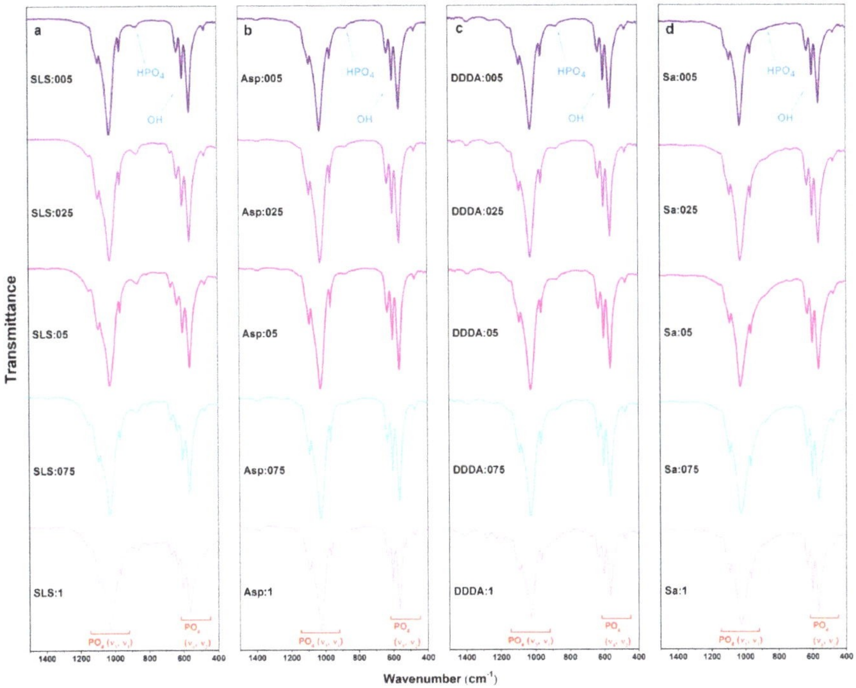

Figure 4. FTIR spectra of the samples prepared using various concentrations of SLS (**a**), DL-aspartic acid (**b**), dodecanedioic acid (**c**), and suberic acid (**d**) after a hydrothermal treatment at 200 °C for 5 h.

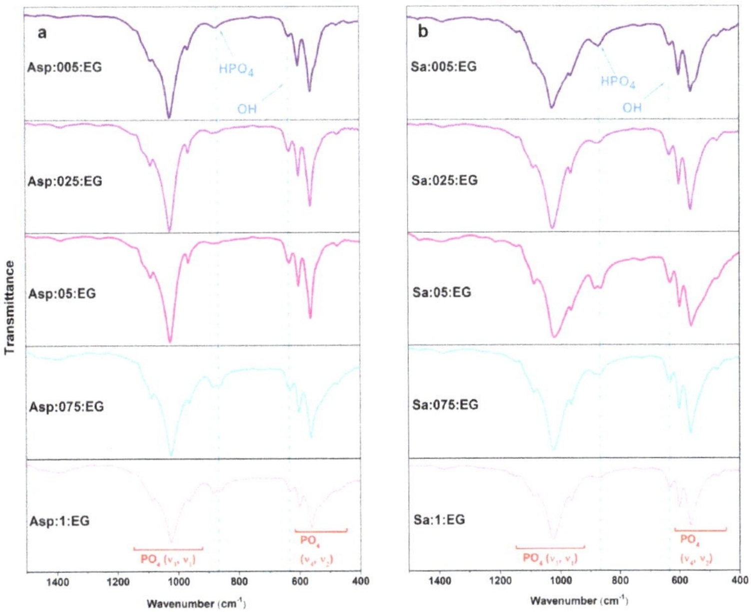

Figure 5. FTIR spectra of the samples prepared using various concentrations of DL-aspartic (**a**) and suberic (**b**) acids in water–ethylene glycol mixture (40:60, v/v) after a solvothermal treatment at 200 °C for 5 h.

The spectral range of 1500–400 cm^{-1} was chosen as representative, because in this range the main absorption bands assigned to CDHA can be observed. As was expected, all the spectra exhibited the absorption bands characteristic of CDHA [22]. The absorption bands of phosphate group in CDHA structure were observed at ca. 560 and 603 cm^{-1} (ν_4), 1020 and 1090 cm^{-1} (ν_3), 960 cm^{-1} (ν_1) which were assigned to O–P–O bending, asymmetric P–O stretching, and symmetric P–O stretching vibrations, respectively. The O–P–O bending vibration mode (ν_2) was observed at 470 cm^{-1} and the band at ca. 630 cm^{-1} was assigned to the hydroxyl group [22]. The absorption band centered at around 870 cm^{-1} was ascribed to the P–O(H) stretching mode of the HPO_4^{2-} group, which confirms the formation of CDHA, since this group is absent in stoichiometric non-deficient HAp [14]. Other bands of low intensity (Figure 4a,c) can be ascribed to the impurities formed during the synthesis and identified by XRD.

The FTIR spectra of synthesized products in the range from 4000 to 400 cm^{-1} are provided in Figures S3–S8; moreover, the spectra of the used organic additives are shown for comparison. The broader range additionally reveals the presence of absorption band centered at ca. 3570 cm^{-1}, which corresponds to the stretching mode of the hydroxyl group [22]. It is seen that in most cases the absorption bands corresponding to additives are absent in the spectra of fabricated CDHA powders, which suggests that the additives were removed from the surface during the washing procedure or the amount of residual organic species is negligible and cannot be detected by FTIR spectroscopy. The only exceptions are the products synthesized with DDDA (Figure S5). The absorption band of DDDA at 1690 cm^{-1} is visible in the spectra of samples synthesized with concentrations of 0.05 mol/L and higher.

The particle size and morphology of HAp are important characteristics of this material due to several reasons [23]. For instance, Dey et al. [24] studied the influence of HAp particle size, morphology, and crystallinity on proliferation of HCT116 colon cancer cells. It was demonstrated that decreasing the HAp powder crystallite size significantly increases the cell inhibition. Another study investigated the effect of nano-HAp particles of different sizes on the proliferation of odontoblast-like MDPC-23 cells comparing them with conventional HAp [25]. The results revealed that nano-HAp expressed an obvious growth-promoting effect. Wen et al. [26] showed that the larger specific surface area associated with the smaller particle size was beneficial for the drug-loading properties of HAp.

The morphology of the samples was found to be dependent both on the nature and concentration of the organic additive. When a small amount of SLS (0.005 mol/L) was used in the hydrothermal reaction, the plate-like crystals dominated in the sample; however, a small number of rods were also seen (Figure 6a). After increasing the amount of organic compound (0.05–0.1 mol/L), the morphology of the CDHA powders changed considerably and rod-shaped assemblies started to form (Figure 6e,i). The huge plate seen in Figure 6i corresponds to $CaSO_4$. The use of low concentration of DL-aspartic acid resulted in the formation of micrometric plate-like particles (Figure 6b), while increasing the amount of acid first led to the reduction of particle size (Figure 6f) and further formation of some rods (Figure 6j). The effect of DDDA was similar to that of aspartic acid and the morphology evolution from plate-like particles to the mixture of plates and rods was observed with an increase in additive concentration. The influence of suberic acid on the morphology of CDHA samples was found to be minor, and plate-like particles were obtained regardless of the concentration of acid. We assume that the effect on the sample morphology might become prominent if the concentrations of suberic acid were further increased.

The SEM images of the samples synthesized in a mixture of water and ethylene glycol in the presence of DL-aspartic and suberic acids are given in Figure 7. In the case of aspartic acid, the presence of the lowest concentration of additive resulted in the formation of a mixture of relatively long rods and short rods (Figure 7a). Using a higher concentration of aspartic acid obtained powders which consisted of micrometric agglomerates, which, in turn, were composed of smaller plate- and rod-like particles (Figure 7b–e). The size of the particles was significantly smaller compared to those synthesized in the aqueous medium

(Figure 6). In the case of suberic acid, relatively large plate-like particles were obtained regardless of the concentration of additive (Figure 7f–j). The morphology was comparable with that of powders obtained by the hydrothermal synthesis with no additives.

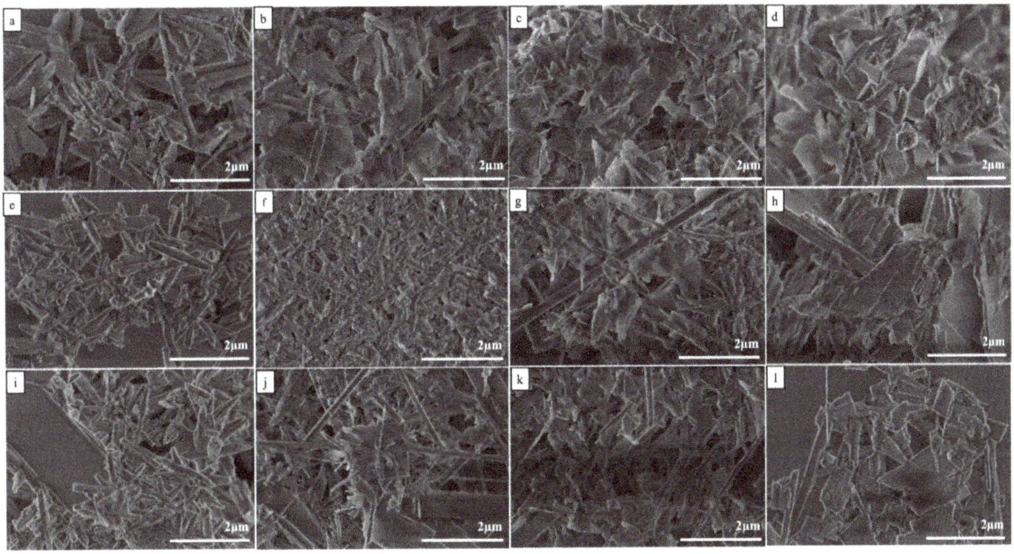

Figure 6. SEM micrographs of the samples after hydrothermal treatment SLS:005 (**a**), SLS:05 (**e**), and SLS:1 (**i**); Asp:005 (**b**), Asp:05 (**f**), and Asp:1 (**j**); DDDA:005 (**c**), DDDA:05 (**g**), and DDDA:1 (**k**); and Sa:005 (**d**), Sa:05 (**h**), and Sa:1 (**l**).

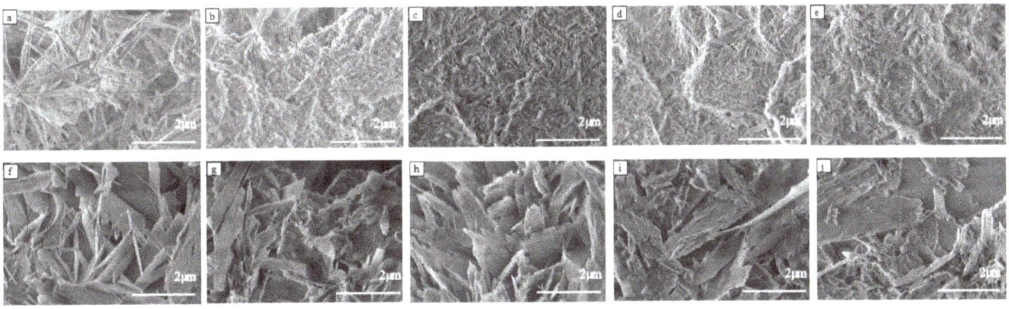

Figure 7. SEM micrographs of the samples after solvothermal treatment under ratio 40:60 of water to ethylene glycol Asp:005:EG (**a**), Asp:025:EG (**b**), Asp:05:EG (**c**), Asp:075:EG (**d**), and Asp:1:EG (**e**); and Sa:005:EG (**f**), Sa:025:EG (**g**), Sa:05:EG (**h**), Sa:075:EG (**i**), and Sa:1:EG (**j**).

To summarize, different impact of additives on the morphology of HAp is probably related to different structures and chemical compositions of organic additives, which leads to a different interaction with HAp particles during the nucleation and crystal growth process.

4. Conclusions

A comprehensive experimental study was carried out to compare the effects of sodium lauryl sulfate and three amino acids (DL-aspartic acid, dodecanedioic acid, and suberic acid) on the formation of calcium-deficient hydroxyapatite via a hydrolysis of α-tricalcium phosphate. It was demonstrated that the phase purity and morphology of the final product strongly depends on the nature and the concentration of organic additives. In aqueous medium at particular concentrations of sodium lauryl sulfate and dodecanedioic acid, the formation of impurities in addition to hydroxyapatite was observed. On the other hand, the use of aspartic and suberic acids did not affect the phase purity. The different impacts of additives on the morphology of hydroxyapatite are probably related to the different structures and chemical compositions of the organic additives, which lea to different interactions with hydroxyapatite particles during the nucleation and crystal growth process. The morphology of the samples prepared in the aqueous medium varied from plate- to rod-like depending on the concentrations of the specific organic additive. The use of a mixture of water and ethylene glycol led to the formation of significantly smaller particles having a shape of rods and narrow plates.

Supplementary Materials: The following supporting information can be downloaded at: https://www.mdpi.com/article/10.3390/cryst13020265/s1, Figure S1: XRD pattern of α-TCP used for the synthesis of CDHA; Figure S2: FTIR spectrum of α-TCP used for the synthesis of CDHA; and Figures S3–S8: FTIR spectra of the samples prepared using various concentrations of SLS, DL-aspartic acid, dodecanedioic acid, and suberic acid in the range from 4000 to 400 cm^{-1}.

Author Contributions: Conceptualization, R.K., A.Z. and A.K.; methodology, R.K.; validation, E.R.-S., J.-C.Y. and A.I.P.; formal analysis, R.K. and A.I.P.; investigation, R.K., E.R.-S., A.I.P. and J.-C.Y.; resources, A.K.; data curation, A.I.P.; writing—original draft preparation, R.K.; writing—review and editing, A.K.; visualization, R.K.; supervision, A.K.; project administration, A.Z.; and funding acquisition, A.Z. All authors have read and agreed to the published version of the manuscript.

Funding: This research was funded by the grant WHITCERAM (No. S-LJB-22-1) from the Research Council of Lithuania.

Institutional Review Board Statement: Not applicable.

Informed Consent Statement: Not applicable.

Data Availability Statement: Not applicable.

Conflicts of Interest: The authors declare no conflict of interest.

References

1. Habraken, W.; Habibovic, P.; Epple, M.; Bohner, M. Calcium phosphates in biomedical applications: Materials for the future? *Mater. Today* **2016**, *19*, 69–87. [CrossRef]
2. Zikrata, O.V.; Larina, O.V.; Valihura, K.V.; Kyriienko, P.I.; Balakin, D.Y.; Khalakhan, I.; Veltruská, K.; Krajnc, A.; Mali, G.; Soloviev, S.O.; et al. Successive Vapor-Phase Guerbet Condensation of Ethanol and 1-Butanol to 2-Ethyl-1-hexanol over Hydroxyapatite Catalysts in a Flow Reactor. *ACS Sustain. Chem. Eng.* **2021**, *9*, 17289–17300. [CrossRef]
3. Goto, T.; Cho, S.H.; Ohtsuki, C.; Sekino, T. Selective adsorption of dyes on TiO_2-modified hydroxyapatite photocatalysts morphologically controlled by solvothermal synthesis. *J. Environ. Chem. Eng.* **2021**, *9*, 105738. [CrossRef]
4. Ivanets, A.; Zarkov, A.; Prozorovich, V.; Venhlinskaya, E.; Radkevich, A.; Yang, J.-C.; Papynov, E.; Yarusova, S.; Kareiva, A. Effect of Mg^{2+}-, Sr^{2+}-, and Fe^{3+}-substitution on ^{85}Sr and ^{60}Co adsorption on amorphous calcium phosphates: Adsorption performance, selectivity, and mechanism. *J. Environ. Chem. Eng.* **2022**, *10*, 107425. [CrossRef]
5. Sinusaite, L.; Antuzevics, A.; Popov, A.I.; Rogulis, U.; Misevicius, M.; Katelnikovas, A.; Kareiva, A.; Zarkov, A. Synthesis and luminescent properties of Mn-doped alpha-tricalcium phosphate. *Ceram. Int.* **2021**, *47*, 5335–5340. [CrossRef]
6. Zhuang, Z.; Fujimi, T.J.; Nakamura, M.; Konishi, T.; Yoshimura, H.; Aizawa, M. Development of a,b-plane-oriented hydroxyapatite ceramics as models for living bones and their cell adhesion behavior. *Acta Biomater.* **2013**, *9*, 6732–6740. [CrossRef]
7. Kawasaki, T.; Takahashi, S.; Ideda, K. Hydroxyapatite high-performance liquid chromatography: Column performance for proteins. *Eur. J. Biochem.* **1985**, *152*, 361–371. [CrossRef]
8. Kawasaki, T.; Ikeda, K.; Takahashi, S.; Kuboki, Y. Further study of hydroxyapatite high-performance liquid chromatography using both proteins and nucleic acids, and a new technique to increase chromatographic efficiency. *Eur. J. Biochem.* **1986**, *155*, 249–257. [CrossRef]
9. Ezerskyte-Miseviciene, A.; Kareiva, A. Everything old is new again: A reinspection of solid-state method for the fabrication of high quality calcium hydroxyapatite bioceramics. *Mendeleev. Commun.* **2019**, *29*, 273–275. [CrossRef]
10. Bogdanoviciene, I.; Beganskiene, A.; Tõnsuaadu, K.; Glaser, J.; Meyer, H.J.; Kareiva, A. Calcium hydroxyapatite, $Ca_{10}(PO_4)_6(OH)_2$ ceramics prepared by aqueous sol–gel processing. *Mater. Res. Bull.* **2006**, *41*, 1754–1762. [CrossRef]
11. Ferraris, S.; Yamaguchi, S.; Barbani, N.; Cazzola, M.; Cristallini, C.; Miola, M.; Vernè, E.; Spriano, S. Bioactive materials: In vitro investigation of different mechanisms of hydroxyapatite precipitation. *Acta Biomater.* **2020**, *102*, 468–480. [CrossRef] [PubMed]

12. Jung, K.-W.; Lee, S.Y.; Choi, J.-W.; Lee, Y.J. A facile one-pot hydrothermal synthesis of hydroxyapatite/biochar nanocomposites: Adsorption behavior and mechanisms for the removal of copper(II) from aqueous media. *Chem. Eng. J.* **2019**, *369*, 529–541. [CrossRef]
13. Goto, T.; Kim, I.Y.; Kikuta, K.; Ohtsuki, C. Hydroxyapatite formation by solvothermal treatment of α-tricalcium phosphate with water–ethanol solution. *Ceram. Int.* **2012**, *38*, 1003–1010. [CrossRef]
14. Sinusaite, L.; Popov, A.; Raudonyte-Svirbutaviciene, E.; Yang, J.-C.; Kareiva, A.; Zarkov, A. Effect of Mn doping on hydrolysis of low-temperature synthesized metastable alpha-tricalcium phosphate. *Ceram. Int.* **2021**, *47*, 12078–12083. [CrossRef]
15. Suchanek, K.; Bartkowiak, A.; Perzanowski, M.; Marszałek, M. From monetite plate to hydroxyapatite nanofibers by monoethanolamine assisted hydrothermal approach. *Sci. Rep.* **2018**, *8*, 15408. [CrossRef]
16. In, Y.; Amornkitbamrung, U.; Hong, M.-H.; Shin, H. On the Crystallization of Hydroxyapatite under Hydrothermal Conditions: Role of Sebacic Acid as an Additive. *ACS Omega* **2020**, *5*, 27204–27210. [CrossRef]
17. Jiang, S.; Cao, Y.; Li, S.; Pang, Y.; Sun, Z. Dual function of poly(acrylic acid) on controlling amorphous mediated hydroxyapatite crystallization. *J. Cryst. Growth* **2021**, *557*, 125991. [CrossRef]
18. Karalkeviciene, R.; Raudonyte-Svirbutaviciene, E.; Gaidukevic, J.; Zarkov, A.; Kareiva, A. Solvothermal Synthesis of Calcium-Deficient Hydroxyapatite via Hydrolysis of α-Tricalcium Phosphate in Different Aqueous-Organic Media. *Crystals* **2022**, *12*, 253. [CrossRef]
19. Sinusaite, L.; Kareiva, A.; Zarkov, A. Thermally Induced Crystallization and Phase Evolution of Amorphous Calcium Phosphate Substituted with Divalent Cations Having Different Sizes. *Cryst. Growth Des.* **2021**, *21*, 1242–1248. [CrossRef]
20. Sinusaite, L.; Grigoraviciute-Puroniene, I.; Popov, A.; Ishikawa, K.; Kareiva, A.; Zarkov, A. Controllable synthesis of tricalcium phosphate (TCP) polymorphs by wet precipitation: Effect of washing procedure. *Ceram. Int.* **2019**, *45*, 12423–12428. [CrossRef]
21. Torres, P.M.C.; Vieira, S.I.; Cerqueira, A.R.; Pina, S.; da Cruz Silva, O.A.B.; Abrantes, J.C.C.; Ferreira, J.M.F. Effects of Mn-doping on the structure and biological properties of β-tricalcium phosphate. *J. Inorg. Biochem.* **2014**, *136*, 57–66. [CrossRef] [PubMed]
22. Koutsopoulos, S. Synthesis and characterization of hydroxyapatite crystals: A review study on the analytical methods. *J. Biomed. Mater. Res.* **2002**, *62*, 600–612. [CrossRef] [PubMed]
23. Zhou, H.; Lee, J. Nanoscale hydroxyapatite particles for bone tissue engineering. *Acta Biomater.* **2011**, *7*, 2769–2781. [CrossRef] [PubMed]
24. Dey, S.; Das, M.; Balla, V.K. Effect of hydroxyapatite particle size, morphology and crystallinity on proliferation of colon cancer HCT116 cells. *Mater. Sci. Eng. C* **2014**, *39*, 336–339. [CrossRef]
25. Li, N.; Wu, G.; Yao, H.; Tang, R.; Gu, X.; Tu, C. Size effect of nano-hydroxyapatite on proliferation of odontoblast-like MDPC-23 cells. *Dent. Mater. J.* **2019**, *38*, 534–539. [CrossRef]
26. Wen, Y.; Li, J.; Lin, H.; Huang, H.; Song, K.; Duan, K.; Guo, T.; Weng, J. Improvement of Drug-Loading Properties of Hydroxyapatite Particles Using Triethylamine as a Capping Agent: A Novel Approach. *Crystals* **2021**, *11*, 703. [CrossRef]

MDPI
St. Alban-Anlage 66
4052 Basel
Switzerland
Tel. +41 61 683 77 34
Fax +41 61 302 89 18
www.mdpi.com

Crystals Editorial Office
E-mail: crystals@mdpi.com
www.mdpi.com/journal/crystals

www.ingramcontent.com/pod-product-compliance
Lightning Source LLC
Chambersburg PA
CBHW061455210926
51850CB00008B/1728
* 9 7 8 3 0 3 6 5 6 9 5 8 1 *